Scanning Electron Microscopy for the Life Sciences

Recent developments in scanning electron microscopy (SEM) have resulted in a wealth of new applications for cell and molecular biology, as well as related biological disciplines. It is now possible to analyze macromolecular complexes within their three-dimensional cellular microenvironment in near native states at high resolution, and to identify specific molecules and their structural and molecular interactions. New approaches include cryo-SEM applications and environmental SEM (ESEM), staining techniques and processing applications combining embedding and resin-extraction for imaging with high-resolution SEM, and advances in immuno-labeling. New developments include helium ion microscopy, automated block-face imaging combined with serial sectioning inside an SEM chamber, and focused ion beam milling (FIB) combined with block-face SEM. With chapters written by experts, this guide gives an overview of SEM and sample processing for SEM, and highlights several advances in cell and molecular biology that have greatly benefited from using conventional, cryo-, immuno-, and high-resolution SEM.

Heide Schatten is Professor at the University of Missouri, Columbia. Her publications include advanced imaging methods, cellular and molecular biology, cancer biology, reproductive biology, microbiology, and space biology. The latter included collaborations with NASA scientists and experiments aboard the Space Shuttle Endeavour to examine the effects of spaceflight on cytoskeletal organization during development. She has received numerous awards including grant awards from NASA, NIH, and NSF. She has published over 200 papers and seven book chapters, and edited several special topics journal issues and eight books, with several more in progress.

Advances in Microscopy and Microanalysis

Books in Series

Published

Heide Schatten, *Scanning Electron Microscopy for the Life Sciences*

Forthcoming

Nigel Browning *et al.*, *Dynamic Transmission Electron Microscopy*
Michael Isaacson, *Microscopic Nanocharacterization of Materials*
Richard Leapman, *Energy Filtered Electron Microscopy and Electron Spectroscopy*

Scanning Electron Microscopy for the Life Sciences

HEIDE SCHATTEN

University of Missouri

CAMBRIDGE
UNIVERSITY PRESS

CAMBRIDGE UNIVERSITY PRESS
Cambridge, New York, Melbourne, Madrid, Cape Town,
Singapore, São Paulo, Delhi, Mexico City

Cambridge University Press
The Edinburgh Building, Cambridge CB2 8RU, UK

Published in the United States of America by Cambridge University Press, New York

www.cambridge.org
Information on this title: www.cambridge.org/9780521195997

First published 2013

Printed and bound in the United Kingdom by the MPG Books Group

A catalog record for this publication is available from the British Library

Library of Congress Cataloging in Publication data
Schatten, Heide.
Scanning electron microscopy for the life sciences / Heide Schatten.
p. cm.
Includes index.
ISBN 978-0-521-19599-7
1. Biology – Methodology. 2. Scanning electron microscopy. I. Title.
QH324.9.X2S33 2013
570.28′25–dc23
2012015496

ISBN 978-0-521-19599-7 Hardback

Contents

Endorsements — *page* vii
List of contributors — ix

1 **The role of scanning electron microscopy in cell and molecular biology: SEM basics, past accomplishments, and new frontiers** 1
Heide Schatten

2 **Corrosion casting technique** 16
Jerzy Walocha, Jan A. Litwin, and Adam J. Miodoński

3 **Revealing the internal structure of cells in three dimensions with scanning electron microscopy** 33
Sol Sepsenwol

4 **Mitochondrial continuous intracellular network-structures visualized with high-resolution field-emission scanning electron microscopy** 50
T. Naguro, H. Nakane, and S. Inaga

5 **Is the scanning mode the future of electron microscopy in cell biology?** 71
Paul Walther, Christopher Schmid, Michaela Sailer, and Katharina Höhn

6 **High-resolution labeling for correlative microscopy** 83
Ralph M. Albrecht, Daryl A. Meyer, and O. E. Olorundare

7 **The use of SEM to explore viral structure and trafficking** 99
Jens M. Holl and Elizabeth R. Wright

8 **High-resolution scanning electron microscopy of the nuclear surface in Herpes Simplex Virus 1 infected cells** 115
Peter Wild, Andres Kaech, and Miriam S. Lucas

9 **Scanning electron microscopy of chromosomes: structural and analytical investigations** 137
Elizabeth Schroeder-Reiter and Gerhard Wanner

10 **A method to visualize the microarchitecture of glycoprotein matrices with scanning electron microscopy** 165
Giuseppe Familiari, Rosemarie Heyn, Luciano Petruzziello, and Michela Relucenti

11 **Scanning electron microscopy of cerebellar intrinsic circuits** 179
Orlando J. Castejón

12 **Application of *in vivo* cryotechnique to living animal organs examined by scanning electron microscopy** 196
Shinichi Ohno, Nobuo Terada, Nobuhiko Ohno, and Yasuhisa Fujii

13 **SEM in dental research** 211
Vladimir Dusevich, Jennifer R. Melander, and J. David Eick

14 **SEM, teeth, and palaeoanthropology: the secret of ancient human diets** 236
Alejandro Romero and Joaquín De Juan

Index 257

The color plates are to be found between pages 116 and 117.

Endorsements

"This book is an excellent exposition of the many-faceted field of biological scanning electron microscopy. A brief introduction to the physics of SEM imaging is followed by an outstanding selection of recent applications, which are written by leaders in their respective fields and which include complete methodological details."

Michael Marko, Wadsworth Center, New York State Department of Health

"This book, *Scanning Electron Microscopy for the Life Sciences*, edited and compiled by Dr. Heide Schatten, comprises an extensive collection of articles demonstrating that the relevance of SEM to biological research is of increasing importance. The book is a very valuable compendium for any researcher interested in the fine structural morphology and chemistry of the cell and its compartments, such as mitochondria, and the nucleus and its contents. This, combined with the wide array of approaches, including recent ones, such as helium ion microscopy and block-face imaging combined with serial sectioning inside the SEM chamber, are all covered in this well-illustrated and also otherwise beautiful produced volume."

Bert Menco, Northwestern University

"*Scanning Electron Microscopy for the Life Sciences* includes an outstanding array of chapters on techniques for sample preparation and SEM imaging for a number of specimen types ranging from entire organs to molecules. Chapters include detailed information on protocols for specimen preparation and data collection, analysis, and presentation that have been applied to specific biological systems. Even though the information presented is specific to the experimental systems used in the laboratories of the authors, the methods and information provided will benefit all who use SEM in their research."

Robert Price, University of South Carolina School of Medicine

"The use of scanning electron microscopy in the life sciences has increased dramatically in the last decade. This has given rise to advances in equipment and development of new techniques. This text provides a needed survey of these advances; displaying the many ways that the beautiful three-dimensional structural detail available with scanning electron microscopy can be exploited. Scanning electron microscopy has long been prized for its ability to visualize high resolution surface detail, but many of the chapters in this book also show how internal detail of cells and tissue can be analyzed by scanning electron microscopy. This is an excellent introductory text for those who want to incorporate scanning electron microscopy in their repertoire. However, the well-written descriptions of cutting-edge techniques and the many 'tips and tricks' provided by expert authors insure that even experienced scanning electron microscopists will find the book valuable."

W. Gray Jerome, Vanderbilt University Medical Center

"This book is an excellent source of information about recent advances in the field of SEM for the life sciences and will assist microscopists in gaining a greater depth of understanding."

Cynthia S. Goldsmith, Centers for Disease Control and Prevention (CDC)

"Authoritative review of modern biological SEM methods. Advanced specimen preparation and imaging methods reveal fine details not observable by other means."

Charles Lyman, Lehigh University

Contributors

Ralph M. Albrecht
University of Wisconsin-Madison

Orlando J. Castejón
Universidad del Zulia, Venezuela

Joaquin De Juan
Universidad de Alicante, Spain

Vladimir Dusevich
University of Missouri – Kansas City

J. David Eick
University of Missouri – Kansas City

Giuseppe Familiari
Sapienza University of Rome, Italy

Yasuhisa Fujii
University of Yamanashi, Japan

Rosemarie Heyn
Sapienza University of Rome, Italy

Katharina Höhn
University of Heidelberg, Germany

Jens M. Holl
Emory University School of Medicine, Atlanta

S. Inaga
Tottori University, Japan

Andres Kaech
University of Zürich

Jan A. Litwin
Jagiellonian University Medical College, Krakow, Poland

Miriam S. Lucas
Swiss Federal Institute of Technology, Zürich

Jennifer R. Melander
University of Missouri – Kansas City

Daryl A. Meyer
University of Wisconsin-Madison

Adam J. Miodonski
Jagiellonian University Medical College, Krakow, Poland

T. Naguro
Tottori University, Japan

H. Nakane
Tottori University, Japan

Shinichi Ohno
University of Yamanashi, Japan

Nobuhiko Ohno
University of Yamanashi, Japan

O. E. Olorundare
University of Wisconsin-Madison and University of Ilorin, Nigeria

Luciano Petruzziello
Sapienza University of Rome, Italy

Michela Relucenti
Sapienza University of Rome, Italy

Alejandro Romero
Universidad de Alicante, Spain

Michaela Sailer
Universität Ulm, Germany

Heide Schatten
University of Missouri-Columbia

Christopher Schmid
Max-Planck-Institute for Molecular Physiology, Germany

Elizabeth Schroeder-Reiter
Ludwig-Maximilians-Universität München, Germany

Sol Sepsenwol
University of Wisconsin, Stevens Point

Nobuo Terada
University of Yamanashi, Japan

Jerzy Walocha
Jagiellonian University Medical College, Krakow

Paul Walther
Universität Ulm, Germany

Gerhard Wanner
Ludwig-Maximilians-Universität München, Germany

Peter Wild
Institute of Veterinary Anatomy and Virology, University of Zürich

Elizabeth R. Wright
Emory University School of Medicine, Atlanta

1 The role of scanning electron microscopy in cell and molecular biology: SEM basics, past accomplishments, and new frontiers

Heide Schatten

1.1 Introduction

New developments in scanning electron microscopy (SEM) have resulted in a wealth of new applications for cell and molecular biology as well as related biological disciplines. New instrument developments coupled with new sample preparation techniques have been key factors in the increasing popularity of this versatile research tool. The desire to view biological material in native states at high resolution has stimulated new approaches to cryo-SEM applications and environmental SEM (ESEM). In addition, new staining techniques and novel processing applications that combine embedding and resin-extraction for imaging with high-resolution SEM has allowed new insights into structure–function relationships. Advances in immuno-labeling have further enabled the identification of specific molecules and their location within the cellular microenvironment. It is now possible to analyze macromolecular complexes within their three-dimensional cellular environment in near native states, and this in many cases has provided advantages over two-dimensional imaging with transmission electron microscopy (TEM). New instrument developments include helium ion microscopy that allows imaging greater details of cellular components; new technology approaches include automated block-face imaging combined with serial sectioning inside an SEM chamber that in recent years has increasingly been utilized for a variety of biological applications. Focused ion beam milling (FIB) combined with block-face SEM is among the newer approaches to analyze cellular structure in three dimensions. The present chapter introduces the basic features of SEM and sample processing for SEM, and highlights several advances in cell and molecular biology that have greatly benefited from using conventional, cryo-, immuno-, and high-resolution SEM.

Scanning Electron Microscopy for the Life Sciences, edited by H. Schatten. Published by Cambridge University Press

1.2 The SEM as a versatile instrument for biological applications

As has been highlighted in previous review papers and books, the SEM is known for its versatility, allowing imaging and analysis of large and small sample sizes and of a diversity of specimens in multiple biological disciplines. Numerous articles are available on SEM instrumentation, modes of operation, imaging capabilities, and resolution (reviewed in Pawley, 2008; Schatten, 2008, 2011); new books are also available that have addressed different aspects of SEM utilization (Schatten and Pawley, 2008; several others are reviewed by Hawkes, 2009). In addition, recent special topics issues of microscopy journals focused on SEM have been devoted to specific biological and material science applications, demonstrating the increased need for more specific information for the increased number of researchers applying SEM to biomedicine and the basic sciences. In this section, the SEM is briefly introduced and the importance of sample preparation for biomedicine and biology is highlighted and detailed for routine sample preparation as well as for several specific applications. Examples of sample preparations that have been designed for specific cellular and molecular investigations are presented in the individual chapters of this book.

For general information a schematic diagram, Figure 1.1, displays the basic components of a conventional SEM.

Images in the SEM are generated by probing the specimen with a focused high-energy beam of electrons that is scanned across the specimen in a raster scan pattern. The electron beam interacts with the specimen surface, and interaction of the beam electrons with the sample atoms produces signals that contain information about the specimen's surface topography and characteristic features. However, internal cellular structures can also be visualized by using preparation methods that "peel" off the regular surface layers and turn internal structures into surfaces that can then be viewed with SEM providing information on surface and internal structures of intracellular components. In addition, isolated cellular components can be visualized clearly by SEM. Such applications are included in Section 1.3 and are detailed for specific applications in several chapters of this book.

The incident electron beam interacting with the specimen produces emission of low-energy (<50 eV) secondary electrons (SE), back-scattered electrons (BSEs), light emission (cathodoluminescence), characteristic X-ray emission, specimen current, and transmitted electrons and others as displayed in Figure 1.2 (color plate). For routine SEM imaging an electron gun with a tungsten filament cathode or a lanthanum hexaboride (LaB_6) cathode is used while a field emission gun (FEG) is used for more detailed SEM imaging (reviewed by Pawley, 2008; Schatten, 2008, 2011). Specific detectors are used to generate information from the specimen: typically an Everhart–Thornley detector is used for SEs, a type of scintillation-photomultiplier system, while a dedicated detector of either a scintillation or semiconductor type is used for BSE detection. For routine SEM imaging a secondary electron detector is used for conventional imaging. This imaging mode may allow significant advantages over TEM, as the

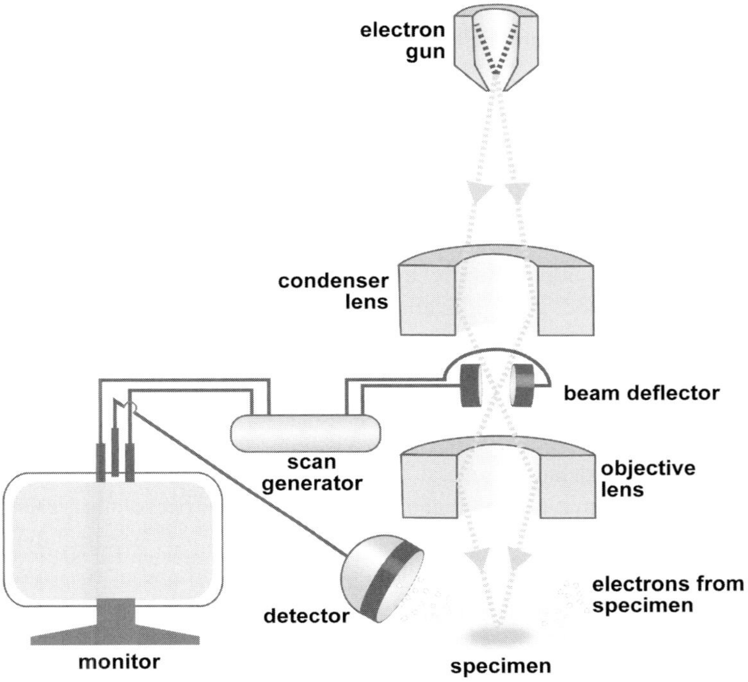

Figure 1.1 The basic components of a conventional SEM.

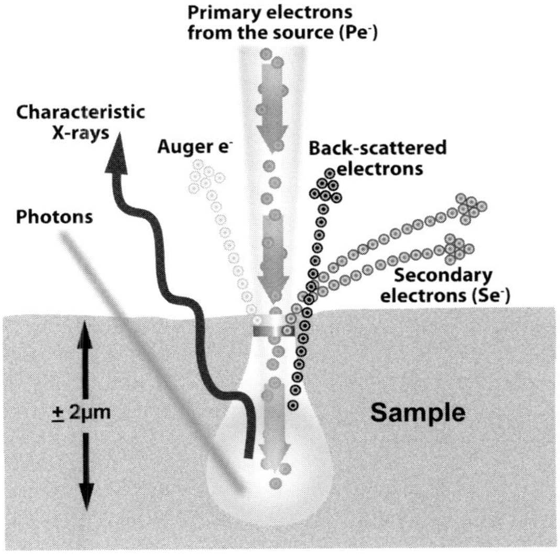

Figure 1.2 The incident electron beam interacting with the specimen produces emission of low-energy secondary electrons (SE), back-scattered electrons (BSEs), light emission (cathodoluminescence), characteristic X-ray emission, specimen current and transmitted electrons, and others as shown. (See plate section for color version.)

depth of field generates images that can readily be interpreted by the brain as three-dimensional representation. BSEs are beam electrons that are reflected from the sample by elastic scattering. The BSE signal intensity is related to the atomic number of the specimen and can therefore provide information about the different elements contained in the sample, which is oftentimes applied for imaging colloidal gold immunolabels of *c*. 5–10 nm in diameter. While characteristic X-rays for elemental analysis are used for biological applications, this form of analytical imaging is currently more frequently utilized in the material sciences to identify the composition of elements in a sample. However, new developments are in progress that may be amenable to biological applications and find new utilization in biology (Newbury, 2008). In this book, elemental analysis is described for the characterization of dental material (Dusevich *et al.*, Chapter 13 of this book). Characteristic X-rays are emitted when an inner shell electron is removed from the sample by beam interaction, which causes a higher energy electron to fill the inner shell and release characteristic energy.

As discussed in detail in several chapters of this book, newer variations of SEMs include the ESEM that allows imaging of relatively unprocessed samples contained in low vacuum or gas. While this mode of imaging is not entirely practical for all biological samples it is excellent for biomaterials and several other biological samples as demonstrated by Dusevich *et al.* (Chapter 13 of this book). Most samples viewed in conventional SEM do require processing, which routinely includes chemical fixation with glutaraldehyde or formaldehyde to stabilize the specimen's mobile macromolecular structure by chemical cross-linking of proteins and osmium tetroxide to stabilize lipids. Cryofixation is being used to preserve structures in their close to native states, which can be achieved with liquid nitrogen or liquid helium temperatures, as described below and detailed in several chapters in this book.

For chemically fixed samples, dehydration follows to replace water with organic solvents such as ethanol (or acetone) in incremental steps that gradually include increased ratios of alcohol (or acetone) to water up to 100% alcohol steps. It is critically important to dehydrate samples fully without leaving water residues to avoid sample preparation artifacts. The preferred choice for sample drying is the critical point procedure, but there are alternatives if a critical point dryer is not available. The dried sample is then mounted on a specimen holder (also called specimen stub). A last step before sample analysis with SEM includes conductive coating of the sample to prevent accumulation of static electric fields which may be caused by electron irradiation during imaging.

1.3 General sample preparation for SEM

The four steps for sample preparation include a) fixation, b) dehydration, c) critical point drying, and d) coating. All four steps can vary significantly and require modifications for specific applications. Adequate sample preparation is critically important to maintain structural integrity and obtain reliable information on cellular components and molecular composition. Poor and inadequate specimen preparation undoubtedly causes artifacts and may yield wrong information, which in some cases has caused confusion and serious concern in the literature (reviewed by Heuser, 2003). A great variety of sample preparation techniques and methods is available, and these have been elaborated by various investigators for specific research questions, some of which are presented in specific chapters of this book and others have been reviewed recently (Schatten, 2011). For routine SEM applications, the most common specimen preparation techniques are discussed below. Specialized sample preparation techniques are discussed in several of the other chapters of this book.

1.3.1 Conventional sample preparation

A most commonly used protocol includes fixing the biological sample with aldehydes such as 2% paraformaldehyde/2% glutaraldehyde in 0.1 M phosphate buffer (PBS) followed by PBS rinses, post-fixation in 1% osmium tetroxide in 0.1% PBS, rinses in PBS, dehydration in increasing series of ethanol or acetone, critical point drying, mounting on aluminum stubs, and coating (reviewed in Schatten, 2011). The times for the specific protocol steps vary, depending on sample size and sample characteristics. For optimal results the investigator is referred to research papers in the specific area of interest or to specific *Methods* books series such as *Methods in Cell Biology* that contain detailed recipes with valuable notes sections addressing potential problems, hints, alternative approaches, and other most valuable information shared by researchers in their fields. General processing information for SEM is given below.

Chemical fixation is most frequently used for SEM sample preparation, but chemical fixation may destroy the structural integrity of certain structures or molecules of interest. For example, because glutaraldehyde cross-links the free amino groups of polypeptides and amino acids and inactivates enzymes it is inadequate for enzyme localization and related studies. In this case and several others, alternative preparation methods may yield better and more reliable results.

Dehydration is necessary to prepare samples for viewing in the SEM vacuum. As mentioned above for fixation, dehydration may be damaging to biological structures, as the shape of a macromolecule or membrane is produced and maintained by its interactions with water. This interaction may be destroyed by removing water during the dehydration process. Shrinkage of biological material has been observed and may need to be calculated for accurate measurements after preparing soft tissue for SEM (Boyde and Maconnachie, 1979, 1981). However, large macromolecules and their associated or covalently attached structural components are preserved and not affected by dehydration (Ris, 1985, 1988, 1990, 1991).

Critical point drying (CPD) is a most reliable method for drying samples in preparation for viewing in the SEM vacuum. However, several precautions are important to avoid artifacts (Ris, 1985), as traces of water contaminating the intermediate liquid used for drying (ethanol or acetone) or in the liquid CO_2 transition fluid may distort ultrastructure and induce considerable artifacts (Ris, 1985). Artifacts introduced by residual water after CPD had caused historic debates when new structures termed microtrabeculae (Porter and Stearns, 1981) were clearly identified as preparation artifacts (Ris, 1985; reviewed in Heuser, 2003). These studies and numerous others have highlighted the importance of proper sample preparation as well as critical evaluation of instrumentation and accessory equipment. It further underlines the importance of critical sample evaluation and data interpretation. Thorough CPD is important for producing artifact-free specimens.

Coating of a specimen is an important aspect in sample preparation, as most biological structures have insulating characteristics and need to be made more conductive by applying a thin layer of metal to reduce the effects of charging (sample glaring; discussed in Hermann and Müller, 1991; Hermann *et al.*, 1996). Heavy metals or heavy metal compounds are used as coating materials and include gold, gold/palladium, platinum, tungsten, graphite, and others (Walther, 2008) that are deposited either by high-vacuum evaporation or by low-vacuum sputter coating of the sample. Sample conductivity may also be increased by using the OTO (osmium, thiocarbohydrazide, osmium) staining method (Seligman *et al.*, 1966; Malick *et al.*, 1975; Familiari *et al.*, Chapter 10 of this book). Determining the correct amount of coating that eliminates charging but allows reliable viewing of biological structures without obscuring the areas of interest is critically important.

Coating also improves contrast; coating thickness may vary with different samples (reviewed in Schatten, 2011).

Optimizing the coating thickness and applying specialized coating techniques (Walther, 2008) may be required for delicate biological structures, especially when viewing with low-voltage field emission SEM (LVFESEM) for which coating with 1–2 nm gold, palladium, or platinum applied by sputter-coating has been optimal for a large variety of samples (reviewed in Schatten, 2008). Coating may affect biologically relevant resolution (reviewed in Pawley, 2008). If charging problems are still encountered, alternative coating methods may need to be considered such as those developed by Walther and Hentschel (1989), who introduced a double coating technique by which the sample is first coated with a thin layer of heavy metal (platinum–carbon or tungsten with an average thickness of 2–3 nm) followed by a 5–10 nm carbon layer (Walther, 2008). Specific coating procedures have been described and discussed in the literature (Peters, 1980; 1982; 1985; 1986a; 1986; 1988; Peters and Fox, 1990; Walther, 2008; reviewed in Schatten, 2008), which includes Pt, W, and Ta by DC-ion sputtering to view cells growing on EM grids (Lindroth *et al.*, 1988; Bell *et al.*, 1989; Lindroth and Sundgren, 1989).

For optimal analysis of cellular and molecular components, appropriate sample preparation and the investigator's expertise with specific biological samples are among the most important criteria for reliable results using SEM. More difficult preparations are oftentimes encountered with plant samples or microbiology specimens including bacteria or parasites for which new preparation techniques may need to be developed. An example for the development of such techniques has been described for *Toxoplasma* (Schatten and Ris, 2002, 2004; Schatten *et al.*, 2003). In *Toxoplasma*, actin visualization had been a problem and several hurdles had to be overcome through step-wise and complementary approaches to determine localization of the fixation-sensitive actin-like fibers. Dobrowolski *et al.* (1997) had used cryo-fixation to determine immunolocalization of actin molecules, which established the presence of actin immunogenicity underneath the *Toxoplasma* surface. Subsequent experiments were performed peeling off the outer surface layer by quick treatment with detergent followed by cytoskeletal stabilization and fixation that revealed actin-like fibers underneath the surface (Schatten *et al.*, 2003). This example demonstrates that specific biological expertise is important to reveal structure–function relationships using SEM; new preparation methods may need to be designed paying attention to the specific biological characteristics and dynamics that require specific biological knowledge to preserve delicate structures that may respond differently to different chemicals used in the preparation protocols. The importance of biological expertise to obtain optimal information is demonstrated in numerous examples. The different requirements to preserve different structures reliably had already been recognized in the early pioneer days of electron microscopy when most samples were fixed in the cold, unknowingly destroying the cold-sensitive microtubule fibers. Correlative microscopy (also see Albrecht *et al.*, Chapter 6 of this book) is oftentimes needed to obtain accurate information for biological material. When it was recognized that cold-fixation indeed destroyed microtubules and the debate was settled, other debates emerged that questioned the characteristics of microtubules that in cross-sectioned EM samples were featured as short stubs, while immunofluorescence microscopy with anti-tubulin antibodies revealed long microtubules that had not previously been shown with TEM. These historic examples clearly show that sample preparation and interpretation of results are highly important and may require several approaches for reliable identification of biological material.

For plant material, optimal SEM preparation techniques are still being elaborated, as plant cells are more difficult to prepare for SEM because of tissue rigidity resulting from polysaccharide-containing cell walls and the large vacuole spaces within cells (Cox *et al.*, 2008). In several cases, use of protoplasts has been the choice for plant material, as protoplasts can be analyzed after removal of the cell wall containing large amounts of cellulose that hinders optimal processing. Isolation and detergent extractions of plant material have resulted in stunning data for cellular components, as seen in Chapter 9 of this book by Schroeder-Reiter and Wanner, which displays details of plant chromatin. Specific processing for microorganisms has been described in excellent detail by Erlandsen (2008).

1.3.2 Freezing methods

Ultra-rapid freezing is frequently used to preserve molecules in a more native state compared to chemical fixation. Ultra-rapid freezing demands avoiding the formation of damaging ice crystals in cells or tissue, which is accomplished by freezing at a rate of 10^4–10^5 degrees C/second. At this cooling rate, cellular water becomes vitrified rather than forming ice crystals. Several freezing methods are readily available, some of which are described in the specific chapters of this book. The basic freezing methods include the following:

Plunge freezing allows an average depth of vitrification of *c.* 1–2 μm with minimal ice crystal artifacts, which is achieved by plunging a specimen into a liquid cryogen such as supercooled liquid nitrogen or supercooled ethane or propane.

Slam freezing (cold metal block freezing) allows an average depth of vitrification of *c*. 10–15 μm with minimal ice crystal artifacts, which is achieved by slamming a specimen onto a copper or silver block that has been chilled to −196 to −269 °C with liquid nitrogen or liquid helium.

Propane jet freezing allows an average depth of vitrification of *c*. 40 μm with minimal ice crystal artifacts, which is achieved by sandwiching a 200–500 μm thick specimen between two metal plates that are clamped into a device that directs jets of liquid propane cooled with liquid nitrogen against both sides of the specimen plates.

High-pressure freezing allows an average depth of vitrification of *c*. 500 μm with minimal ice crystal artifacts. This modification of propane jet freezing or liquid nitrogen freezing is achieved by pressurizing the specimen to 2100 atmospheres to suppress or reduce growth of ice crystals at the moderate freezing rates that can be achieved in the depth of the sample. High pressure lowers the freezing point of water as well as the rate of ice crystal formation.

These basic freezing methods and several modifications have been applied with great success to a variety of specimens (reviewed in Schatten, 2011). Freezing followed by freeze drying is among the methods of choice for many applications in cell biology (Pawley and Ris, 1987) and freezing followed by freeze-substitution has gained increasing popularity (Erlandsen, 2008) for the superior ultrastructural preservation of cellular components and structures. Direct observation of frozen specimens (Pawley *et al*., 1991) has provided resolution above 3 nm, and freeze–fracture (Haggis and Pawley, 1988) has been applied successfully to visualize and analyze intracellular structures (reviewed in Pawley, 2008). Analysis of intracellular structure has further been accomplished by dry fracture of tissue culture cells achieved by touching intact cells to the surface of adhesive tape (Lim *et al*., 1987; Ris, 1988, 1989; Ris and Pawley, 1989; Sepsenwol, Chapter 3 of this book), allowing excellent insights into intracellular structure. Viewing of incorporated labels that decorate internal cell structure can also be accomplished with this method.

Among the advantages of cryofixation over chemical fixation is the arrest of cells in a "life-like" state; cryo-immobilization takes only milliseconds compared to chemical fixation, which may take seconds, or even longer. Freeze-substitution in acetone, methanol, or other solvents is frequently used for subsequent processing and permanent fixation.

In addition to freezing alone, several combination methods have also been used for specific biological applications including cryo-SEM of chemically fixed cells, as described by Erlandsen (2008). Other investigators have used chemical fixation with very low concentrations of glutaraldehyde (0.1–1.0%) for 10 to 15 min to stabilize macromolecules prior to cryo-immobilization (Centonze and Chen, 1995; Chen *et al*., 1995). Such fixation approaches preserved macromolecular complexes excellently and revealed remarkable detail of actin filaments (Erlandsen, 2008) after coating with chromium, allowing clear visualization of the helical twists of two polypeptide chains in the filament and 5 nm subunits (reviewed in Schatten, 2011). In other studies using cryo-methods, biological resolution of 2–3 nm could be achieved when viewing macromolecular complexes with high-resolution in-lens cryo-SEM (Erlandsen *et al.*, 2001), revealing details of the glycocalyx on the extracellular surface of human platelets that were labeled with three colloidal-gold markers to detect all three cell-adhesion molecules in the glycocalyx. This study used plunge-freezing into propane chilled with liquid nitrogen followed by partial freeze-drying at −85 °C and the double-coating method developed by Walther *et al.* (1995). This coating method involves cryo-coating by evaporation of 2 nm TaW at 45° through electron-beam deposition and 7–10 nm carbon at 90 °C. The double-coating technique has also been used in numerous other applications including double-layer coating of yeast cells after high-pressure freezing and freeze–fracture (reviewed in Erlandsen, 2008). All these examples clearly demonstrate that combination methods of various complexities and high-resolution SEM coupled with specific expertise of biological structure can be superior over other imaging methods and can reveal unique three-dimensional information.

If an SEM is equipped with a cold stage for cryo-microscopy, additional preparation techniques are available including cryo-fracture under vacuum, sputter coating and transfer to the SEM cryo-stage while still frozen. This method is particularly useful for imaging and analysis of temperature-sensitive cellular components including fats. Direct viewing of frozen specimens in the SEM by using a cold stage had become possible when side-entry eucentric goniometer stages were developed that could accept high-stability, cryotransfer stage rods. Uncoated and slightly coated specimens have been viewed in this mode (Herter, 1991; Pawley *et al.*, 1991; Müller *et al.*, 1992; Boyde, 2008; Walther, 2008), allowing detailed three-dimensional information to be obtained, as for example by stereo-imaging of frozen-hydrated mitochondria. Combination methods include freeze–fracture followed by the thaw–fix technique developed by Haggis (Haggis, 1987; Haggis and Pawley, 1988), involving fresh-freezing in propane, freeze–fracture and thawing into fixative, critical point drying and ion-beam-sputter coating with Pt before imaging. These examples also demonstrate the usefulness of stereo-imaging that allows better understanding of complex structural interactions of cellular components (reviewed in Pawley, 2008).

1.4 High-resolution low-voltage SEM and combination methods

Several chapters in this book describe use of high-resolution low-voltage field emission SEM (HRLVFESEM) with great success to view and analyze isolated structures or delicate internal cellular components. These applications have greatly benefited from the development of field-emission sources that has allowed formation of an intense beam of low-voltage electrons with small beam diameter (reviewed in Albrecht and Meyer, 2008).

HRLVFESEM has increasingly found new applications in cell and molecular biology for the study of structure–function relationships on three-dimensional levels. In addition, HRLVFESEM has also been an indispensable approach to image and analyze isolated structures that previously had been analyzed mainly by TEM negative staining. These new applications utilizing low-voltage electron microscopy take advantage of accelerating voltages at or below 5 keV. With these new capabilities combined with improved sample processing and sample coating as described in several chapters of this book, modern SEMs can achieve resolution for biological material down to 2–5 nm, a level previously only possible with TEM (reviewed in Schatten, 2011). As shown in subsequent chapters in this book, detailed analysis of chromosomes, cytoskeletal components, viruses, and other biological material has been performed using HRLVFESEM and revealed new detailed information in three dimensions that is superior to data obtained with TEM.

New applications with cryo-SEM including cryo-microtomy of cryo-immobilized plant and animal cells (Nusse and Van Aelsi, 1999; Walther and Müller, 1999) have yielded new information on internal cellular structures. In these specific applications the surface of the tissue block is examined rather than sections by cryo-SEM. Furthermore, a cryo-dual beam instrument has been utilized that incorporates both focusing electrons (SEM) and focusing ion beam (FIB) columns. Such applications have been employed by Mulders (2003) to analyze biological samples including yeast, bacteria, and gut epithelial cells.

New combination methods have been developed that will be addressed below. These new applications and others offer new approaches to identify biological components reliably inside cells, as described in excellent detail in several chapters of this book. A variety of different methods may need to be employed for optimal information, taking into consideration that some biological specimens are more fragile and complex than others, requiring more complex specimen preparation and processing.

In the author's lab, FESEM has been used to image delicate mitotic spindles and sperm asters yielding new information on cytoskeletal interactions in three dimensions (reviewed in Pawley, 2008; Schatten, 2008, 2011). In addition, the technique has allowed analysis of isolated centrosomes in three-dimensional configuration (Thompson-Coffe *et al.*, 1996).

1.5 New developments and future perspectives

One of the goals for new instrument and sample preparation development is to achieve higher resolution and imaging of samples in more native states. Such approaches have been pursued in recent years by Boyde (2008) and by researchers designing various types of microfluidic chamber that can be placed inside an SEM (Thiberge *et al.*, 2004; Boyde, 2008). The design and testing of microfluidic chambers is intensively being pursued by several groups with applications for transmission EM and SEM (Thiberge *et al.*, 2004; Klein *et al.*, 2011).

A most impressive development first presented by Denk and Horstmann (2004) introduced automated block face imaging combined with serial sectioning inside the chamber of an SEM. This development required several technical modifications that have been described in detail (Denk and Horstmann, 2004). With this new technology development the authors were able to trace even the thinnest axons to identify synapses in nerve tissue. The authors reported several hundred sections of 50–70 nm thickness that will lead to further developments to reconstruct large areas of neuronal tissue. Building on these developments, Knott *et al.* employed combination methods using light and electron microscopy. Serial section SEM of adult brain tissue using focused ion beam milling allowed visualization of the ultrastructure (Knott *et al.*, 2008). Several combination methods for live cell analysis using light microscopy followed by analysis with SEM have further been employed for nerve cells (Knott *et al.*, 2009). These studies included identification of live structures of interest with confocal microscopy followed by analysis with focused ion beam/scanning electron microscopy (FIB/SEM) and serial block face/scanning electron microscopy (SBF/SEM). Such studies and others offer new approaches aimed at studying live-like events at high resolution, and are currently being pursued with predicted success (Denk *et al.*, 2012). When coupled with new method developments, such as the recently reported novel genetically encoded tag for correlated light and electron microscopy (Shu *et al.*, 2011), it can further be predicted that we have entered a new area of discoveries on ultrastructural/functional levels that will have been made possible as a result of these new instrument/method developments.

Among the newest instrument developments is scanning helium ion microscopy (SHIM or HeIM), offering new advantages over conventional SEM (reviewed in Morgan *et al.*, 2006) including higher resolution and increased brightness. The high source brightness and short wavelength of the helium ions yield qualitative data that provide sharp images on a wide range of materials. As the secondary electron yield is quite high, it allows for imaging with currents as low as 1 femtoamp. A surface resolution of 0.24 nm has been demonstrated. The detectors provide information-rich images that reveal topographic, material, crystallographic, and electrical properties of the sample. In contrast to other ion beams, there is no discernible sample damage due to the relatively low mass of the helium ion. Since 2007 this technology has been commercialized and instruments have been utilized successfully.

Taken altogether, since its introduction as a research tool for the life sciences (reviewed by Pawley, 2008), the SEM has enjoyed enormous utilization in cell and molecular biology, facilitated by new sample preparation methods, advances in instrument development and new technological approaches that have generated novel two- and three-dimensional information for a variety of biological specimens. All chapters in this book are written by scientists with specific expertise in their respective fields of science and specific expertise in SEM methodology that can be applied to various other areas of interest in cell and molecular biology.

1.6 Acknowledgments

Donald Connor's professional help with the illustrations and Howard A. Wilson's professional help with the presentation of several images in the book are gratefully acknowledged.

1.7 References

Albrecht, R. and Meyer, D. (2008). Molecular labeling for correlative microscopy: LM, LVSEM, TEM, EF-TEM and HVEM. In: *Biological Low-Voltage Scanning Electron Microscopy*. Edited by H. Schatten and J. Pawley. New York, Springer, pp. 171–196.

Bell, P. B., Lindroth, M., and Fredriksson, B. A. (1989). Problems associated with the preparation of cytoskeletons for high resolution electron microscopy. *Scanning Microsc.*, Supplement **3**, 117–135.

Boyde, A. and Maconnachie, E. (1979). Volume changes during preparation of mouse embryonic tissue for scanning electron microscopy. *Scanning*, **2**, 149–163.

Boyde, A. and Maconnachie, E. (1981). Morphological correlations with dimensional change during SEM specimen preparation. *Scanning Electron Microsc.*, **IV**, 27–34.

Boyde, A. (2008). Low kV and video-rate, beam-tilt stereo for viewing live-time experiments in the SEM. In: *Biological Low-Voltage Scanning Electron Microscopy.* Edited by H. Schatten and J. Pawley. New York, Springer, pp. 197–214.

Centonze, V. E. and Chen, Y. (1995). Visualization of individual reovirus particles by low-temperature high-resolution scanning electron microscopy. *J. Struct. Biol.*, **115**, 215–225.

Chen, Y., *et al.* (1995). Imaging of cytoskeletal elements by low-temperature high-resolution scanning electron microscopy. *J. Microsc.*, **179**, 67–76.

Cox, G., Vesk, P., Dibbayawan, T., Baskin, T. I., and Vesk, M. (2008). High-resolution and low-voltage SEM of plant cells. In: *Biological Low-Voltage Scanning Electron Microscopy.* Edited by H. Schatten and J. Pawley. New York, Springer, pp. 229–244.

Denk, W. and Horstmann, H. (2004). Serial block-face scanning electron microscopy to reconstruct three-dimensional tissue nanostructure. *PLoS Biology*, **2**(11), e329.

Denk, W., Briggman, K. L., and Helmstaedter, M. (2012). Structural neurobiology: missing link to a mechanistic understanding of neural computation. *Nat. Rev. Neuroscience*, **13**(5), 351–358.

Dobrowolski, J. M., Niesman, I. R., and Sibley, D. L. (1997). Actin in the parasite *Toxoplasma gondii* is encoded by a single copy gene, ACT1, and exists primarily in a globular form. *Cell Motility and the Cytoskeleton*, **37**, 253–262.

Erlandsen, S. L. *et al.* (2001). High resolution cryo-FESEM and detection of individual cell adhesion molecules by stereo-imaging in the glycocalyx of human platelets: Immunogold localization of P-selectin (CD62P), integrin GpIIb/IIIa (CD41/CD61), and GpI-IX (CD42a, b). *J. Histochem. Cytochem.*, **49**, 809–819.

Erlandsen, S. L. (2008). Cryo-SEM of chemically fixed animal cells. In: *Biological Low-Voltage Scanning Electron Microscopy.* Edited by H. Schatten and J. Pawley. New York, Springer, pp. 215–228.

Knott, G. W., Marchman, H., Wall, D., and Lich, B. (2008). Serial section scanning electron microscopy of adult brain tissue using focused ion beam milling. *Neurosci.*, **28**(12), 2959–2964.

Knott, G. W., Holtmaat, A., Trachtenberg, J. T., Svoboda, K., and Welker, E. (2009). A protocol for preparing GFP-labeled neurons previously imaged *in vivo* and in slice preparations for light and electron microscopic analysis. *Nat. Protocol.*, **4**(8), 1145–1156.

Haggis, G. H. (1987). Freeze-fracture of cell nuclei for high-resolution SEM and deep-etch TEM. *Proc. Electron Microsc. Soc. Am.*, **45**, 560–564.

Haggis, G. H. and Pawley, J. B. (1988). Freeze-fracture of 3T3 cells for high resolution scanning electron microscopy. *J. Microsc.*, **150**, 211–218.

Hawkes, P. (2009). Nature's infinite books of secrecy. *Ultramicroscopy*, **109**, 1393–1410.

Hermann, R. and Müller, M. (1991). Prerequisites of high resolution scanning electron microscopy. *Scan. Electron Microsc.*, **5**, 653–664.

Hermann, R., Walther, P., and Müller, M. (1996). Immunogold labeling in scanning electron microscopy. *Histochem. Cell Biol.*, **106**, 31–39.

Herter, P., *et al.* (1991). High-resolution scanning electron microscopy of inner surfaces and fracture faces of kidney tissue using cryo-preparation methods. *J. Microsc.*, **161**, 375–385.

Heuser, J. (2003). Whatever happened to the 'microtrabecular concept'? *Biol. Cell*, **94**, 561–596.
Klein, K. L., Anderson, I. M., and de Jonge, N. (2011). Transmission electron microscopy with a liquid flow cell. *J. Microsc.*, **242**(2), 117–123.
Lindroth, M., Bell, P. B., and Fredriksson, B. A. (1988). Comparison of the effects of critical point drying and freeze-drying on cytoskeletons; and microtubules. *J. Microsc.*, **151**(2), 103–114.
Lindroth, M. and Sundgren, J. E. (1989). Ion beam-sputtered and magnetron-sputtered thin films on cytoskeletons: A high resolution TEM study. *Scanning I*, **1**, 243–253.
Lim, S. S., Ris, H., and Schnasse, B. (1987). Pigment granules in goldfish xanthophores are attached to intermediate filaments. *J. Cell Biol.*, **105**, 37a.
Malick, L., Wilson, E., Richard, B., and Stetson, D. (1975). Modified thiocarbohydrazide procedure for scanning electron microscopy: routine use for normal, pathological, or experimental tissues. *Biotech. Histochem.*, **50**(4), 265–269.
Morgan, J., Notte, J., Hill, R., and Ward, B. (2006). An introduction to the helium ion microscope. *Microscopy Today*, **14**(4), 24–31.
Müller, M. (1992). The integrating power of cryo-fixation based electron microscopy in biology. *Acta Microscopica*, **1**, 37–44.
Mulders, H. (2003). The use of a SEM/FIB dual beam applied to biological samples. *GIT Imag. Microsc.*, **2**, 8–10.
Newbury, D. (2008). Developments in instrumentation for microanalysis in low-voltage scanning electron microscopy. In: *Biological Low-Voltage Scanning Electron Microscopy*. Edited by H. Schatten and J. Pawley. New York, Springer, pp. 263–304.
Nusse, J. and Van Aelsi, A. C. (1999). Cryo-planning for cryo-scanning electron microscopy. *Scanning*, **21**, 372–378.
Pawley, J. B. and Ris, H. (1987). Structure of the cytoplasmic filament system in freeze-dried whole mounts viewed by HVEM. *J. Microsc.*, **13**, 319–332.
Pawley, J. B., Walther, P., Shih, S. J., and Malecki, M. (1991). Early results using high resolution low voltage low temperature SEM. *J. Microsc.*, **162**, 327–335.
Pawley, J. B. (2008). LVSEM for biology. In: *Biological Low-Voltage Scanning Electron Microscopy*. Edited by H. Schatten and J. Pawley. New York, Springer, pp. 27–106.
Peters, K. R. (1980). Penning sputtering of ultra-thin metal films for high resolution electron microscopy. *Scanning Electron Microscopy 1980*. Edited by O. Johari and I. Corvin. Chicago 1, SEM Inc., pp. 143–154.
Peters, K. R. (1982). Conditions required for high quality high magnification images in secondary electron scanning electron microscopy. *Scanning Electron Microscopy 1982*. Edited by O. Johari and I. Corvin. Chicago IV, SEM Inc., 1359–1372.
Peters, K. R. (1985). Working at higher magnifications in scanning electron microscopy with secondary and backscattered electrons on metal coated biological specimens and imaging macromolecular cell membrane structures. *Scanning Electron Microscopy 1985*. Edited by O. Johari and I. Corvin. Chicago IV, SEM Inc., 1519–1544.
Peters, K. R. (1986a). Rationale for the application of thin, continuous metal films in high magnification electron microscopy. *J. Microsc.*, **142**, 25–34.
Peters, K. R. (1986b). Metal coating thickness and image quality in scanning electron microscopy. *Proc. EMSA*, **44**, 664–667.
Peters, K. R. (1988). Current state of biological high resolution scanning electron microscopy. *Proc. EMSA*, **46**, 180–181.
Peters, K. R. and Fox, M. D. (ed.) (1990). Ultra-high resolution cinematic digital 3D imaging of the cell surface by field emission scanning electron microscopy. *Proc. XIIth ICEM Mtg.*, **1**, 12–13.

Porter, K. R. and Stearns, M. E. (1981). Stereomicroscopy of whole cells. *Meth. Cell Biol.*, **22**, 53–75.

Ris, H. (1985). The cytoplasmic filament system in critical point-dried whole mounts and plastic-embedded sections. *J. Cell Biol.*, **100**, 1474–1487.

Ris, H. (1988). Application of LVSEM in the analysis of complex intracellular structures. *ProcEMSA,* **46**, 212–213.

Ris, H. (1989). Three-dimensional imaging of cell ultrastructure with high resolution low voltage SEM. Inst. Phys. Conf. Ser. 98,657462.

Ris, H. and Pawley, J. B. (1989). Analysis of complex three-dimensional structures involved in dynamic processes by high voltage electron microscopy and low voltage high resolution scanning electron microscopy. In: *Microscopy of Subcellular Dynamics*. Edited by H. Pattner. Boca Raton, FL, CRC Press, pp. 309–323.

Ris, H. (1990). Application of low voltage high resolution SEM in the study of complex intra-cellular structures. *Proc Xllth ICEM Mtg*., Seattle, 18–19.

Ris, H. (1991). The three-dimensional structure of the nuclear pore complex as seen by high voltage electron microscopy and high resolution low voltage scanning electron microscopy. *EMSA Bull.*, **21-1**, 54–56.

Schatten, H. and Ris, H. (2002). Unconventional specimen preparation techniques using high resolution low voltage field emission scanning electron microscopy to study cell motility, host cell invasion, and internal structures in *Toxoplasma gondii. Microsc. Microanal.*, **8**, 94–103.

Schatten, H., Sibley. D., and Ris, H. (2003). Structural evidence for actin filaments in *Toxoplasma gondii* using high resolution low voltage field emission scanning electron microscopy. *Microsc. Microanal.*, **9**, 330–335.

Schatten, H. and Ris, H. (2004). Three-dimensional imaging of *Toxoplasma gondii*-host cell membrane interactions. *Microsc. Microanal.*, **10**, 580–585.

Schatten, H. (2008). High-resolution, low voltage, field-emission scanning electron microscopy (HRLVFESEM) applications for cell biology and specimen preparation protocols. In: *Biological Low-Voltage Scanning Electron Microscopy*. Edited by H. Schatten and J. Pawley. New York, Springer, pp. 145–169.

Schatten, H. and Pawley, J. (ed.) (2008). *Biological Low-Voltage Scanning Electron Microscopy*. New York, Springer.

Schatten, H. (2011). Low voltage high resolution SEM (LVHRSEM) for biological structural and molecular analysis. Special issue: Biospecimens for high resolution. *Micron*, **42**(2), 175–85.

Seligman, A. M., Wasserkrug, H. L., and Hanker, J. S. (1966). A new staining method for enhancing contrast of lipid-containing membranes and droplets in osmium tetroxide-fixed tissue with osmiophilic thiocarbohydrazide (TCH). *J. Cell Biol.*, **30**(2), 424–432.

Shu, X., Lev-Ram, V., Deerinck, T. J., *et al.* (2012). A genetically encoded tag for correlated light and electron microscopy of intact cells, tissues, and organisms. *PLoS Biology*, **9**(4): e1001041.

Thiberge, S., Nechushtan, A., Sprinzak, D., *et al.* (2004). Scanning electron microscopy of cells and tissues under fully hydrated conditions. *Proc. Nat. Acad. Sci. USA*, **101**(10), 3346–3351.

Thompson-Coffe, C., Coffe, G., Schatten, H., Mazia, D., and Schatten, G. (1996). Cold-treated centrosomes: isolation of the centrosomes from mitotic sea urchin eggs, production of an anticentrosomal antibody, and novel ultrastructural imaging. *Cell Motil. Cytoskeleton*, **33**, 197–207.

Walther, P. and Hentschel, J. (1989). Improved representation of cell surface structures by freeze substitution and backscattered electron imaging. *Scanning Microsc.*, **3**, Supplement 3, 201–211.

Walther, P. *et al.* (1995). Double layer coating for high-resolution low temperature SEM. *J. Microsc.*, **179**, 229–237.

Walther, P. and Müller, M. (1999). Biological ultrastructure as revealed by high-resolution cryo-SEM of blockfaces after cryo-sectioning. *J. Microsc.*, **196**(3), 279–287.

Walther, P. (2008). High-resolution cryoscanning electron microscopy of biological samples. In: *Biological Low-Voltage Scanning Electron Microscopy*. Edited by H. Schatten and J. Pawley. New York, Springer, pp. 245–262.

2 Corrosion casting technique

Jerzy Walocha, Jan A. Litwin, and Adam J. Miodoński

2.1 Introduction

Among various techniques of biological material preparation for scanning electron microscopy, corrosion casting – sometimes also referred to as microcorrosion casting – is the method of choice for studying three-dimensional topography of microvascular systems, offering spatial continuity of the cast vessels combined with high resolution and quasi-3-D quality of SEM images (Figure 2.1). In essence, the technique requires injection of the studied microvasculature with solidifiable medium and subsequent corrosion (maceration) of the tissue. The resulting cast is cleaned of possible tissue/reagent debris and examined in SEM. Apart from revealing the topographical pattern of microvasculature, the casts can also provide information concerning luminal differentiations, such as bulging endothelial cell nuclei, fenestrated capillary areas, and valves or local constrictions reflecting the presence of contracted smooth muscle cells or pericyte processes, since the cast surface replicates in detail the luminal surface of blood vessels. In theory, corrosion casting is suitable to replicate all luminal systems or spaces of the organism, but its applications to studies of organs/systems other than blood or lymphatic vessels encounter significant technical difficulties and have been very rare.

2.2 History

Although corrosion casting is now regarded as one of the material preparation techniques for SEM, it has its roots in gross anatomy. Invention of a casting substance, capable of filling major body spaces and blood vessels and subsequent hardening *in situ*, resistant to mechanical and chemical damage and thus appropriate to create durable specimens that can be presented to students, had been an anatomist's dream for many years. The idea of using corrosion casts for gross anatomical studies is a few hundred years old. Leonardo da Vinci (1452–1519) made wax casts of brain ventricles and heart chambers of humans. Jan Swammerdam (1637–1680) is commonly mentioned as the inventor of solidifying injection mass (although he also used melted wax) and syringe to perform injections of blood vessels. At the turn of the seventeenth and eighteenth century and later, different casting media were tried: Gottfried Bidloo in 1685 used melted metal to inject trachea

Scanning Electron Microscopy for the Life Sciences, edited by H. Schatten. Published by Cambridge University Press

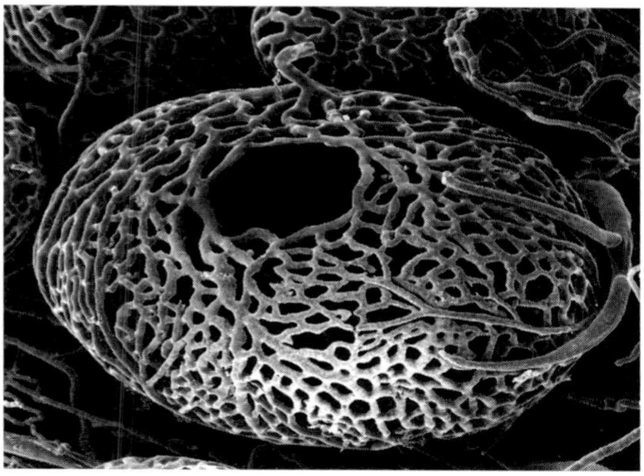

Figure 2.1 Microvasculature of poison gland in the skin of salamander. Reprinted with kind permission from Springer Science+Business Media, from: Miodoński, A. and Jasiński, A. (1979) Scanning electron microscopy of microcorrosion casts of the vascular bed in the skin of the spotted salamander, *Salamandra salamandra L. Cell Tissue Res.*, **196**, 153–62.

and bronchi, 70 years later Johannes N. Lieberkühn employed a mixture of 1/10 natural resin and 9/10 turpentine and was the first to produce successfully injected microvessels of the gastrointestinal tract mucosa.

The corrosion casting era was opened by Ruysch (1725) and Lieberkühn (1748) who injected human organs with casting medium and then used insect larvae to corrode the injected specimens. By the end of the nineteenth century Hyrtl, Schiefferdecker, Teichmann, Hoyer Sr. and Jr., Gerlach, Voigt, Storch, and Kadyi continued the systemic anatomical studies using different kinds of casting media such as gelatin, celloidin, cellulose, or modified glazier's putty (Kuś, 1969). The latter medium was employed by Teichmann, one of the most outstanding pioneers of casting techniques, famous for his unique cast specimens presently exhibited in the museum of Chair of Anatomy, Jagiellonian University, Krakow, Poland.

The currently used casting media – synthetic resins – had their precursors in the first half of the twentieth century. In 1935, Schummer introduced a polymerizing resin called Plastoid and injected testis and ureter (Schummer, 1935). In 1936, Narat *et al.* used for the first time vinyl-polychloride for injection of placental, renal, and splenic blood vessels. Significant technical progress was achieved in the 1950s and later, when synthetic resins were introduced as casting media: acrylic resin (Taniguchi *et al.* in 1952); polyester resin (Aleksandrowicz and Łoziński in 1959); vinyl chloride (Goetzen in 1966); mixture of methylmethacrylates; methylmethacrylate (Murakami in 1971); Batson No. 17 (Nopanitaya *et al.* in 1979); Araldite CY 223; Tardoplast (Amselgruber and König in 1987), (Lametschwandtner *et al.*, 1990).

Another type of casting medium tested in the 1970s was latex and silicone rubber (Cementex, Microfil). However, rubber casts showed extreme fragility and easily disintegrated during corrosion. They required freeze-drying or critical point drying to maintain their three-dimensional arrangement and did not replicate luminal surface microstructures consistently.

Recently, a new polyurethane-based PU4ii resin has been proposed as an optimal casting medium (Krucker *et al.*, 2006).

The corrosion media used to produce the casts included various alkalis (NaOH, KOH, sodium hypochloride) and acids (HCl, H_2SO_4, HNO_3, bichromate sulfuric acid, chromium trioxide, and formic acid), although it has been shown that some of them can cause severe damage to the casts and that different casting substances require different corrosion media to prevent such damage (Hodde *et al.*, 1990).

Forty years ago, Janice Nowell and her coworkers for the first time used a combination of corrosion casting and observation of the casts in SEM to study the structure of avian lungs (Nowell *et al.*, 1970). One year later, Murakami (1971) applied that approach to the study of vascular networks. Since then, over 2000 papers presenting scanning electron microscopy of corrosion casts in animals, humans, and even invertebrates have been published. The data provided by this technique were helpful not only in evaluating microanatomy of the studied systems, but also in elucidating some aspects of development, differentiation, aging, and pathological processes.

2.3 Corrosion casting procedure

The corrosion casting/SEM technique consists of several stages:

- washing of the space to be cast in order to remove its content (e.g. blood)
- injection of casting medium
- hardening of casting medium
- corrosion (maceration)
- dissection (optional)
- cleaning and drying
- mounting and conductive treatment
- SEM examination
- quantitative analysis of SEM images (optional).

Since the vast majority of corrosion casting/SEM studies has been performed on blood vessels, the technical aspects are discussed in this context.

In small animals, a cast can be made of the entire vascular bed. Larger animals require casting of selected vascular systems, usually those of organs chosen for the study, by injecting the casting medium via their main supplying vessels and allowing outflow of the medium via the main draining vessels. This can be done either *in situ* or after removal of the organ. It is often difficult to obtain such casts of an acceptable quality. Larger volumes of rinsing and casting media have to be used, there is a risk of vasoconstriction and intravascular coagulation before the casting step leading to incomplete replication, and furthermore, casts can be easily damaged during specimen handling, especially when cast dissection is necessary. The former risk can be minimized if organs are perfused *in situ* in an anesthetized animal and if the perfusion solution contains anticoagulants. Biomedical research on humans is particularly demanding, since casting has to be carried out in material obtained post-operatively or on autopsies. Fresh post-operative specimens of human organs can only rarely be used because whole organs are required for resin injection, while at least fragments of the organ must be collected for histopathological examination. Corrosion casting of autopsy material bears a risk of poor tissue preservation and endothelial cell necrosis. For obvious reasons, the completeness of replication and number of artifacts depend on the time elapsing between the interruption of the circulation and filling of vessels with the casting medium. Several authors have reported successful post-mortem casting and SEM analysis of human microvascular systems (Karaganov *et al.*, 1981; Banya *et al.*, 1989; Murakami *et al.*, 1992; Walocha *et al.*, 2003; 2012). They were able to obtain acceptable casts from organs collected upon autopsy within 24 hours after death. Acceptable post-mortem vascular casts have also been obtained in large animals (Martin-Orti *et al.*, 1999).

There are two main obstacles that can hamper the complete filling of the vascular system with casting medium: blood clots and vasoconstriction caused by metabolic or neurogenic factors. Therefore, the first step of the procedure – perfusion of the vessels with a washing solution is crucial for optimal replication. The solution (PBS, Tyrode's solution, Ringer's solution) should contain anticoagulant (heparin) and a spasmolytic agent (lidocaine, papaverine). Subsequent perfusion with a low-concentration fixative, e.g. formaldehyde or glutaraldehyde (vascular fixation) is optional, but fixation of the vascular walls is believed to increase their stability and to prevent ruptures leading to extravasation of the casting medium or dilatation of the vessels under its pressure.

The properties of the casting medium are another key factor for high-quality casts. An ideal casting medium:

- has relatively low viscosity (as close to that of body fluids as possible) to ensure perfect filling of the smallest spaces
- is chemically and physiologically neutral in the system to be cast
- polymerizes within an appropriately short time (3–15 minutes)
- does not shrink/deform during polymerization and drying
- allows microdissection of the cast without breaking or deformation
- is fully resistant to corroding reagents (cast shows no surface damage)
- resists electron bombardment during SEM examination
- allows replication of minute details of cast surfaces
- reveals no toxicity.

The casting media currently in use are not perfect – some of the above listed criteria are only partially fulfilled (e.g. there seems to be an inverse correlation between viscosity of the liquid resin and its shrinkage rate during polymerization) – but still they provide satisfactory replication at magnifications offered by SEM. The most widely employed media are methylmethacrylates: Mercox, Batson's no. 17, and Technovit 8001. The casting kit usually includes the resin, polymerization catalyst (plasticizer), and sometimes polymerization promoter (initiator/accelerator).

Perfusion of specimens with casting medium should be performed under controlled pressure, since too low a pressure usually leads to incomplete replication, while too high a pressure, albeit minimizing that hazard, can induce deformation and damage of the vascular walls resulting in casting artifacts such as bulges and extravasations. Manual perfusion is the simplest, but it cannot be standardized and most authors use specially designed mechanical injection/perfusion devices, which allow injection of preprogrammed volumes of casting medium per minute at controlled injection pressure.

Polymerization of the resin is possible at room temperature, but it is usually carried out at elevated temperature by placing the specimen in a water bath at 40 to 60 °C for a few hours. Such treatment accelerates polymerization and allows obtaining less fragile casts.

Corrosion of tissues surrounding the cast is mostly performed in solutions of sodium or potassium hydroxide at room temperature or at 37–39 °C. Although a wide range of hydroxide concentrations (15–60%) has been used in corrosion casting studies, it seems that the optimal concentration is 5–20%, since concentrations of 40% and more can inhibit the maceration process by saponification of proteins (Hodde *et al.*, 1990). Corrosion is a time-consuming process and in larger specimens can take a few days. It can be accelerated by relatively frequent changes of the hydroxide followed by washes with gently running warm tap water.

Small casts can be directly cleaned and mounted for observation in SEM. In most cases, however, the size of cast obtained from an injected organ requires its dissection, not only to cope with the space available in the SEM specimen chamber, but also to reveal deeper regions of the cast. Dissection should not alter natural shape and three-dimensional topography of the cast. It can be carried out before or after corrosion. Dissection performed before corrosion has an advantage of the cast being supported by the surrounding tissues with distinguishable anatomical landmarks. Dissection after corrosion requires particular precision and delicacy – it can be performed with the use of a microtome blade, microsurgical tools or by low-power laser beam, although the latter can produce thermal artifacts at the dissection plane. When highly precise dissection of small areas is needed, it can be done under a stereomicroscope on a dried and mounted cast, in a warm alcohol bath, or even using a micromanipulator associated with SEM (Lametschwandtner *et al.*, 1990). However, the risk of cast damage is lower when prior to dissection the cast is stabilized by embedding in a solidifying medium, such as water-soluble wax (e.g. Aquax – Miodoński *et al.*, 1980) or a mixture of polyethylene glycols (Walocha *et al.*, 2002), which do not interfere chemically with the casting medium.

After dissection, the surface of the cast should be cleaned to remove possible tissue/reagent debris. This can be achieved by immersing the cast in 5–10% trichloroacetic acid, 2% HCl or 2–3% formic acid, although other cleaning media, such as HCl-collagenase solution or alcohol, have also been used. This step is completed by a wash in tap water followed by distilled water, which removes all salts that might later crystallize on the drying replica.

Casts can be dried in air, although tension exerted on the cast during evaporation of water increases the risk of local deformations (Lametschwandtner *et al.*, 1990); hence a short rinse in ethanol is recommended to minimize that effect. Freeze-drying or, less commonly used, critical point drying are useful options.

The casting media are not electron conductive, hence the casts have to be mounted on a metallic support by a conductive mediator and then metal-coated. Silver tape or adhesive tape coated with silver colloid have been used for mounting, although in the latter case the colloid can be adsorbed by adjacent regions of the cast, hampering SEM examination.

"Conductive bridges" introduced by Lametschwandtner *et al.* (1980) reduce specimen charging during SEM examination and allow a relatively safe remounting of the cast. Such bridges are usually made of fine copper wires attached with silver glue to various regions of the cast on one end and to the support on the other. Their number, depending on the size of the cast, ranges from one for small casts (using a modified support and relatively thick wire, the cast can be rotated to allow SEM examination of its "bottom") to about twenty for very large casts. Recently, Verli *et al.* (2007) used carbon tape to connect the wires to the mounting platform.

The casts are usually coated with gold, often with an underlying layer of carbon, although other metals, such as chromium, platinum, palladium, and osmium have also been used (Belz and Auchterlonie, 1995). Coating can be performed by sputtering or evaporation. Sputtering seems to be safer, since the heat associated with evaporation can damage the cast. If the cast is large, coating should be repeated several times, to allow penetration of the metal to the deepest regions of the specimen. With time, fissures and cracks can appear in the coating layer, so the casts should be examined in the SEM relatively shortly after coating. After longer storage, casts may require recoating.

2.4 Microvascular casts – interpretation of SEM images

The quasi-3-D appearance of cast microvascular systems in SEM facilitates their analysis based on continuity of successive vessel types (arteries – arterioles – capillaries – venules – veins). However, casts do not allow the investigator to identify direction of blood flow, thus discrimination between arteries/arterioles and veins/venules is crucial for interpretation of the vascular bed (Figure 2.2). It was Miodoński *et al.* (1976) who first noticed that nuclear imprints of arteries were different from those of veins. The surfaces of arterial casts exhibit ovoid or fusiform nuclear imprints, with their long axes oriented along the long axis of the vessel. The imprints, quite regularly distributed, appear as sharply demarcated depressions. In the veins, imprints are roundish, shallower, less sharply outlined, and less regularly distributed on the cast surface (Figure 2.3). In capillary casts, the nuclear imprints are less distinct or even absent (the surface can look smooth). The shape of endothelial cells can also be observed, since the intercellular borders are visible as delicate furrows. In the arteries, endothelial cells are elongated and rhomboidal, oriented according to the long axis of the vessel, whereas in veins the cells are more polygonal. Later studies have confirmed these observations (Gnepp and Green, 1979; Nopanitaya *et al.*, 1979). The branching pattern can also be helpful in distinguishing arterioles from venules: an arteriolar branching pattern is mostly symmetrical, with branches showing a similar diameter, whereas a venular branching pattern is in many cases asymmetrical, with the venules receiving tributaries nearly as large as the draining venule or as small as capillaries (Hodde *et al.*, 1977; Christofferson and Nilson, 1988).

Apart from discrimination between arteries/arterioles and veins/venules, the identification of different microvessel types according to classification based on their wall structure – arterioles, terminal arterioles (metaarterioles), capillaries, postcapillary venules, collecting

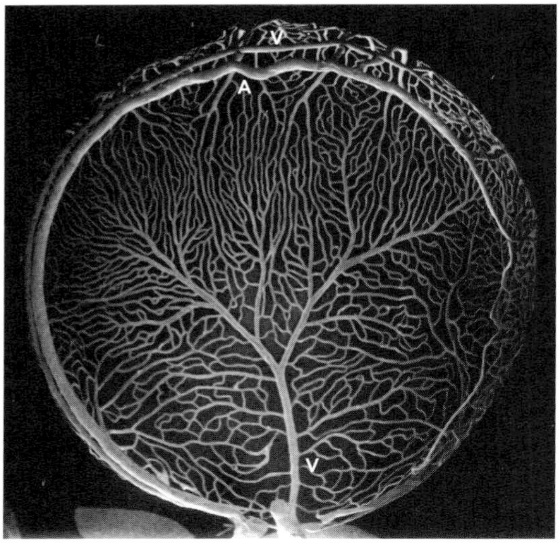

Figure 2.2 An example of a closed microvascular system. Differentiation between artery (A) and vein (V) is necessary to determine direction of blood flow. Reprinted with kind permission from Springer Science+Business Media, from: Miodoński, A.J. and Bär, T. (1987) The superficial vascular hyaloid system in eye of frogs, *Rana temporaria* and *Rana esculenta*. Scanning electron-microscopic study of vascular corrosion casts. *Cell Tissue Res.*, **250**, 465–473.

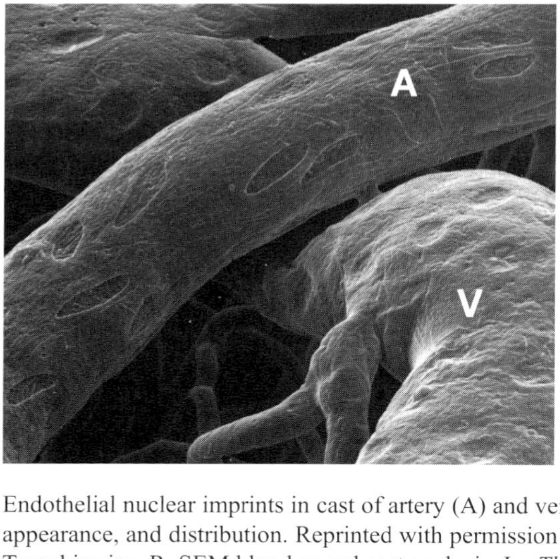

Figure 2.3 Endothelial nuclear imprints in cast of artery (A) and vein (V). Note differences in their shape, appearance, and distribution. Reprinted with permission from Miodoński, A., Kuś, J., and Tyrankiewicz, R. SEM blood vessel cast analysis. In: *Three-dimensional Microanatomy of Cells and Tissue Surfaces* (ed. Allen, D.J., DiDio, L.J.A., and Motta, P.M.), Amsterdam, Elsevier, 1981, pp. 71–87.

venules, muscular venules, and small collecting veins – is practically limited in the casts to estimation of the vessel diameter and should be treated with caution.

The casts also visualize details of luminal topography of blood vessels reflecting some structures associated with the vascular wall. Venous valves appear in the casts as slight

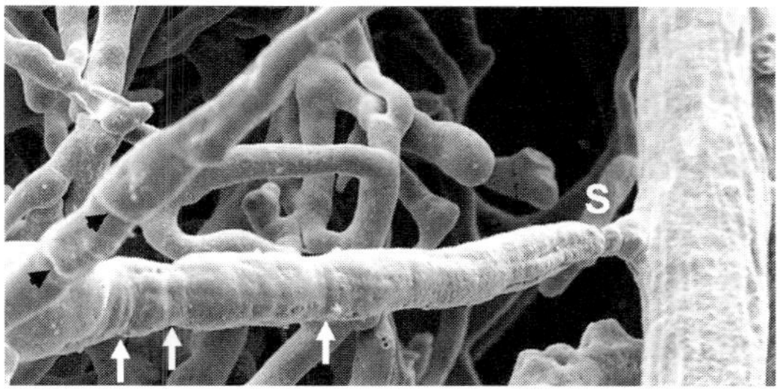

Figure 2.4 Arteriole with a sphincter (S) at the site of its origin and with shallower circular imprints of smooth muscle cells (white arrows). In a nearby capillary, two narrow imprints of pericyte processes can be seen (black arrowheads).

expansions at valve sinuses and deep slits at the sites of valve leaflets. Valvular malformations, e.g. leaflets shared by adjacent valves, can also be easily observed (Hossler and West, 1988).

Continuous capillaries can be distinguished from fenestrated ones: casts of the former are straight and uniform in diameter, whereas casts of the latter are undulating and show eccentric smooth-surfaced dilatations alternating with narrower segments. Although fenestrations themselves are too small to be replicated, comparative SEM and TEM analysis has revealed that the bulging cast segments corresponded to the fenestrated areas of capillaries (Aharinejad and Böck, 1994).

Thin, annular or semilunar grooves observed on the cast surfaces of capillaries and postcapillary venules reflect local constrictions of the capillary/postcapillary wall induced by pericyte processes located on the outer surface of the endothelial lining (Aharinejad and Böck, 1992). Accordingly, wider annular constrictions observed in arteriolar or venular casts correspond to contracted smooth muscle cells, and if the constriction is deep, it suggests the presence of a sphincter (Aharinejad *et al.*, 1992) or intra-arterial cushion (Matsuura and Yamamoto, 1988) (Figure 2.4).

Scanning electron microscopy of corrosion casts can also reveal some structural features of the capillary bed indicative of angiogenesis. Short, blind "capillary sprouts" have been commonly interpreted as marks of ongoing angiogenesis, although incomplete replication should also be taken under consideration in such cases. Small, smooth "holes" ranging from 0.5 to 2 μm in diameter (Figure 2.5) are characteristic of another angiogenic process, the intussusceptive capillary growth (Zagórska-Świeży *et al.*, 2008), characterized by formation of transcapillary tissue pillars (Burri *et al.*, 2004).

2.5 Quantitative analysis of vascular corrosion casts in SEM

An early and simple method of quantification was weighing the cast of a vascular system fully replicated with resin of a known density – it permitted calculation of the vascular bed volume (Weiger *et al.*, 1986).

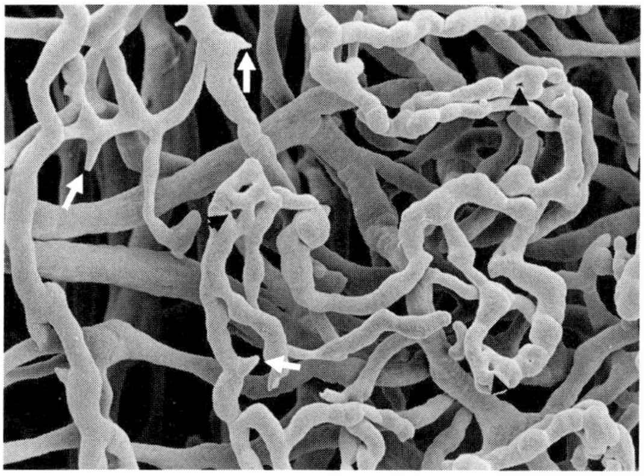

Figure 2.5 Capillary network with features of angiogenesis: blind capillary sprouts (white arrows) and holes indicative of intussusceptive capillary growth (black arrowheads).

Morphometry, in the last decades assisted by computerized image analysis, enables measurements of various vascular system parameters such as interbranch and intervascular distances, branching angles, vessel lengths and diameters – combination of these parameters quantitatively defines three-dimensional (3-D) structure of the vascular bed.

For a long time the measurements had been carried out on two-dimensional SEM micrographs and translation of results into 3-D was problematic. Minnich *et al.* (1999) presented a method for accurate dimensional and angular measurements of miscrostructures inspected in SEM. Their approach included collection of stereopaired digital images from SEM and creation of 3-D image representations, as well as setting of measuring points in the images, computation of precise space coordinates from the corresponding point coordinates, vector equation determination of distances and angles between the consecutive corresponding points, and processing of data by analysis software.

This method was further developed to quantify bifurcation indices, area ratios, and asymmetry ratios, and to test the bifurcations for the optimality principles. The obtained data allowed drawing some functional conclusions concerning bifurcation pattern in the studied vascular system, which favored minimum pumping power and minimum volume rather than minimum surface and minimum drag (Lametschwandtner *et al.*, 2005). Manelli *et al.* (2007) described an automatic 3-D reconstruction of vascular casts directly from stereo-images and developed software that performed morphometric measurements on such 3-D constructs.

Recently, "true" 3-D visualization and quantification of corrosion cast microvascular networks was achieved by using microcomputed tomography (micro-CT) instead of SEM images. The resolution of micro-CT images is inferior to that of SEM, but fully sufficient for studying the spatial distribution of microvessels. This approach made possible a highly precise reconstruction of the microvascular topography and its multiparameter analysis (Heinzer *et al.*, 2006; Atwood *et al.*, 2010).

2.6 Corrosion casting studies of microvascular systems

Studies of vascular systems using corrosion casting and SEM have included three major fields: architecture of normal vascular systems, architecture of developing vascular systems, and vascular systems in experimental and clinical pathology.

An impressive – nearly complete – list of papers published in the first two decades of vascular corrosion casting/SEM research was presented in the fundamental review by Lametschwandtner *et al.* (1990). Since then, hundreds of studies using that technique have appeared in the literature – the interested reader can access them via the Pubmed website.

Normal mature and developing vascular systems have been investigated in a wide spectrum of animal species and in humans. Human developmental studies are relatively rare because of difficulties in collection and processing of fetal material, although Miodoński's group published a series of papers presenting developing vasculature of various organs in the second trimester of fetal life (e.g. Zagórska-Świeży *et al.*, 2008). Corrosion casting has also been employed occasionally to study lymphatic vessels and lymphatic spaces (Castenholz and Castenholz, 1996; Okada *et al.*, 2002; Schraufnagel *et al.*, 2003).

In biomedical research, SEM studies of vascular corrosion casts have included such topics as reconstructive medicine and vascular changes in a variety of pathological processes, with special emphasis on formation and remodeling of microvascular networks in tumors (Skinner *et al.*, 1995).

The technique has provided valuable information concerning vascularization of various types of flaps used in plastic/reconstructive surgery (skin flaps, musculocutaneous perforator flaps, mucoperiosteal flaps, venous flaps) (Bergeron *et al.*, 2006).

Dilatation of capillary plexuses was observed in inflammation (Ravnic *et al.*, 2007). In cerebral ischemia, casting revealed avascular areas, arteriolar vasospasm, interrupted arteriolar branches and thin, "stringy" capillaries (Ohtake *et al.*, 2004). Choroidal and retinal vasculature of hypertensive rats showed tortuosity, irregularity, and narrowing of arteries and capillaries (Bhutto and Amemiya, 2002). In cholesterolemia, alterations observed in retinal vessels included straight, string-like capillaries as well as elongation and straightening of precapillary arterioles (Yamakawa *et al.*, 2001). Numerous anomalies of retinal vessels were found in diabetes: arteries were tortuous, narrow, and irregular in diameter, precapillary arterioles formed hairpin loops, venules were sparse and capillaries showed undulations, loops, and general narrowing with local dilatations described as microaneurysms (Bhutto *et al.*, 2002). Loss of typical capillary pattern and decrease in the number of feeding arterioles in the optic nerve, as well avascular areas in choroid and retina were demonstrated in glaucomatous eyes (Zhao and Cioffi, 2000). In cirrhotic liver, the distance between pre- and post-sinusoidal vessels was reduced and newly formed vessels appeared in the hepatic tissue, some of them connected pre- and post-sinusoidal vessels bypassing the sinusoids, others formed perinodular plexuses (Gaudio *et al.*, 1993). Disarranged microvascular patterns with features of regression and formation of dense plexuses around the cysts were observed in polycystic kidney disease (Wei *et al.*, 2006). In a mouse model of Alzheimer's disease, brain microvessels showed alterations even before the formation of amyloid plaques and at more advanced stages of the disease areas occupied by plaques were avascular (Meyer *et al.*, 2008).

Differentiated microvascular patterns were revealed by corrosion casting and SEM in tumors, both transplanted into animals and surgically removed from patients or collected at autopsy. In the former, a relatively simple angioarchitecture was usually observed, with numerous angiogenic capillary sprouts and variable capillary diameter as well as intervessel and interbranch diameters (Grunt *et al.*, 1986; Konerding *et al.*, 1999; Tsunenari *et al.*, 2002; Sangiorgi *et al.*, 2006). Human tumors (hepatocellular carcinoma, malignant melanoma, colorectal cancer, bladder cancer, renal clear cell carcinoma, larynx cancer, leiomyomas) injected with casting media after resection showed more complex vascular architecture, which was rather specific for the tumor type (Konerding *et al.*, 2001), but shared such common features as increased and/or irregular capillary diameter and signs of active angiogenesis. Some tumors revealed an outer vascular "coat" or "capsule" with very high vessel density and less vascularized central areas (Grunt *et al.*, 1986; Bugajski *et al.*, 1989; Walocha *et al.*, 2003). Unusual capillary patterns – flattened "sheets", multiple, sometimes tortuous loops, glomerular or basket-like arrays (Figure 2.6) – were frequently observed in various tumor types (Miodoński *et al.*, 1980, 1998; Arashiro *et al.*, 1995; Skinner *et al.*, 1995; Arashiro, 2002). Hepatocellular carcinoma nodules were supplied by peripheral vascular plexuses containing branches of both hepatic artery and portal vein and their intrinsic vessels communicated with sinusoids of the surrounding normal hepatic tissue. This multiple blood supply explains survival of cancer cells after arterial embolization (Kita *et al.*, 1991).

In spite of morphological differentiation, quantitative comparison of vascular network parameters in 13 experimental and 3 human tumors of different origin showed a high degree of similarity (Konerding *et al.*, 1995).

2.7 Other applications of corrosion casting/SEM to life sciences

In contrast to innumerable papers devoted to vascular corrosion casts, surprisingly few authors have tried to apply this technique to other systems or organs. The scarcity of such studies has probably resulted from considerable difficulties in injecting "blind" tubular systems such as ducts of glands or respiratory passages. However, some attempts have been successful. Replication of air passages allowed study of the three-dimensional topography of animal airways (Nowell *et al.*, 1970; Nettum, 1996) as well as human respiratory acinus and its development during the fetal period (Dilly, 1984). Corrosion casting combined with SEM revealed the spatial arrangement and microanatomical details of the biliary tree (Gaudio *et al.*, 1988), pancreatic ducts (Ashizawa *et al.*, 1997), and duct system of mammary gland (Schenkman *et al.*, 1985). In the eye, drainage routes for aqueous humor have been replicated (Ujiie and Bill, 1984; Krohn and Bertelsen, 1997). Corrosion casting of a brain ventricle system allowed visualization of not only its spatial topography, but also local variations in the ependymal lining (Liao and Lu, 1993). Injection of kidney with a resin via a ureter produced casts of collecting

Figure 2.6 Extremely dense, sheet-like and glomerular capillary plexuses in exophytic portions of urinary bladder cancer. Reprinted with kind permission from Springer Science+Business Media, from: Miodoński *et al.* (1998).

tubules and Henle's loops of sufficient quality to study their spatial arrangement and dimensions (Magaudda *et al.*, 1990). In the reproductive system, corrosion casting has revealed the successive segments of the efferent ductules (Stoffel *et al.*, 1991) and changes in the patency of the utero-oviductal tract in the course of an estrous cycle (Doboszyńska and Sobotka, 2002). Okada *et al.* (2002) studied macerated compact bone and succeeded in obtaining corrosion casts of not only vascular canals but also minute structures such as bone lacunae and canaliculi, demonstrating age-associated changes in their shape and distribution. Even more impressive results were presented by Meyer (1989), who replicated the insect tracheal system and was able to visualize even the thinnest tracheoles, approx. 70 nm in diameter.

2.8 Concluding remarks

Corrosion casting combined with SEM is now a well-established method used in many areas of life science research, allowing three-dimensional observation of large-scale specimens. Although technically demanding, it does not require sophisticated equipment and can be carried out in any SEM laboratory. Future perspectives seem to be focused on introducing new low-viscosity casting media, which would precisely replicate very small structures present on the luminal surfaces of the cast spaces, such as gold particles used as labels of antibodies or lectins. With the help of high-resolution SEM, this could make possible combination of corrosion casting with immunocytochemistry or lectin histochemistry and broaden the scope of corrosion casting research to studies on distribution of specific molecules (e.g. receptors) on the cast surfaces.

2.9 Acknowledgments

The authors thank J. Urbaniak and K. Zagórska-Świeży for skillful technical assistance.

2.10 References

Aharinejad, S. and Böck, P. (1992). Luminal constrictions on corrosion casts of capillaries and postcapillary venules in rat exocrine pancreas correspond to pericyte processes. *Scanning Microsc.*, **6** (*3*), 877–86.

Aharinejad, S. and Böck, P. (1994). Identification of fenestrated capillary segments in microvascular corrosion casts of the rat exocrine pancreas. *Scanning*, **16**, 209–14.

Aharinejad, S., Böck, P., Lametschwandtner, A., and Firbas, W. (1992). Scanning and transmission electron microscopy of venous sphincters in the rat lung. *Anat. Rec.*, **233**, 555–68.

Aleksandrowicz, R. and Łoziński, J. (1959). Metoda stosowania żywic poliestrowych do sporządzania preparatów i badań anatomicznych [The method of using polyester resin to make preparations and anatomical studies]. *Folia Morphol.*, X.(XVIII), **4**, 445–50 [in Polish].

Amselgruber, W. and König, H. E. (1987). Simplified production of blood vessel corrosion preparations for scanning electron microscopy studies. *Z. Mikrosk. Anat. Forsch.*, **101** (*3*), 523–31.

Arashiro, K. (2002). The tumor vasculature in cutaneous malignant melanoma: scanning electron microscopy of corrosion casts. *Plast. Reconstr. Surg.*, **110**, 717–18.

Arashiro, K., Ohtsuka, H., and Miki, Y. (1995). Three-dimensional architecture of human cutaneous vascular lesions: a scanning electron microscopic study of corrosion casts. *Acta Derm. Venereol.*, **75**, 257–63.

Ashizawa, N., Endoh, H., Hidaka, K., Watanabe, M., and Fukumoto, S. (1997). Three-dimensional structure of the rat pancreatic duct in normal and inflammated pancreas. *Microsc. Res. Tech.*, **37**, 543–56.

Atwood, R. C., Lee, P. D., Konerding, M. A., Rockett, P., and Mitchell, C. A. (2010). Quantisation of microcomputed tomography-imaged ocular microvasculature. *Microcirculation*, **17**, 59–68.

Banya, Y., Ushiki, T., Takagane, H., *et al.* (1989). Two circulatory routes within the human corpus cavernosum penis: a scanning electron microscopic study of corrosion casts. *J. Urol.*, **142**, 879–83.

Belz, G. T. and Auchterlonie, G. J. (1995). An investigation of the use of chromium, platinum and gold coating for scanning electron microscopy of casts of lymphoid tissues. *Micron*, **26**, 141–44.

Bergeron, L., Tang, M., and Morris, S. F. (2006). A review of vascular injection techniques for the study of perforator flaps. *Plast. Reconstr. Surg.*, **117**, 2050–57.

Bhutto, I. A. and Amemiya, T. (2002). Choroidal vasculature changes in spontaneously hypertensive rats – transmission electron microscopy and scanning electron microscopy with casts. *Ophthalmic Res.*, **34**, 54–62.

Bhutto, I. A., Lu, Z. Y., Takami, Y., and Amemiya, T. (2002). Retinal and choroidal vasculature in rats with spontaneous diabetes type 2 treated with the angiotensin-converting enzyme inhibitor cilazapril: corrosion cast and electron-microscopic study. *Ophthalmic Res.*, **34**, 220–31.

Bugajski, A., Nowogrodzka-Zagórska, M., Leńko, J., and Miodoński, A. J. (1989). Angiomorphology of the human renal clear cell carcinoma. A light and scanning electron microscopic study. *Virchows Arch. A. Pathol. Anat. Histopathol.*, **415**, 103–13.

Burri, P. H., Hlushchuk, R., and Djonov, V. (2004). Intussusceptive angiogenesis: its emergence, its characteristics, and its significance. *Dev. Dynam.*, **231**, 474–88.

Castenholz, A. and Castenholz, H. F. (1996). Casting methods of scanning electron microscopy applied to hemal lymph nodes in rats. *Lymphology*, **29**, 95–105.

Christofferson, R. H. and Nilsson, B. O. (1988). Microvascular corrosion casting with analysis in the scanning electron microscope. *Scanning*, **10**, 43–63.

Dilly, S. A. (1984). Scanning electron microscope study of the development of the human respiratory acinus. *Thorax*, **39**, 733–42.

Doboszyńska, T. and Sobotka, A. (2002). The patency of the utero-oviductal tract during the estrous cycle in the pig. *Pol. J. Vet. Sci.*, **5**, 131–37.

Gaudio, E., Pannarale, L., Carpino, F., and Marinozzi, G. (1988). Microcorrosion casting in normal and pathological biliary tree morphology. *Scanning Microsc.*, **2**, 471–75.

Gaudio, E., Pannarale, L., Onori, P., and Riggio, O. (1993). A scanning electron microscopic study of liver microcirculation disarrangement in experimental rat cirrhosis. *Hepatology*, **17**, 477–85.

Gnepp, D. R. and Green, F. H. (1979). Scanning electron microscopy of collecting lymphatic vessels and their comparison to arteries and veins. *Scanning Electron Microsc.*, **3**, 756–62.

Goetzen, B. (1966). The use of latex geon 265 in anatomic studies of blood vessels and ducts. *Folia Morphol. (Warsz.)*, **25** (1), 147–151.

Grunt, T. W., Lametschwandtner, A., Karrer, K., and Staindl, O. (1986). The angioarchitecture of the Lewis lung carcinoma in laboratory mice (a light microscopic and scanning electron microscopic study). *Scanning Electron Microsc.*, **Pt 2**, 557–73.

Heinzer, S., Krucker, T., Stampanoni, M., *et al.* (2006). Hierarchical microimaging for multiscale analysis of large vascular networks. *Neuroimage*, **32**, 626–36.

Hodde, K. C., Miodoński, A., Bakker, C., and Veltman, W. A. M. (1977). SEM of microcorrosion casts with special attention on arteriovenous differences and application to the rat's cochlea. *Scanning Electron Microsc.*, **II**, 477–84.

Hodde, K. C., Steeber, D. A., and Albrecht, R. M. (1990). Advances in corrosion casting methods. *Scanning Electron Microsc.*, **4**, 693–704.

Hossler, F. E. and West, R. F. (1988). Venous valve anatomy and morphometry: studies on the duckling using vascular corrosion casting. *Am. J. Anat.*, **181**, 425–32.

Karaganov, I. L., Mironov, A. A., Mironov, V. A., and Gusev, S. A. (1981). Scanning electron microscopy of corrosion preparations. *Arkh. Anat. Gistol. Embriol.*, **81**, 5–21.

Kita, K., Itoshima, T., and Tsuji, T. (1991). Observation of microvascular casts of human hepatocellular carcinoma by scanning electron microscopy. *Gastroenterol. Jpn.*, **26**, 319–28.

Konerding, M. A., Fait, E., and Gaumann, A. (2001). 3D microvascular architecture of pre-cancerous lesions and invasive carcinomas of the colon. *Br. J. Canc.*, **84**, 1354–62.

Konerding, M. A., Miodoński, A. J., and Lametschwandtner, A. (1995). Microvascular corrosion casting in the study of tumor vascularity: a review. *Scanning Microsc.*, **9**, 1233–44.

Konerding, M. A., Malkusch, W., Klapthor, B., *et al.* (1999). Evidence for characteristic vascular patterns in solid tumours: quantitative studies using corrosion casts. *Br. J. Canc.*, **80**, 724–32.

Krohn, J. and Bertelsen, T. (1997). Corrosion casts of the suprachoroidal space and uveoscleral drainage routes in the human eye. *Acta Ophthalmol. Scand.*, **75**, 32–35.

Krucker, T., Lang, A., and Meyer, E. P. (2006). New polyurethane-based material for vascular corrosion casting with improved physical and imaging characteristics. *Microsc. Res. Tech.*, **69**, 138–47.

Kuś, J. (1969). The history of injection methods in the morphological sciences. *Folia Morphol. (Warsz.)*, **27**, 134–46.

Lametschwandtner, A., Lametschwandtner, U., and Weiger, T. (1990). Scanning electron microscopy of vascular corrosion casts – technique and applications: updated review. *Scanning Microsc.*, **4**, 889–940.

Lametschwandtner, A., Minnich, B., Stöttinger, B., and Krautgartner, W. D. (2005). Analysis of microvascular trees by means of scanning electron microscopy of vascular casts and 3D-morphometry. *Ital. J. Anat. Embryol.*, **110** (*Suppl 1*), 87–95.

Lametschwandtner, A., Miodoński, A., and Simonsberger, P. (1980). On the prevention of specimen charging in scanning electron microscopy of vascular corrosion casts by attaching conductive bridges. *Mikroskopie*, **36**, 270–73.

Liao, K. K. and Lu, K. S. (1993). Cast-model and scanning electron microscopy of the rat brain ventricular system. *Gaoxiong Yi Xue Ke Xue Za Zhi*, **9**, 328–37.

Lieberkühn, J. N. (1748). Sur les moyens propres a découvrir la construction des viscères. *Memoires del Academie Royale des Sciences*, **IV**, Berlin.

Magaudda, L., Cutroneo, G., De Leo, S., *et al.* (1990). Use of synthetic resin casts for the scanning electron microscopic study of the kidney tubule system (in Italian). *Arch. Ital. Anat. Embryol.*, **95**, 87–104.

Manelli, A., Sangiorgi, S., Binaghi, E., and Raspanti, M. (2007). 3D analysis of SEM images of corrosion casting using adaptive stereo matching. *Microsc. Res. Tech.*, **70**, 350–54.

Martin-Orti, R., Stefanov, M., Gaspar, I., Martin, R., and Martin-Alguacil, I. (1999). Effect of anticoagulation and lavage prior to casting of postmortem material with Mercox and Batson 17. *J. Microsc.*, **195** (*Pt 2*), 150–60.

Matsuura, T. and Yamamoto, T. (1988). An electron microscope study of arteriolar branching sites in the normal gastric submucosa of rats and in experimental gastric ulcer. *Virchows Arch. A. Pathol. Anat. Histopathol.*, **413**, 123–31.

Meyer, E. P. (1989). Corrosion casts as a method for investigation of the insect tracheal system. *Cell Tissue Res.*, **256**, 1–6.

Meyer, E. P., Ulmann-Schuler, A., Staufenbiel, M., and Krucker, T. (2008). Altered morphology and 3D architecture of brain vasculature in a mouse model for Alzheimer's disease. *Proc. Nat. Acad. Sci. USA*, **105**, 3587–92.

Minnich, B., Leeb, H., Bernroider, E. W., and Lametschwandtner, A. (1999). Three-dimensional morphometry in scanning electron microscopy: a technique for accurate dimensional and angular measurements of microstructures using stereopaired digitized images and digital image analysis. *J. Microsc.*, **195** (*Pt 1*), 23–33.

Miodoński, A. J., Bugajski, A., Litwin, J. A., and Piasecki, Z. (1998). Vascular architecture of human urinary bladder carcinoma: a SEM study of corrosion casts. *Virchows Arch.*, **433**, 145–51.

Miodoński, A., Hodde, K. C., and Bakker, C. (1976). Rasterelektronenmikroskopie von Plastik-Korrosions-Präparaten: morphologische Unterschiede zwischen Arterien and Venen. *BEDO*, **9**, 435–42.

Miodoński, A., Kuś, J., Olszewski, E., and Tyrankiewicz, R. (1980). Scanning electron microscopic studies on blood vessels in cancer of the larynx. *Arch. Otolaryngol.*, **106**, 321–32.

Murakami, T. (1971). Application of the scanning electron microscope to the study of fine distribution of the blood vessels. *Arch. Histol. Jpn.*, **32**, 445–54.

Murakami, T., Fujita, T., Taguchi, T., Nonaka, Y., and Orita, K. (1992). The blood vascular bed of the human pancreas, with special reference to the insulo-acinar portal system. Scanning electron microscopy of corrosion casts. *Arch. Histol. Cytol.*, **55**, 381–395.

Nettum, J. A. (1996). Combined bronchoalveolar-vascular casting of the canine lung. *Scanning Microsc.*, **10**, 1173–79.

Nopanitaya, W., Aghajanian, J. G., and Gray, L. D. (1979). An improved plastic mixture for corrosion casting of the gastrointestinal microvascular system. *Scanning Electron Microsc.*, **3**, 751–55.

Nowell, J., Pangborn, J., and Tyler, W. S. (1970). Scanning electron microscopy of the avian lung. *Scanning Electron Microsc.*, **1**, 249–256.

Ohtake, M., Morino, S., Kaidoh, T., and Inoué, T. (2004). Three-dimensional structural changes in cerebral microvessels after transient focal cerebral ischemia in rats: scanning electron microscopic study of corrosion casts. *Neuropathology*, **24**, 219–27.

Okada, S., Albrecht, R. M., Aharinejad, S., and Schraufnagel, D. E. (2002). Structural aspects of the lymphocyte traffic in rat submandibular lymph node. *Microsc. and Microanal.*, **8**, 116–33.

Okada, S., Yoshida, S., Ashrafi, S. H., and Schraufnagel, D. E. (2002). The canalicular structure of compact bone in the rat at different ages. *Microsc. Microanal.*, **8**, 104–15.

Ravnic, D. J., Konerding, M. A., Tsuda, A., *et al.* (2007). Structural adaptations in the murine colon microcirculation associated with hapten-induced inflammation. *Gut*, **56**, 518–23.

Ruysch, F. (1725). Opera omnia anatomico-medico-chirurgica. Amsterdam, vol. **1–4**.

Sangiorgi, S., Congiu, T., Manelli, A., Dell'Eva, R., and Noonan, D. M. (2006). The three-dimensional microvascular architecture of the human Kaposi sarcoma implanted in nude mice: a SEM corrosion casting study. *Microvasc. Res.*, **72**, 128–35.

Schenkman, D. I., Berman, D. T., and Albrecht, R. M. (1985). Use of polymer casts or metal particle infusion of ducts to study antigen uptake in the guinea pig mammary gland. *Scanning Electron Microsc.*, **Pt 3**, 1209–14.

Schraufnagel, D. E., Agaram, N. P., Faruqui, A., *et al.* (2003). Pulmonary lymphatics and edema accumulation after brief lung injury. *Am. J. Physiol. Lung Cell Mol. Physiol.*, **284**, L891–897.

Schummer, A. (1935). Ein neues Mittel („Plastoid") und Verfahren zur Herstellung korrosionsanatomischer Präparate. *Anatomischer Anzeiger*, **81**, 177–201.

Skinner, S. A., Frydman, G. M., and O'Brien, P. E. (1995). Microvascular structure of benign and malignant tumors of the colon in humans. *Dig. Dis. Sci.*, **40**, 373–84.

Stoffel, M., Friess, A. E., and Kohler, T. (1991). Efferent ductules of the boar – a morphological study. *Acta Anat. (Basel)*, **142**, 272–80.

Taniguchi, Y., Ohta, Y., and Tajiri, S. (1952). New improved method for injection of acrylic resin. *Okaj. Folia Anat. Jpn.*, **24**, (4), 259–267.

Tsunenari, I., Yamate, J., and Sakuma, S. (2002). Three-dimensional angioarchitecture in transplantable rat fibrosarcomas. *J. Comp. Pathol.*, **126**, 66–70.

Ujiie, K. and Bill, A. (1984). The drainage routes for aqueous humor in monkeys as revealed by scanning electron microscopy of corrosion casts. *Scanning Electron Microsc.*, **Pt 2**, 849–56.

Verli, F. D., Rossi-Schneider, T. R., Schneider, F. L., Yurgel, L. S., and de Souza, M. A. (2007). Vascular corrosion casting technique steps. *Scanning*, **29**, 128–32.

Walocha, J. A., Litwin, J. A., Bereza, T., Klimek-Piotrowska, W., and Miodoński, A. J. (2012). Vascular architecture of human uterine cervix visualized by corrosion casting and scanning electron microscopy. *Human Reproduction*, **27**, 727–732.

Walocha, J. A., Litwin, J. A., and Miodoński, A. J. (2003). Vascular system of intramural leiomyomata revealed by corrosion casting and scanning electron microscopy. *Hum. Reprod.*, **18**, 1088–93.

Walocha, J. A., Miodoński, A. J., Nowogrodzka-Zagórska, M., Kuciel, R., and Gorczyca, J. (2002). Application of a mixture of glycol polyethylenes for the preparation of microcorrosion casts – an observation. *Folia Morphol. (Warsz.)*, **61**, 313–16.

Wei, W., Popov, V., Walocha, J. A., Wen, J., and Bello-Reuss, E. (2006). Evidence of angiogenesis and microvascular regression in autosomal-dominant polycystic kidney disease kidneys: a corrosion cast study. *Kidney Int.*, **70**, 1261–68.

Weiger, T., Lametschwandtner, A., and Stockmayer, P. (1986). Technical parameters of plastics (Mercox CL-2B and various methylmethacrylates) used in scanning electron microscopy of vascular corrosion casts. *Scanning Electron Microsc.*, **Pt 1**, 243–52.

Yamakawa, K., Bhutto, I. A., Lu, Z., Watanabe, Y., and Amemiya, T. (2001). Retinal vascular changes in rats with inherited hypercholesterolemia – corrosion cast demonstration. *Curr. Eye Res.*, **22**, 258–65.

Zagórska-Świeży, K., Litwin, J. A., Gorczyca, J., Pityński, K., and Miodoński, A. J. (2008). The microvascular architecture of the choroid plexus in fetal human brain lateral ventricle: a scanning electron microscopy study of corrosion casts. *J. Anat.*, **213**, 259–65.

Zhao, D. Y. and Cioffi, G. A. (2000). Anterior optic nerve microvascular changes in human glaucomatous optic neuropathy. *Eye (Lond.)*, **14** (*Pt 3B*), 445–49.

3 Revealing the internal structure of cells in three dimensions with scanning electron microscopy

Sol Sepsenwol

3.1 Introduction

Scanning electron microscopy (SEM) is ideally suited to imaging structures with complex topology. The limitation of SEM is that it can only give information about the surface layer of the sample. In biology, many structures of interest are internal, so that the problem is to open up the cell or extracellular structures, preferably without disturbing their normal relationships in the intact cell. With the advent of high-resolution, low-voltage, field emission scanning electron microscopy (HRSEM) and special preparation techniques, it is possible to view both internal and external macromolecular structures in three dimensions, thereby restoring depth information lacking in conventional microscopic studies. HRSEM has important advantages over transmission electron microscopy (TEM) methods used to visualize three-dimensional (3-D) structures at high resolution. TEM is limited by sample thickness ($\leq$ 1.0 micron), even at high accelerating voltages, and by beam damage to delicate structures, such as cytoskeletal fibers, especially when unsupported by plastic embedment. The deep etch freeze–fracture metal replica techniques for TEM (Heuser, 1976; Henser and Salpeter, 1979) work beautifully to circumvent some of these thickness limitations[1] but require, among other things, a cryogenic evaporative metal coating system not available in most EM labs. Cryo-TEM tomography likewise can create detailed views of cells to a depth of a few hundred nm. HRSEM is not so limited and does not require the computational reconstruction of tomography. Newer HRSEM can produce nanometer scale resolution, 3-D images at accelerating voltages below 1 kV (Ris, 1991; Lück *et al.*, 2010). The relatively simple methods described here for direct visualization and immunolabeling of the internal structures of cells or extracellular structures at high resolution in three dimensions should be generally useful in exploring complex internal cell architectures. Most of the techniques below were developed to study intracellular ultrastructure of cells either in monolayers on substrate or in suspension, but can be readily adapted to tissue

[1] See the Heuser Lab website (http://www.heuserlab.wustl.edu) for a dazzling gallery of deep-etch TEM images, many unpublished.

Note: seven of the images in this chapter are presented in two-color anaglyph stereo in the color plate section and should be viewed with red–green decoding glasses (red – left eye). Two-color glasses are readily available from EM supply houses and stereo specialty houses.

Scanning Electron Microscopy for the Life Sciences, edited by H. Schatten. Published by Cambridge samples. These techniques are not restricted to HRSEM, but may also be used with lower resolution, tungsten filament SEMs. They are: (a) detergent extraction, (b) cleaving on scored glass, (c) tape ripping, (d) agar/alginate string fracturing, (e) polylysine wet ripping and immunolabeling, (f) de-embedding of semi-thick plastic sections, (g) sectioned fresh material on a freezing stage, and (h) high-pressure freezing and cryo-SEM.

3.2 Notes on fixation and dehydration of samples

3.2.1 Fixation

Fixation, of course, is critical to reliable preservation of cell ultrastructure and also affects immunolabeling. In the writer's lab, samples are usually fixed in freshly made 1–2% EM grade glutaraldehyde in compatible salt solution containing 50 mM HEPES buffered between pH 6.8 and 7.8. For most mammalian cells, this can be based on a pH 7.2–7.4 HEPES saline buffer with added calcium, magnesium, and other salts. Sometimes 0.025% saponin is added to the fixative to permeabilize cells for more rapid fixation of internal structures. Cells are usually post-fixed with freshly made 0.1% osmium tetroxide in the same buffer to provide rigidity to the structures, as well as an enhanced secondary electron signal. The light osmium post-fixation will not obscure back-scatter detection of colloidal gold immunolabeling, given the newer high-sensitivity BSE detectors. For structures known to be labile during fixation at 0–25 °C, flash freezing and freeze substitution osmium fixation can be used (Buser and Walther, 2008).

3.2.2 Dehydration

Meticulous dehydration is as essential to reliable preservation of internal structures as fixation, since vanishingly small amounts of water can cause adherence between adjacent protein structures, leading to artifacts that may be beautiful, but biologically irrelevant (Small, 1981). Good dehydration can be achieved by attention to the following: (a) ethanol dehydration times should be longer than 10 min per change with continuous, gentle agitation, (b) de-fined alumina dehydrating agents (molecular sieve) should be used in the absolute ethanol stock bottle, (c) exchanges of liquid CO_2 during critical point drying (CPD) should be extensive with agitation, (d) a molecular sieve trap should be used in the liquid CO_2 line, (e) chambers or racks should allow free flow of liquid CO_2 over the sample (Sepsenwol, 2005). In the writer's experience, other solvent dehydration systems such as hexamethyldisilazine (HMDS) or Pel-Dri® do not produce results comparable to CPD for high-resolution work.

3.3 Detergent treatment to expose internal cell structure demembranation during fixation

This technique has been used to investigate internal cytoskeletal architecture in cells attached to substrate by treating the cells with a nonionic detergent (usually

Triton X-100) in a buffer known to stabilize actin and intermediate filaments (Schliwa and van Blerkom, 1981). While detergent treatment has its own peculiar artifacts, one is desirable – the removal of the overlying plasma membrane and most of the cytoplasmic organelles and unstable polymers, which can obscure the overall superstructure of the assembled cytoskeleton and organelles associated with the attached surface of the cell. In so doing, it creates a simplified view of the cytoskeleton, which may be useful for understanding the much more complex, dynamic cytoskeleton *in situ*. The author uses fixative and detergent together, 1% glutaraldehyde containing 0.1% Triton X-100 detergent in a HEPES isotonic saline buffer, followed by buffer rinses and 0.1% osmium post-fixation in the same buffer. Because the amount of disruption of the detergent is sensitive to shear forces, time, and fixative, it is hard to create consistent results, even within a single sample. Detergent fixation with high voltage, 1000 kV, transmission EM (HVEM) was originally used to demonstrate a system of fiber complexes in crawling sperm of the nematode, *Ascaris suum* (Sepsenwol *et al.*, 1989). Figure 3.1 (and color plate) compares the results of detergent treatment as imaged by HVEM and by HRSEM. The HVEM image (Figure 3.1A) shows the

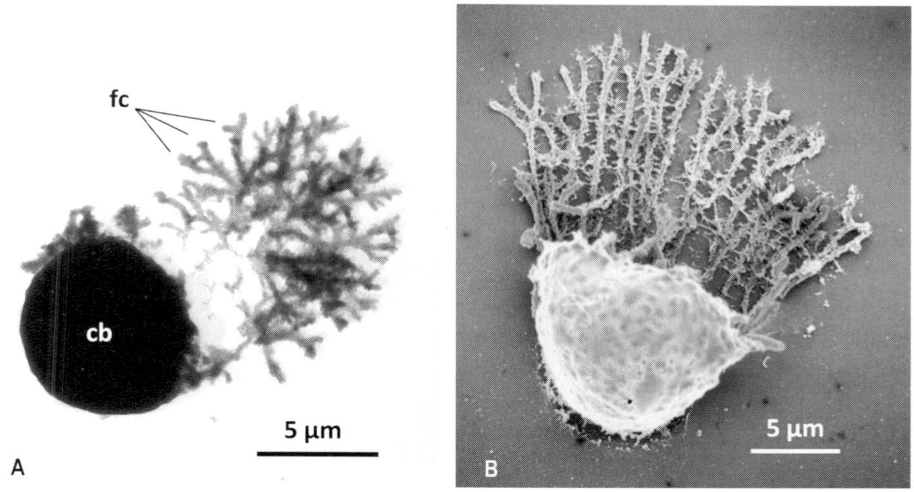

Figure 3.1 Comparison of high-voltage TEM (HVEM) with high-resolution SEM of triton-treated, fixed, crawling sperm of the nematode intestinal parasite, *Ascaris suum*. (A) Activated sperm cell crawling on coated grid, fixed and CPD. The pseudopod has been de-membranated, revealing the "bottlebrush" arrangement of major sperm protein filaments into fiber complexes (*fc*) that represent its machine for crawling. Because of thickness limitations of TEM, most of the cytoskeleton was removed, so that the full array of complexes cannot be seen. Accelerating voltage (Vacc) = 1000 kV; in stereo, left eye red. (B) This cell is from the same HVEM preparation, but Pt-coated and viewed with HRSEM. The branching and composition of the fiber complexes are clearer, as is the loss of all but the adherent branches of the cytoskeleton caused by the detergent treatment. It also demonstrates the integrity of the cell–body plasma membrane, which is resistant to detergent disruption, and some of its unique surface features. Vacc = 1.5 kV; in stereo, left eye red. (See color plate section for two-color stereo 3-D version.)

basic fiber complex arrangement: complexes branching from their origins at the leading edge of the pseudopod. Most of the features of the cell body, about 5 μm thick, are not visible with HVEM. The HRSEM of the same preparation (Figure 3.1B) shows not only the loss of pseudopod components, including parts of the cytoskeleton itself, but also details of the cell body and its unique membrane. While detergent extraction may be useful for simplifying the basic structures left behind, inevitably other methods are necessary that better preserve *in vivo* architecture and its special relationships to other parts of the cell.

3.4 Cleaved glass fracturing

This simple method provides a side view of attached cells and is useful for studying cell-to-substrate or cell-to-cell contacts. Occasionally, it can provide sections through cells to show internal components associated with attachment. Cells are cultured on finely scored glass, the glass is then broken along the scores and mounted edge on for SEM observation. Specifically, acid cleaned #2 thickness coverslips are lightly scored with a diamond scribe into 1 mm lanes, sterilized, placed in a culture dish, overlaid with cell suspension, and cultured. Slips with attached cells are glutaraldehyde- and osmium-fixed, CPD, then cleaved into 1 mm strips with fine forceps. The strips are mounted on edge to double-sided conductive tape on a conductive base and sputter coated with Pt (see inset to Figure 3.2). This produces useful side views of the cells and their special types of contact with the substrates. When cells attach across score lines, cleaving creates a section through the cell that allows a detailed view of

Figure 3.2 Cleaving attached cells on glass. *Ascaris* sperm crawling on scored glass are fixed, CPD, Pt-coated, and mounted edge on. Crawling *Ascaris* sperm, shown in side view 15 min after activation, is fully attached to glass. The individual points of attachment, the villipodia, spread out along the bottom to form a continuous sheet of contact. This cell has been cleaved along the glass score, showing that the fibrous cytoskeleton of the pseudopod underlies the cell body so that the cell body rides above the pseudopod, rather than behind it. The cell in the background is forming a pseudopod and has not yet attached to the glass. Vacc = 1.5 kV. (Inset) A 1 mm coverslip strip mounted on edge on a silicon chip with conductive tape.

substrate attachment and cytoskeletal elements associated with it (see Figure 3.2). It is also possible to fix cell suspensions in glutaraldehyde and attach them to poly-D-lysine coated, scored coverslips for the same effect using the "wet-ripping" method described below.

3.5 Tape-ripping of fixed, dried cells

Tape-ripping is an old technique originally used by botanists to create peels of fresh or fossilized leaf epidermis for light microscopic observation. It works well to generate many internal views of cells attached to a substrate, often including whole, rather than sectioned, structures. Cells attached to glass are glutaraldehyde fixed, post-fixed in osmium, then CPD. A thin strip of adhesive is applied to the edge of a #2 coverslip from a nonconductive adhesive tab cut in half to provide a straight, thin edge of adhesive (Figure 3.3A (and color plate)). Under a dissecting microscope, the adhesive edge of the coverslip (the "ripper") is laid over an area of attached cells. Cells are ripped open by gently pressing over the adhesive strip with fine forceps until tiny areas of contact can be seen (Figure 3.3B). When the adhesive slip is lifted, one has created complementary samples, the "ripped" and the "ripper" slips (Figure 3.3C). The process may be repeated using the ripper slip as the sample and another tape strip to produce a "double-ripped" sample. Both glass and adhesive strips are mounted on conductive tape, Pt sputter-coated, and examined by SEM. Figures 3.3D and E show tops of cells adhering to the ripper slip. They show the dynamic region between cell body and pseudopod and unique cytoskeletal associations with the pseudopod membrane, the villipodia. This simple method may also be used to rip open fixed, dried, firmly attached samples already mounted on stubs.

3.6 Wet-ripping fixed cells for immunolabeling

The wet-ripping technique uses light fixation and a method of opening the cells without destroying immunoreactivity. Adherent cells on glass are first fixed, then torn open with a PDL-coated coverslip. Fixation before ripping preserves *in situ* relationships between plasma membrane, cytoskeleton, and organelles. In more detail, cells attached to the substrate are fixed as before, but with as low as possible a concentration of glutaraldehyde to preserve ultrastructure, but prevent denaturation of potential antigens. In the author's experience, 0.5–1.0% glutaraldehyde in buffer for 45 min at room temperature (RT) has been sufficient, but this requires some experimentation. The coverslip is washed in buffer and prepared for ripping.

Preparation of PDL coverslips: a cleaned coverslip was incubated 20 min at RT in a 1% poly-D-lysine solution (PDL) in water to create a sticky surface. To rip the cells, the edge of a PDL-coated slip was placed over the attached cells, gently weighted for about 20 min, then lifted off the sample. As with the tape-ripping technique, the samples contain complementary ripped and undisturbed cells. Figure 3.4A (and color plate) shows the tops of a number of ripped cells on the PDL ripper slip.

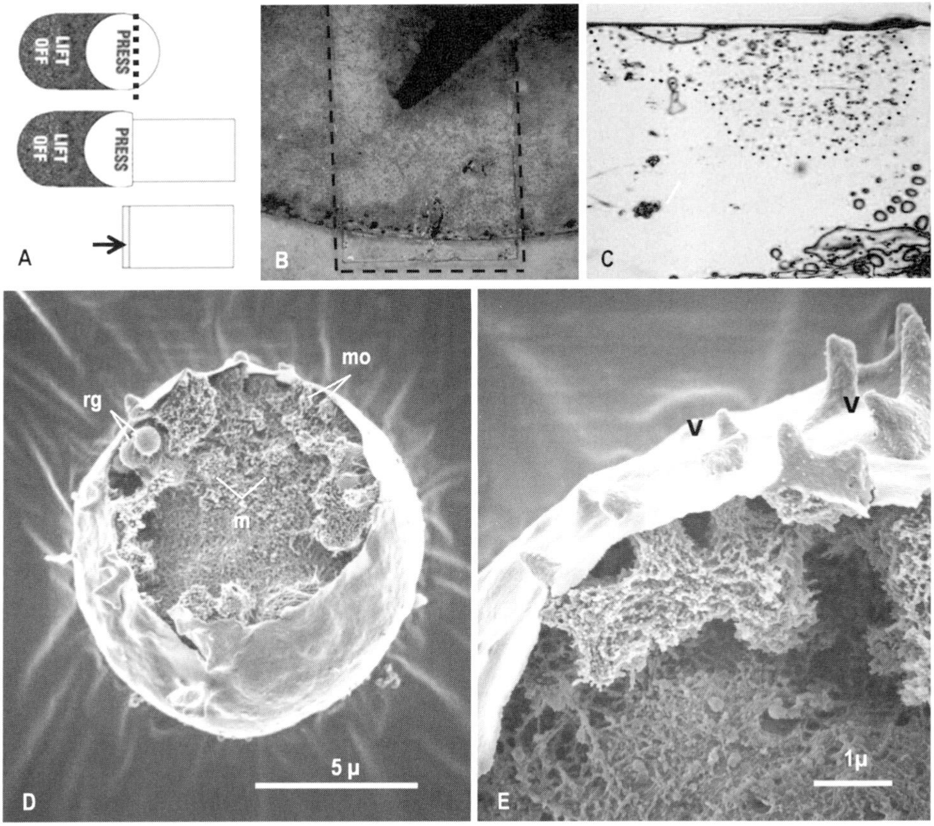

Figure 3.3 Tape-ripping. Adherent *Ascaris* sperm are fixed, CPD, then torn open with double-stick tape and both surfaces Pt-coated. (A) Thin strip of adhesive from an adhesive mounting tab is applied to the end of a cut coverslip. (B) Small area of adhesive strip (dotted outline) is gently pressed against cells on glass with forceps. (C) After adhesive is pulled away, numerous cell fragments are seen on ripper strip adhesive, shown in D and E. (D) *Ascaris* sperm cell body where it joins the pseudopod. The smaller corpuscles at the interface are mitochondria, the larger spherical particles are refringent storage granules. (E) Showing the relationship between the external villipodia (*v*) which adhere to the substrate and the cytoskeleton that forms them. Vacc = 1.5 kV; in stereo, left eye red. (See color plate section for two-color stereo 3-D version.)

Immunolabeling: the wet-ripped cells are immunolabeled with antibody gold by incubating 60 min with a blocking buffer consisting of 1% BSA + 0.1% Tween-20 detergent in HEPES saline buffer, then 60 min with primary antibody in the same buffer, followed by washing with blocking buffer. Second antibody of anti-species – Ig 10–20 nm gold – labeled antibody follows the same procedure as for primary labeling, followed by extensive washing with buffer. Cells are post-fixed with 0.1% osmium tetroxide in buffer, dehydrated, and CPD. It is also possible to double-label by repeating the procedure with a second primary antibody followed by a different sized anti-Ig colloidal gold label. Complementary slips are mounted on conductive tape and sputter-coated with Pt. By

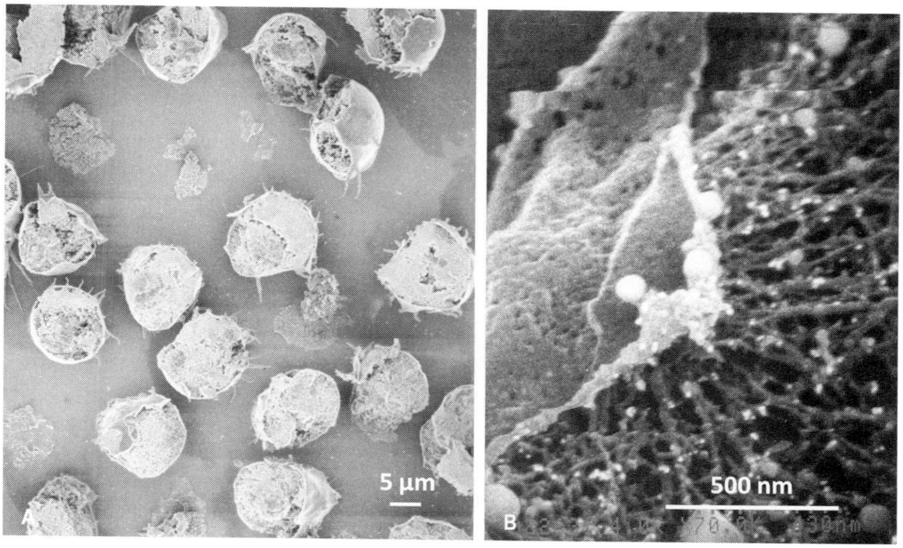

Figure 3.4 Wet-ripping and immunolabeling. *Ascaris* sperm cells attached to glass are lightly fixed, then ripped open by adhesion to an overlying coverslip treated with poly-D-lysine (PDL). The ripped cells were rinsed in blocking buffer, and incubated with primary antibody against nematode *major sperm protein*, the major constituent of the fibrous cytoskeleton of the pseudopodium. Cells were then incubated with anti-species Ig adsorbed to 10 nm colloidal gold. (A) A view of the PDL "ripper" slip showing the bases of numerous cells which have been torn away from the substrate, with many views of pseudopod cytoskeleton and cell body. Vacc = 1.5 kV. (B) Anti-MSP-gold labeling of the fiber complexes associated with pseudopod membrane, imaged with the back-scatter electron detector. Bright dots are colloidal gold. Note that the label is localized to filaments and the interior of the pseudopod membrane. Vacc = 4 kV, back-scatter mode; in stereo, left eye red. (See color plate section for two-color stereo 3-D version.)

using a high-resolution YAG-type back-scatter electron detector (Autrata *et al.*, 1991) at modest accelerating voltages, it is possible simultaneously to image both gold label and the sample ultrastructure (Figure 3.4B). This method, in combination with stereo presentation, can provide much more information about the 3-D distribution of an antigen than can be provided from sectioned material or light microscopy. The wet-ripping method can be modified for cells or tissues not already attached to a substrate by transferring cells or tissues still in glutaraldehyde fixative to PDL-coated coverslips, and incubating at least 60 min to cross-link free aldhehydes of fixative with those of the polylysine. The PDL "ripper" slip is then applied as described.

3.7 Agar – or alginate – string fracturing of embedded cells

This technique is most useful for splitting open unattached cells or particle suspensions. Fixed cells are embedded in a gel, the gel suspension dehydrated to ethanol, frozen in ethanol, then fractured with a blade to cleave the embedded cells. It takes advantage of the depth of field of the SEM to generate beautiful, highly detailed cross-sections of the suspended cells, while still displaying many of the 3-D features of the embedded sample, something not possible with transmission microscope imaging (Figure 3.5 (and color plate)).

Making agar strings: cells are fixed in glutaraldehyde and post-fixed in osmium as described. While still in the distilled water rinse following osmium post-fixation, the cells

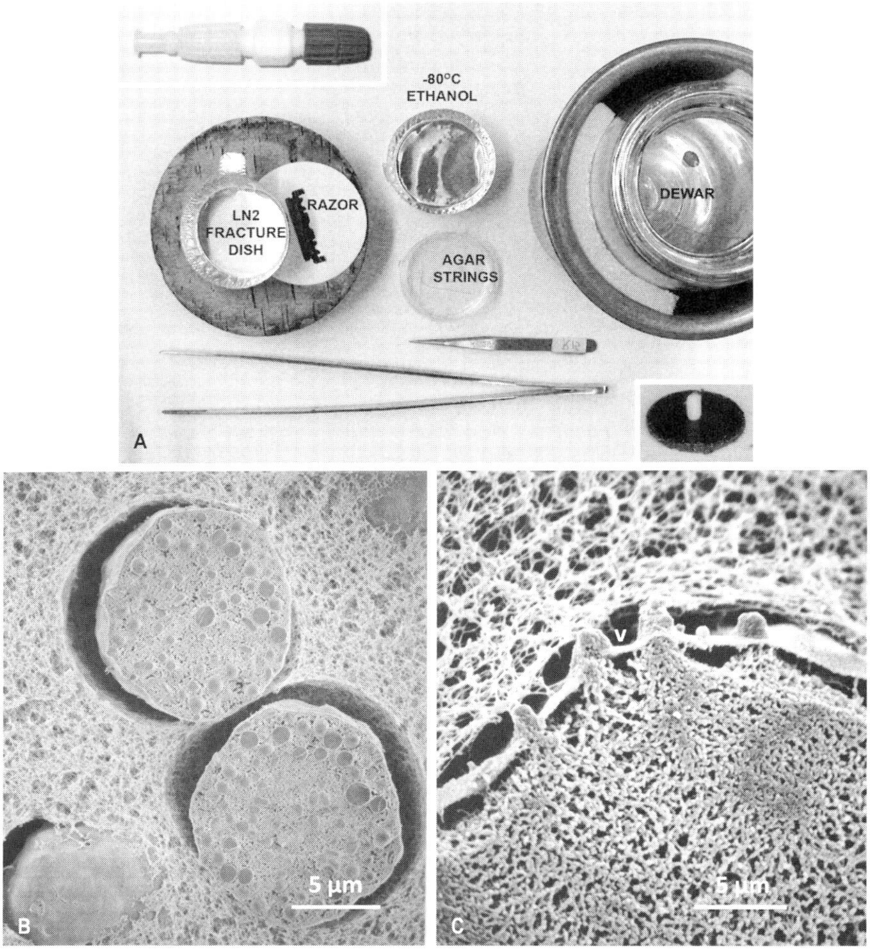

Figure 3.5 Fractured agar strings. *Ascaris* sperm are fixed, rinsed in buffer and embedded in liquefied agar, chilled, extruded into strings, dehydrated to absolute ethanol, frozen in liquid nitrogen and fractured, CPD, then mounted on conductive tape, fractured-end-up, presenting numerous three-dimensional sections. (A) Setup for fracturing agar strings. Upper inset: capillary tube ejector. Lower inset: a fractured stub mounted on a conductive adhesive disc for use in an in-lens HRSEM. (B) Two sections through the cell bodies of sperm cells. Agar strands retract with drying, leaving a view of the cell's exterior. Vacc = 1.5 kV; (C) A fracture section through a pseudopod showing some of the fine structure of the cytoskeleton's association with the pseudopod membrane, the villipodia (*v*). Vacc = 1.5 kV; in stereo, left eye red. Compare to tape-ripped villipodia in Figure 3.3B. (See color plate section for two-color stereo 3-D version.)

are resuspended with gentle pipetting in 50–100 μl 1.5% purified agar in water at 50 °C. The suspension is drawn into an uncoated 50 μl capillary tube using a Bio-Rad mini tube-gel ejector fitted to a 1 ml syringe (Figure 3.5A, upper inset). The agar-filled capillary tube is then gelled on ice for 20 min. The agar gel is extruded as strings from the capillary into water or buffer using the Bio-Rad adapter and syringe.

Making alginate strings: Alginic acid may be used in place of agar and has several advantages. It gels instantly in the presence of calcium ion at room temperature and so avoids exposing cells to elevated temperature (Ogneva and Neronov, 1995). Alginate forms a more robust, stable gel around the sample than agar which may be needed to fracture small, tough cells or particles like bacteria. A 1% solution of alginic acid made in incubation buffer is used to suspend cells, which are drawn into a glass capillary as in the agar technique above. A string is formed by gently ejecting the alginate suspension into phosphate-free buffer containing 75 mM $CaCl_2$, while pulling the capillary away from the extruding string. The gelled alginate strings are stable in buffer indefinitely. Dehydration times for alginate gels are at least twice those for agar (see below).

Freeze – fracture of agar or alginate strings: Figure 3.5A shows the setup for fracturing strings, consisting of

1 wide mouth 1 l Dewar, half-filled with LN2
1 cork or polystyrene foam base
1 metal block, cooled to −80 °C, or slab of dry ice
several aluminum weighing dishes
1 long tongs
1 med forceps
several double-edge, cleaned razor blades, broken in half.

Agar or alginate strings are dehydrated in graded ethanols to 100%. Ten min dehydration steps are adequate for agar strings, 20 min steps or longer are required for alginate strings. Strings in 100% ethanol in an aluminum weighing dish are chilled 5 min on the –80 °C metal block or on a block of dry ice. Chilled strings are transferred from ethanol to a weighing dish half-filled with liquid nitrogen and allowed to freeze (a few seconds).

Another weighing dish is used with the tongs as a dipper to replenish the LN2 in the fracturing dish. The razor and tips of the medium forceps are chilled in LN2. The razor, held in the forceps, is positioned over the frozen string and pressed with a finger to fracture the string into 2 mm segments. Unlike water, ethanol does not form sharp crystals when it freezes, and so does not damage the cells. The original ends are discarded so that all ends are fresh-fractured. Fractured segments are transferred to –80 °C ethanol, then allowed to warm to RT. Once warmed, segments are transferred to two changes of fresh, desiccated 100% ethanol and CPD. Dried segments are cut in half and each piece mounted fractured end up on conductive carbon tape and Pt sputter-coated (Figure 3.5A, lower inset). The gel matrix pulls away from the cells/particles during the dehydration so that the outside of the cell and its section are visible at the same time, allowing one to determine more accurately where in the cell the fracture occurred (Figure 3.5B). Figure 3.5C shows not only the high level of detail that is preserved with this technique, but also the valuable 3-D relationships of the cytoskeletal fibers that form the villipodia, which attach the pseudopod for crawling.

3.8 De-embedding of semi-thick plastic sections

This technique is quite versatile in that it can provide cross-sections of cells, tissues or extracellular structures while preserving some of the three-dimensional information of the sample (Figure 3.6 (and color plate)). There are two parts to the technique: (a) preparation of glass substrate to create a strongly adherent surface for the tissue – usually already osmium-fixed – in the plastic sections, and (b) removal of the plastic from the sections. Here follows a summary of the technique used by the author, based on the technique of Ris and Malecki (1993), which was modified from the procedure of Iwadare *et al.* (1990):

Preparation of coverslips: coverslips (22 × 50 mm, #2 thickness) are cleaned with a mild abrasive powder like Bon Ami® until water sheets evenly over the glass. Slips are rinsed well with distilled water and cut into 5 × 11 mm strips. Glass strips are attached to a strip of Post-It® adhesive for ease of handling, glow-discharged for about 3 min (or processed in a sputter coater using the etch function), then carbon-coated to create a light-brown layer. The strips are then treated with molecular-sieve-dried absolute acetone for 2 hr, dried in a 200 °C oven for 1 hr, and allowed to cool. Strips are derivatized by placing them in separate 1.5 ml microcentrifuge vials, covering with a 4% solution of 3-aminopropyltriethoxysilane in dried absolute acetone after the method of Buechi and Bachi (1979), then incubated overnight at 50 °C. Silane-acetone is aspirated and replaced with dried, absolute acetone. Acetone is replaced with freshly made 1% glutaraldehyde in distilled water and incubated 1 hr at 4 °C. Glutaraldehyde is aspirated and replaced with distilled water. Treated glass strips may be stored at 4 °C for up to one week.

Embedding and thick-sectioning of samples: specimens are fixed and embedded according to standard EM procedures, using an epoxy embedding medium which preserves dimensional stability. These include Epon 812®, Araldite 502®, Eponate 12®, Poly/Bed 812® or mixtures thereof. A ribbon of semi-thick sections, 200 to 300 nm (blue to green interference colors), is floated onto water and expanded until flat with xylene vapor or heat. A silane-treated glass strip, carbon side up, is slipped under the ribbon of sections and slowly withdrawn at a shallow angle to the surface, while the ribbon is positioned longitudinally at one end with a hairbrush.[2] The sections are allowed to air dry on the strips, and their positions marked with a diamond scribe.

Plastic extraction: 5 ml of a crown ether-dimethylsulfoxide mixture is made up fresh. Sections on strips are placed in separate 1.5 ml microcentrifuge tubes. At 1 min intervals, 750 μl of extraction mix is added to each tube, which is inverted four times initially and at 1 min intervals for 6 min. At 6 min, slips are transferred to distilled water in new microcentrifuge vials and inverted twice to inactivate the extraction medium. Sections on strips in individual drying chambers are dehydrated to absolute ethanol in 10 min steps,

[2] Usually made with an eyelash hair glued with nail polish to the end of a tapered wooden applicator stick, a hairbrush is best made with Dalmatian dog hair.

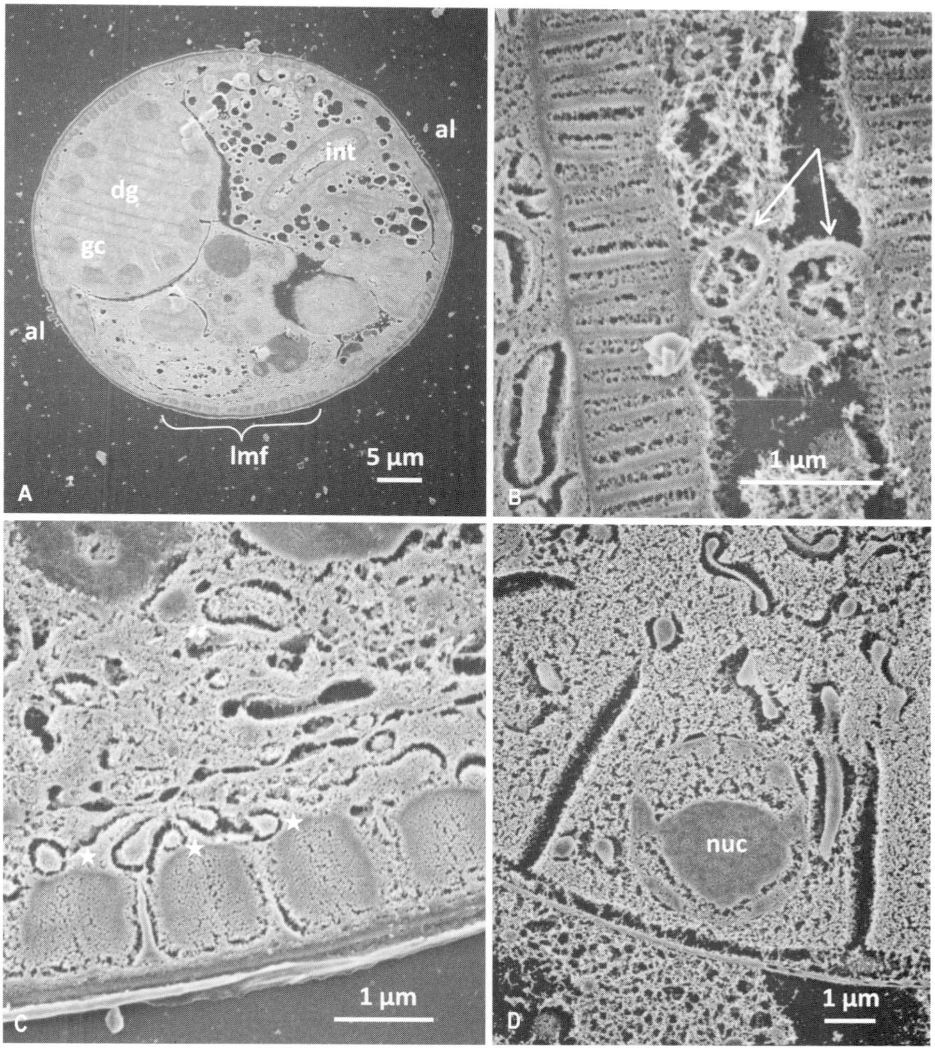

Figure 3.6 De-embedding of plastic sections. Hermaphrodite *Caenorhabditis elegans* were fixed by freeze-substitution and embedded in an Epon–Araldite mixture. Blocks were cut in 300–500 nm thick sections, which were transferred to silanized glass strips. Plastic was removed as described and the strip Pt-coated. (A) Low-power cross section through distal third of worm, showing features comparable to TEM micrographs of *C. elegans*. Major features: intestine (*int*), ventral longitudinal muscle fibers (*lmf*), alae (*al*) extensions from cuticle, distal gonad (*dg*) of syncitial germ cells (*gc*). (B) An enlarged view of the intestinal wall and lumen, showing the carcasses of two *E. coli* (arrows). (C) Enlarged view of ventral longitudinal muscle cells and muscle arms (*stars*) directed toward their synapse with the ventral nerve cord. (D) Section of a germ cell in the distal gonad. While limited by section thickness, stereo enhances some of the tubular elements that join the cells with the center of the gonad, the rachis. Nucleus, *nuc*. calibration bar = 1 µm. In stereo, left eye red. All micrographs, Vacc = 1.5 kV. (See color plate section for two-color stereo 3-D version.)

followed by CPD as usual and Pt sputter coating. Using HRSEM, one can generate excellent images of a great variety of sections of cells or tissue (Figures 3.6A–C) that compare well with TEM, with the added advantage of being able to recreate depth information missing in other sectioned material (Figure 3.6D (and color plate)). This is especially valuable in capturing the three-dimensional structures of submicron specimens that are too large or too rare for thin sections and TEM.

3.9 Use of an SEM freezing stage for examining internal structure in unfixed, wet samples

Most SEMs come with the option of sample chambers that can operate in a variable pressure or environmental mode, that is, at a relatively low vacuum for observing undried samples. For very wet samples, whose internal structures we may wish to image, water vapor pressure may exceed the maximum chamber pressure for imaging. Freezing the sample can reduce this. Relatively inexpensive after-market, Peltier-cooled freezing stages are available for the SEM and can be valuable adjuncts for examination of internal structures of fresh, unfixed, undried material at magnifications under 5000×. It works especially well with plant material. Most cooling stages are water-cooled and can reach −25 to −50 °C (see Figure 3.7A). Because freezing on the stage is relatively slow (seconds), there is some ultrastructural damage due to water ice-crystal formation, but the results can be nonetheless valuable. Figures 3.7B and C show a composite image and detail of a leaf of the aquatic plant, *Elodea canadensis*, cut with a double-edge razor, then mounted cut side up and frozen in place on the freezing stage. Because of the relatively fast sublimation of water from the frozen sample in the SEM at the modestly low temperatures of the stage, working time is limited to 20–40 min before cells collapse. For stage freezing samples, 0.5 M sucrose is an effective, slow sublimating freezing medium. Samples can be manipulated under a dissecting microscope if they are first mounted on 10–15 mm discs punched from disposable aluminum weighing pans and the discs mounted to the freezing stage with a drop of 0.5 M sucrose. To avoid the ultrastructural damage inevitable with slow freezing, one must use the much more expensive alternatives of rapid freezing and cryo-SEM described below.

3.10 High-pressure freezing and preparation for cryo-SEM

The following techniques employ specialized instrumentation to image the interior of cells in the unfixed, frozen state and originated with methods developed by Heuser and others for TEM. Other methods of freezing and techniques peculiar to HRSEM are presented in great detail by Walther (2007) and are only presented in outline form here. The advantages of directly observing frozen samples with HRSEM are that it presents the cell's 3-D ultrastructure undisturbed by fixation and without the need to create a shadowed metal surface replica, as with the deep etch technique for TEM. The disadvantage is that the sample is destroyed once allowed to thaw. While expensive,

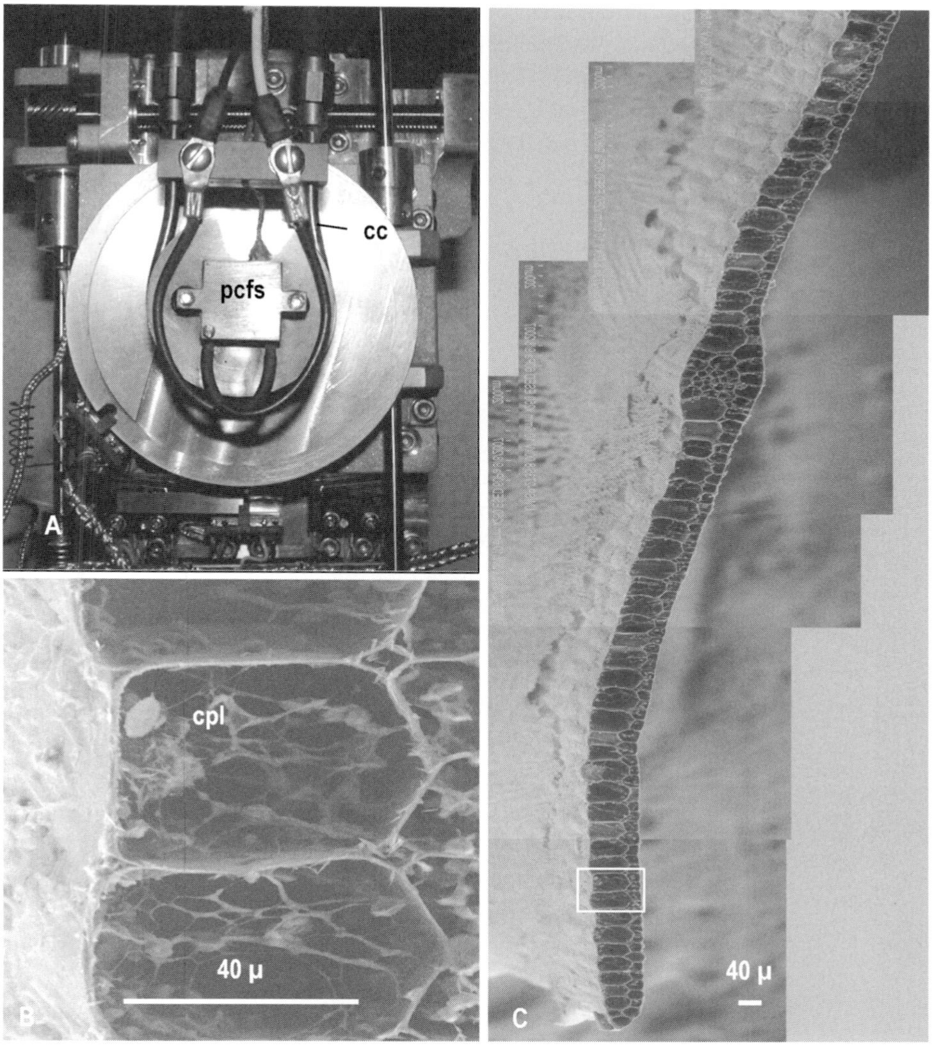

Figure 3.7 Examining fresh, sectioned, frozen material with a Peltier-cooled freezing stage. (A) An EF Fullam Peltier-cooled stage (*pcfs*, Ted Pella, Inc., Redding, CA) mounted in a Hitachi S–3400 variable pressure SEM. The cooling coil (*cc*) receives 10 °C water from an external chiller-circulator. Since the freezing stage is fixed, samples are prepared on thin aluminum discs that are attached to the stage with a drop of freezing medium. (B) Enlarged portion of (C), razor-sectioned, frozen *Elodea Canadensis* showing cell wall and a number of intact chloroplasts (*cpl*) in the parenchymal cells. (C) Composite micrograph of the edge-on view of a 1 mm wide strip of cross-sectioned *Elodea* leaf, with upper-surface parenchymal cells to the left. It is possible to see the frozen 0.5 M sucrose covering the lower portion of the leaf strip. Vacc = 8 kV; environmental secondary detector; chamber pressure = 60 Pa.

cryo-SEM is an excellent check on how well fixation and dehydration steps used in other procedures preserve ultrastructure; it is not uncommon to see labile structures in the frozen state not seen in fixed preparations. Figures 3.8A and B (and color plate) are examples of flash-frozen, unfixed fresh bull sperm briefly treated with Triton X-100. The method follows that developed by Walther *et al.* (1992): live cells are transferred to a gold planchette, covered with a drop of hexadecane, transferred to a Balzers high-pressure LN2 freezing system and flash-frozen in liquid nitrogen under *c.* 5.5 MPa pressure. The frozen cells in the planchette are transferred to an LN2-cooled specimen holder, fractured under LN2 using LN2-cooled pointed forceps, and inserted into a cryo-freeze-etch, sputter-coating apparatus. Frozen samples are allowed to sublimate about 10 min at −115 °C, are coated with a thin layer of Pt, and the LN2-cooled sample holder is transferred to the SEM. Specimen contamination is reduced with an LN2 cold finger above the holder. The slight shear forces of pipetting have dislodged some tails from heads so that we can see both intact sperm (Figure 3.8A) and exposed internal structures (Figure 3.8B) in the same preparation. In addition to features that can be seen in conventionally fixed sperm, such as the basal plate, connecting piece, and oblique proximal centriole, cryo-SEM reveals nanometer scale surface features of the outer membrane of the head in the frozen state that are not seen in conventionally fixed sperm cells (Figure 3.8B) – all of which can be presented in 3-D stereo.

3.11 Use of stereo 3-D presentation for SEM images

The ability to create true stereo 3-D images for the display of ultrastructure is one of the most powerful tools available to SEM users. Often the term "three-dimensional" refers to a 2-D representation of the object with shading and surface features that imply depth. The 2-D SEM image can be very satisfying in that it does not require mental reconstruction of serial sections to imagine the overall surface topology of the sample. On the other hand, when a true stereo 3-D image is recreated in the visual system, one comprehends the range of information that cannot be inferred from 2-D representation. Because the SEM can so easily create a 3-D image by tilting, there really is no reason to compromise the specimen's depth information. Inexpensive digital storage, easy-to-use stereo software and widely available presentation technologies make the use of stereo 3-D an essential part of scanning electron microscopy. In practice, one makes a tilt–stereo pair, usually the first perpendicular to the beam (0° tilt) and the second, 6–10° tilted to the first, with the centers of the two images being coincident. Being able to display the first stored image while positioning the live image makes centering images simple. A variety of software packages can be used to align the two digital images exactly and assemble them into stereo format. The simplest stereo format is side-by-side, cross-eyed stereo, which requires no special viewing tools, but practiced eyes. Another format uses alternate raster lines for left and right video images (interlacing), the left and right images being alternately projected to each eye with synchronized shutter glasses. Monitors with twice the usual pixel density for this type of 3-D

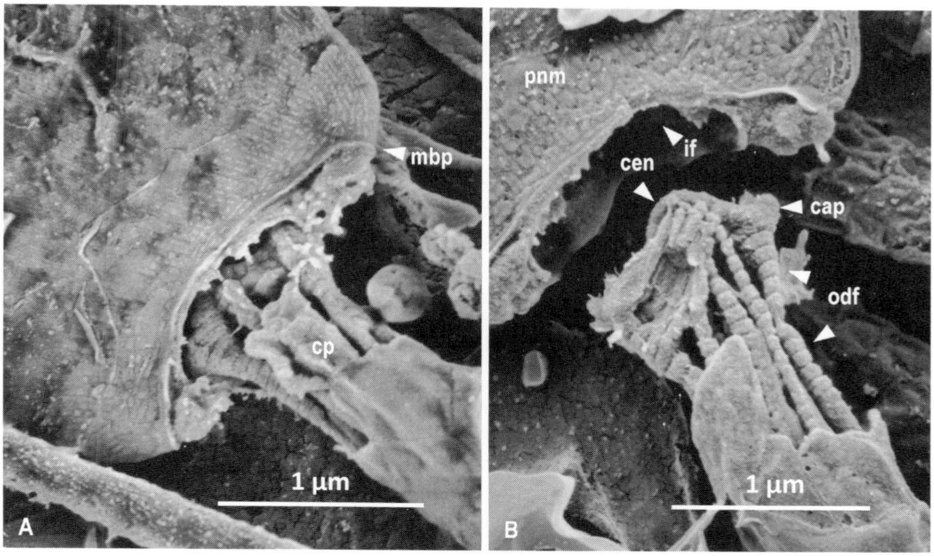

Figure 3.8 Cryo-SEM of de-membranated bull sperm. Fresh bull sperm (American Breeders Service, DeForest, WI, USA) was briefly incubated with Triton X-100 detergent to expose sperm tail elements without damage to other structures, then flash-frozen under high-pressure LN2, etched to expose cells in ice, coated while frozen with Pt, and observed frozen with an LN2-cooled specimen holder. The membrane of the neck area containing the connecting piece appears to be most vulnerable to the brief detergent treatment. (A) The intact connecting piece (*cp*) is seen in its normal relationship to the margin of the basal plate (*mbp*), with centriole and outer dense fibers just visible inside. (B) The components of the connecting piece have been pulled away from the implantation fossa (*if*), showing the proximal centriole (*cen*), at an angle to the axis of the tail, the striated outer dense fibers (*odf*) and capitulum (*cap*). The proximal centriole nucleates the first mitotic spindle in the zygote in some species. The striking embossing on the post-nuclear outer membrane (*pnm*) is not seen in chemically fixed sperm and corresponds to hexagonally packed surface glycoproteins known to reside there. Vacc = 1.5 kV; in stereo, left eye red. (See color plate section for two-color stereo 3-D version.)

presentation without loss of detail are now available. A third more elaborate format, similar to commercial 3-D movie making technology, employs polarization to separate and decode images. It requires two aligned video projectors, whose right–left images are plane-polarized 90° to each other with filters, then superimposed and projected onto a special lenticular screen that does not interfere with the polarization. The left–right images are decoded with polarizer glasses. The limitations of each of these formats based on equipment requirements or size of audience are obvious. The author's laboratory uses the simplest and most versatile method of presenting stereo 3-D, the two-color anaglyph, in which the right and left eye images are color coded, superimposed digitally and decoded with two-color glasses. Since SEM images are grayscale, color coding does not create color ambiguities. The most popular two-color formats are red–blue, red–cyan, and red–green, with red representing the left eye view.

Red–cyan provides the highest image contrast and red–green glasses produce the brightest stereo experience whether used with red–blue, red–cyan, or red–green anaglyphs. Convenience is high and costs are low. Pasteboard color decoding glasses are inexpensive, take up little space (about 5 cm high for a stack of 100 glasses), and they are commonly available at EM and specialty suppliers. The color anaglyph can be presented to a few people as a color print or digital image on a monitor. For a large audience, it can be viewed by video projection on a normal, bright, glass-beaded screen in a dimmed room. In this author's opinion, the stereo processing software *3D Stereo Image Factory Plus*©[3] is the best found to date for creating anaglyph stereo from SEM image files of up to 10 MB each in TIF, JPG, or BMP formats. The package has convenient alignment and centering tools and options for anaglyph color adjustment and for saving stereo pairs in other stereo formats (including interlaced format). Although no longer supported by its author, *3D Stereo Image Factory Plus* will run on all Microsoft PC operating systems after 1999, including Windows 7®.

3.12 Acknowledgments

This chapter is dedicated to the late Hans Ris (1914–2004), cell biology pioneer, literal visionary of cell structure and imaging, tireless researcher, inspiring mentor and collaborator, and patriarch of a great international family of scientists to which I was privileged to belong. My thanks to James Pawley, Ya Chen, and David Wokosin at the University of Wisconsin's Integrated Microscopy Resource for instrumentation assistance, to Tom Roberts, Department of Biology, Florida State University, for the anti-MSP primary antibody used in Figure 3.4B, and to John White, UW Madison, for the freeze-substituted *C. elegans*. Some of this work was supported by NIH AREA grant R15 GM42081 – 01, NSF – CCLI grant 9051267 and University of Wisconsin, Stevens Point, UPDC intramural funding.

3.13 References

Autrata, R., Hermann, R., and Mueller, M. (1991). An efficient BSE single-crystal detector for SEM. *Scanning*, **14**, 127–135.

Buechi, M. and Bachi, T. (1979). Immunofluorescence and electron microscopy of the cytoplasmic surface of the human erythrocyte membrane and its interaction with Sendai virus. *Journal of Cell Biology*, **83**, 338–347.

Buser, C. and Walther, P. (2008). Freeze-substitution: the addition of water to polar solvents enhances the retention of structure and acts at temperatures around −60 °C. *Journal of Microscopy*, **230**, 268–277.

[3] *3D Stereo Image Factory Plus*, ver. 2.5 was created by D. Harvey and last updated in 1999. It can be downloaded from a variety of sites on the internet. Detailed instructions for using this software with electron micrographs are available from the author.

Heuser, J. E. (1976). Morphology of synaptic vesicle discharge and reformation at the frog neuromuscular junction. In: *The Motor Innervation of Muscle*. Edited by S. Thesleff. London, Academic Press, pp. 51–115.

Heuser, J. E. and Salpeter, S. R. (1979). Organization of acetylcholine receptors in quick-frozen, deep-etched, and rotary-replicated Torpedo postsynaptic membrane. *Journal of Cell Biology*, **82**, 150–173.

Iwadare, T., Harada, E., Yoshino, S., and Arai, T. (1990). A solution for removal of resin from epoxy sections. *Stain Technology*, **65**, 205–209.

Lück, S., Sailer, M., Schmidt, V., Beil, M., and Walther, P. (2010). Three-dimensional analysis of intermediate filament networks using SEM-tomography. *Journal of Microscopy*, **239** (*Pt 1*), 1–16.

Ogneva, V. and Neronov, A. (1995). A method for preparation of immobilized cells and tissues for light and electron microscopy studies. *Microscopy Research and Technique*, **30**, 265–267.

Ris, H. (1991). The three-dimensional structure of the nuclear pore complex as seen by high voltage electron microscopy and high resolution, low voltage scanning electron microscopy. *EMSA Bulletin*, **21**, 54–56.

Ris, H. and Malecki., M. (1993). High-resolution field emission scanning electron microscope imaging of internal cell structures after Epon extraction from sections: a new approach to correlative ultrastructural and immunocytochemical studies. *Journal of Structural Biology*, **111**, 148–157.

Schliwa, M. and van Blerkom, J. (1981). Structural interaction of cytoskeletal components. *Journal of Cell Biology*, **90**, 222–235.

Sepsenwol, S. (2005). Specimen capsules for critical-point drying. *Microscopy Today*, **13** (1), 45.

Sepsenwol, S., Ris, H., and Roberts, T. M. (1989). A unique cytoskeleton associated with crawling in the amoeboid sperm of the nematode, *Ascaris suum*. *Journal of Cell Biology*, **108**, 55–66.

Small, J. V. (1981). Organization of actin in the leading edge of cultured cells: influence of osmium tetroxide and dehydration on the ultrastructure of actin meshworks. *Journal of Cell Biology*, **91**, 695–705.

Walther, P. (2007). High resolution cryoscanning electron microscopy of biological samples. In: *Biological Low-Voltage Scanning Electron Microscopy*. Edited by H. Schatten and J. B. Pawley. New York, Springer, pp. 245–261.

Walther, P., Chen, Y., Pech, L. L., and Pawley, J. B. (1992). High resolution scanning electron microscopy of frozen-hydrated cells. *Journal of Microscopy*, **168**, 169–180.

4 Mitochondrial continuous intracellular network-structures visualized with high-resolution field-emission scanning electron microscopy

T. Naguro, H. Nakane, and S. Inaga

4.1 Introduction

Since the discovery that mitochondria form a single continuous network within yeast cells, reported by Hoffmann and Avers (1973), many papers have reported the existence of similar mitochondrial networks in various kinds of cells of multiple organisms.

For most of these studies originating during the 1970s and the 1980s TEM was used and three-dimensional images of mitochondrial networks were made by skillful reconstructions using elaborate serial-thin sectioning methods. At the time the materials used for investigations were confined to protists, algae, and some cultured mammalian cells (reviewed by Bereiter-Hahn, 1990). From the beginning of the 1990s studies on the morphology of mitochondria rapidly advanced using, in particular, confocal-light microscopy on cultured cells and unicellular organisms. Such investigations allowed morphologists to gain direct live images of mitochondrial dynamics in living cells (e.g. Bereiter-Hahn and Vöth, 1994; Dlasková *et al.*, 2010). In addition, the discovery of various molecular components regulating morphological dynamics and the organization of the mitochondria in cells provided some of the most exciting breakthroughs on mitochondrial research in the last decade (Benard and Karbowski, 2009). On the basis of such ample evidence it is now assumed that mitochondria are dynamic singular organelles that continuously repeat fusion and fission (Legros *et al.*, 2002; Chen *et al.*, 2003; Karbowski and Youle, 2003; Mattenberger *et al.*, 2003; Osteryoung and Nunnari, 2003; Rube and Van der Bliek, 2004; Jendrach *et al.*, 2005; Malka *et al.*, 2005; Okamoto and Shaw, 2005; Anesti and Scorrano, 2006; Chan, 2006; McBride *et al.*, 2006: Cerveny *et al.*, 2007; Hoppins *et al.*, 2007; Twig *et al.*, 2008), i.e. a continuous network (a giant mitochondrion) that transiently manifests itself as a reticulum or chondriome (Bereiter-Hahn and Jendrach, 2010).

The functional significance of mitochondrial-network formation is being studied in a variety of cultured cells and unicellular organisms (Bereiter-Hahn and Jendrach, 2010), however, there are no reports on mitochondrial networks in neurons *in situ* (*in vivo* cells within the tissue).

Scanning Electron Microscopy for the Life Sciences, edited by H. Schatten. Published by Cambridge University Press

Here we report on the existence of a single mitochondrion as a continuous network in one such neuron, the granule cell in the rat's olfactory bulb. We demonstrate the three-dimensional overall structure using HR-FE-SEM, followed by discussion of the functional significance of the mitochondrial network in these granule cells. To our knowledge, this is the first report to provide evidence for the existence of a mitochondrial network in mammalian nerve cells *in situ*.

4.2 Mitochondria inside the cell body (perikaryon) of granule cells

Granule cells in the mammalian olfactory bulb possess one large dendrite and a few short neurites but have no axons. The common appearance of granule cells prepared by the osmium maceration method, called the AODO method (see Section 4.14), is shown in Figure 4.1 (color plate). Intracellular organelles of the cell such as the nucleus, endoplasmic reticulum, Golgi apparatus, and mitochondria are clearly revealed three-dimensionally, since the cytoplasmic matrix is removed from the cracked surface during the process of maceration. The nucleus, about 5 μm in diameter, occupies a relatively large part of a rather small cell body of about 8 μm across; the cytoplasmic compartment of the granule cell is very small compared to most other neuronal cell types. The identification of mitochondria is easy when they show their characteristic internal structures referred to as cristae (Figure 4.2 (color plate)), but otherwise it is difficult to distinguish mitochondria from other spherical organelles. In Figure 4.1, because the nucleus positioned in the center of the cell body conceals the spatial configuration of mitochondria, mitochondria are seemingly dispersed in the cell body and they appear to be mutually independent, just like the images seen in ultrathin TEM sections. Thus, an important question is whether mitochondria within a cell exist one by one independently or whether they are mutually connected.

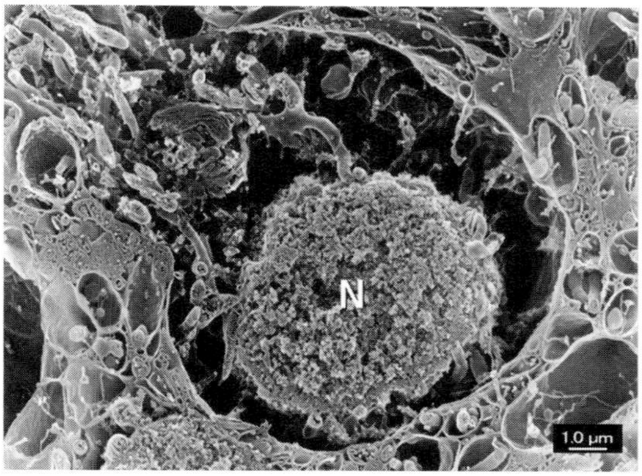

Figure 4.1 HR-FE-SEM image of a cracked granule cell in the rat's olfactory bulb. The nucleus (N), situated at the center of the small cell body, occupies a large area of the central region and conceals the spatial configuration of cell organelles from the view. This granule cell has one apical dendrite (upper left corner). ER (magenta), Golgi apparatus (green), mitochondrial outer membrane (blue), mitochondrial cristae (yellow). (See plate section for color version.)

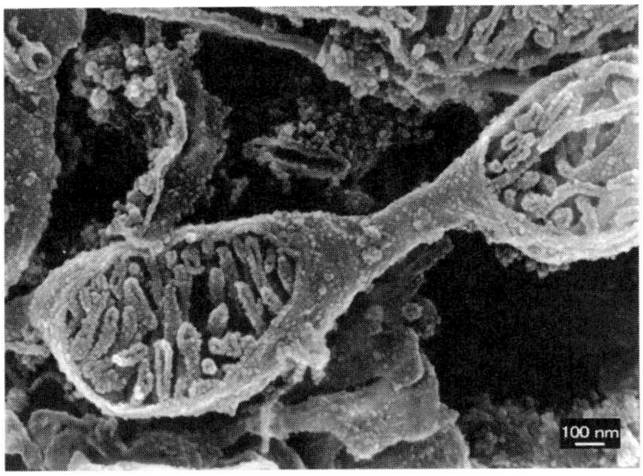

Figure 4.2 High-magnification view of mitochondrial network Type-4 in an olfactory-bulb granule-cell body. Mitochondrial cristae (yellow) are well preserved and do not show any deformation. Golgi apparatus (green), ER (magenta), mitochondrial outer membrane (blue). (See plate section for color version.)

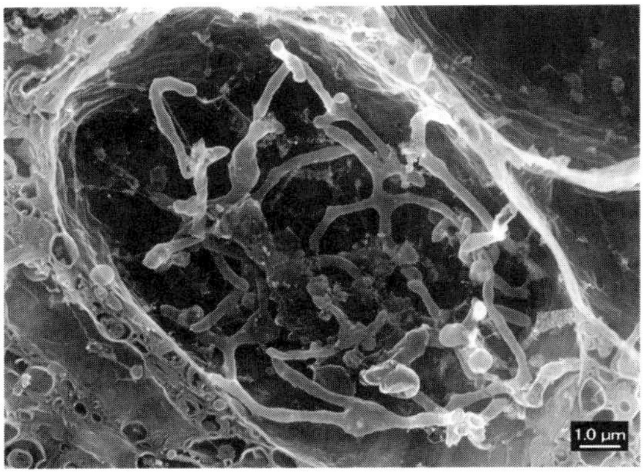

Figure 4.3 Low-magnification view of mitochondrial network Type-1. A large mitochondrion (blue) appears to be in the form of a single continuous network structure extending throughout the cell body. The widths of tubular components of a mitochondrial network are fairly uniform in size, about 250–300 nm. Golgi apparatus (green), ER (magenta), mitochondrial cristae (yellow). (See plate section for color version.)

Cells whose nuclei were removed during specimen preparation give a clear answer to this question (Figures 4.3–4.8 (color plates)). Mitochondria in cells without a nucleus show a unique spatial architecture. They appear to form a continuous network distributed throughout the perikaryon. Indeed, almost every mitochondrial network consists of a single mitochondrion within a cell. Although the mitochondrial network in each cell is highly variable in its spatial configuration, the most typical network is composed of branching and anastomosing tubules about 250–300 nm in width (Figure 4.3). Figures 4.3–4.8 show that the various forms of mitochondrial networks can be roughly classified by their morphological appearances into four groups.

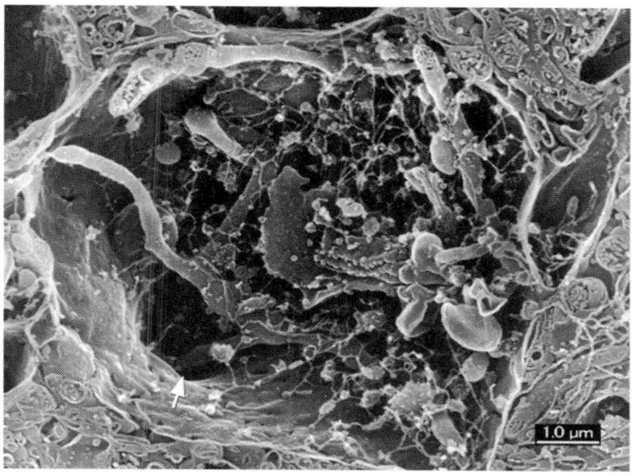

Figure 4.4 Mitochondrial network Type-1 in an olfactory-bulb granule-cell body. A large hole in the bottom left corner represents an entrance to the apical dendrite. Note an elongated part of the mitochondrial network in the cell body that enters into the dendrite (arrow). Multipartite rER forms vesicles or irregular cisterns and is always connected with filamentous sER, around 25 nm in diameter. They form an additional (additional to the mitochondrial network) network system joining together in the cell body. Golgi apparatus (green), ER (magenta), mitochondrial outer membrane (blue), mitochondrial cristae (yellow). (See plate section for color version.)

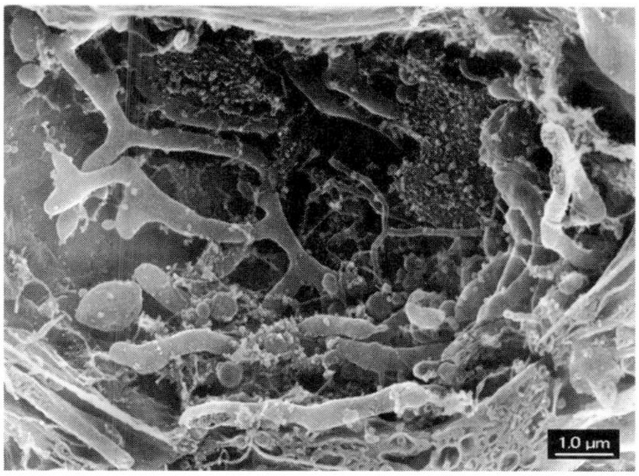

Figure 4.5 Mitochondrial network Type-2 in an olfactory-bulb granule-cell body. The network is composed of two parts different in width. Filamentous parts, around 75 nm in diameter across, originate from the tips of tubular parts of normal size about 250–300 nm. The rER shows a wide variation in shape and size, while the sER appears as a filamentous structure of about 25 nm. Golgi apparatus (green), ER (magenta), mitochondrial outer membrane (blue), mitochondrial cristae (yellow). (See plate section for color version.)

4.3 Classification of mitochondrial networks

4.3.1 Type-1 (Figures 4.3 and 4.4)

As described above, this type of mitochondrial network is the typical form and a rather simple configuration as compared to other types. This network is composed of several cylindrical tubules that branch and anastomose and look like a dermal capillary network.

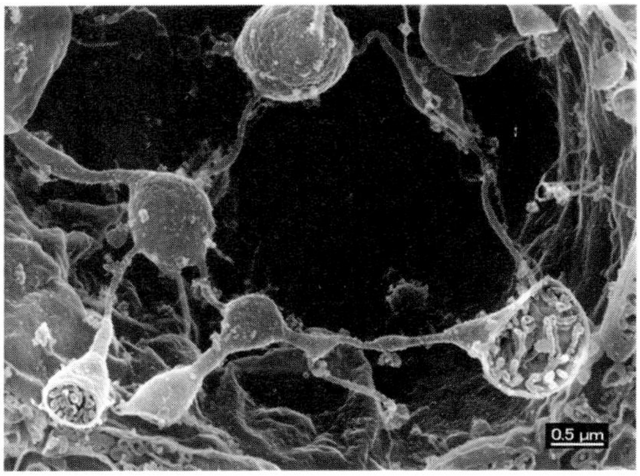

Figure 4.6 Mitochondrial network Type-3 in an olfactory-bulb granule-cell body: a necklace-like loop, consisting of globular parts and filamentous parts. Note the mitochondrial cristae (yellow) in globular parts. They are well preserved without any deformation. Mitochondrial outer membrane (blue), ER (magenta). (See plate section for color version.)

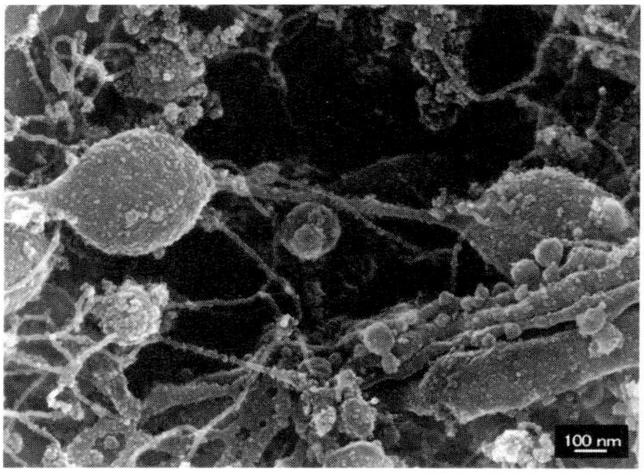

Figure 4.7 Higher magnification image of mitochondrial network Type-3 in an olfactory-bulb granule-cell body. A long filamentous mitochondrial network connecting two globular parts is 45 nm in width. This size is almost twice that of a filamentous tubule of sER that runs parallel to the long axis of the filamentous part of a mitochondrion. Golgi apparatus (green), mitochondrial outer membrane (blue), ER (magenta). (See plate section for color version.)

The majority of cylindrical tubules, apart from branching portions, are relatively uniform in width, about 250–300 nm across. Thus, the distinctive feature of this type is that the network is composed only of tubules of almost the same size.

4.3.2 Type-2 (Figure 4.5)

This type of mitochondrial network is rarely observed. At first glance it seems that there is no difference in form between this type and Type-1, but after more thorough

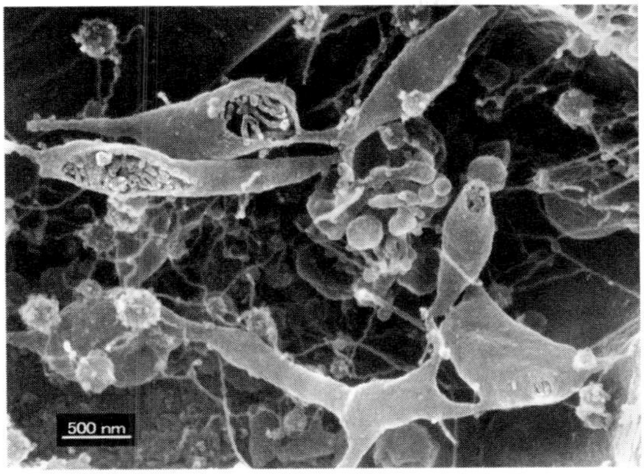

Figure 4.8 Mitochondrial network Type-4 in an olfactory-bulb granule-cell body. The HR-FE-SEM image shows the structural complexities of this mitochondrial network of this Type-4. Due to a wide variety of structural appearances, it is difficult exactly to describe the morphological features of this type of mitochondrial network. Golgi apparatus (green), mitochondrial outer membrane (blue), ER (magenta). (See plate section for color version.)

observations, clear differences can be seen in its constituent parts. The mitochondrial network of Type-2 is composed of two kinds of cylindrical tubule, those like Type-1, about 250–300 nm across, and filamentous tubules around 75 nm in width. Normal-sized tubules are always predominant whereas small numbers of filamentous tubules are sparsely distributed within the mitochondrial network. The filamentous tubules are always connected to tubules of normal size and do not exist independently. One edge of a filamentous tubule is connected with a tip of a normal size tubule, whereas the other edge is "floating" free showing no connection with other tubules. In other words, two kinds of tubule of different sizes are in fact a continuous structure varying in diameter. Some tubules of normal size have filamentous parts at their tips. There is a distinctive structure at the transition where a tubule of normal size changes to a small size. At the transition part the tubule of normal size suddenly decreases its size, resulting in an abrupt transition to a filamentous tubule. The filamentous tubules vary in length and sometimes branch.

4.3.3 Type-3 (Figures 4.6 and 4.7)

This type of mitochondrial network is rarer than that of Type-1 and/or Type-4. Because this type of mitochondrial network shows a peculiar form, it is very easy to distinguish from other types. The mitochondrial network is composed of two different parts: globular parts around 1.0 μm in diameter and filamentous parts about 50–100 nm in width. Several globular parts are connected to each other by the long filamentous parts, together forming a necklace-like structure (Figure 4.6) altogether resembling "beads on a string" (Figure 4.7). The filamentous parts of this type of mitochondrial network have the smallest width of all types of mitochondrial network described here, being 45 nm across

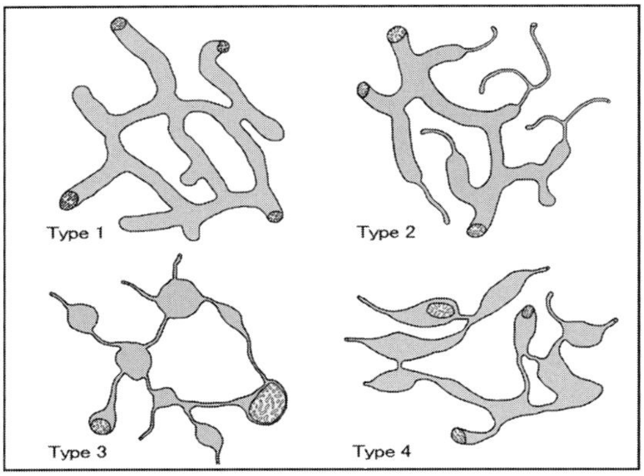

Figure 4.9 Types of mitochondrial network. The mitochondrial network of Type-1 is composed of tubular parts rather uniform in size. Type-2 is composed of two kinds of tubular part differing in diameter. Type-3 is composed of two parts, globular and filamentous. Type-4 is composed of irregularly shaped parts. Mitochondrial outer membrane (blue), mitochondrial cristae (yellow). (See plate section for color version.)

(Figure 4.7). The mitochondrial cristae of the globular parts are well organized and have a normal appearance (Figure 4.6).

Although this chapter does not deal with details on mitochondrial cristae morphology, the observation that the mitochondrial cristae in the globular parts are normal in appearance is important for the discussion (see Section 4.8.)

4.3.4 Type-4 (Figure 4.8)

The last type of mitochondrial network shows a distinctive feature setting it apart from the other three. This type is the most complex of the four types and is composed of many parts that vary in shape such as spindles, boxes, spheres, tubules, and filaments as well as some indeterminate forms. They are connected to one another by short filamentous parts.

A summary diagram of the classification set out above is shown in Figure 4.9 (color plate). Although we do not have enough data to indicate statistically the frequency of occurrence of each type, mitochondrial networks of Type-1 and/or Type-4 seem to be predominant relative to Type-2 and Type-3. The possible meaning of the diversity of mitochondrial networks is discussed later (Section 4.8).

4.4 Mitochondria inside dendrites

We have not examined the whole length of very long meandering dendrites. However, using HR-FE-SEM we have obtained a good view of the distribution pattern of mitochondria inside extended parts of such a dendrite that was cracked longitudinally (Figure 4.10 (color plate)). Mitochondria in these dendrites are not of the small vesicular type fragmented into several pieces but are of the long tubular Type-1 aligned parallel to

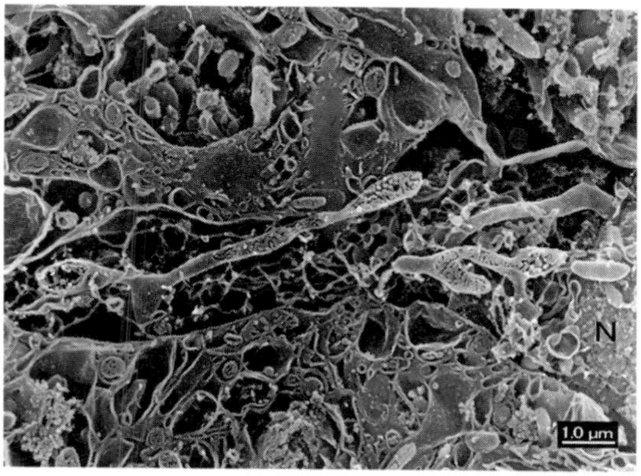

Figure 4.10 Longitudinal section of an apical dendritic trunk of an olfactory-bulb granule cell. Mitochondria in a dendritic trunk display long slender tubules. In this HR-FE-SEM photograph, the direct continuity between mitochondria in the dendrite and the mitochondrial network in a cell body is unclear (but see Figures 4.4 and 4.11). Abundant filamentous sER (magenta) also forms a network structure. Nucleus (N), Golgi apparatus (green), mitochondrial outer membrane (blue), ER (magenta). (See plate section for color version.)

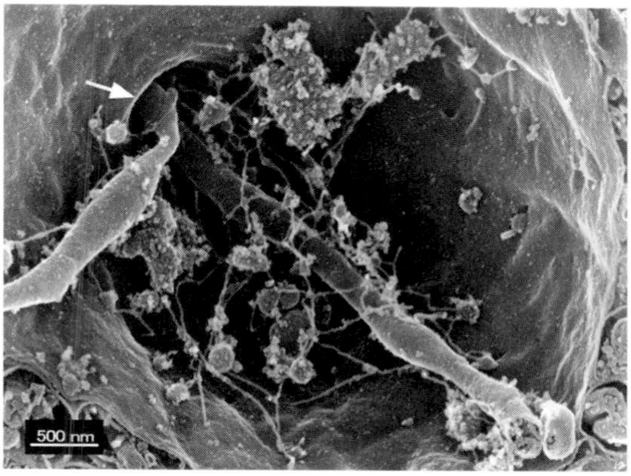

Figure 4.11 Base of an apical dendrite viewed from the cell body. Long tubular mitochondria enter into a dendrite from the cell body passing through a hole (arrow) that borders the dendrite and the cell body. Mitochondrial outer membrane (blue), mitochondrial cristae (yellow), ER (magenta). (See plate section for color version.)

the long axis of the dendrite. Although in Figure 4.10 a long tubular mitochondrion about 8 μm in length is seemingly one piece perhaps of a longer mitochondrion, that appears to have no direct connection to the mitochondrial network of the cell body. Actually, there is continuity between the mitochondrial network in the cell bodies and mitochondria inside dendrites. Indeed, the mitochondrial network inside cell bodies often extends deeply its long tubular part into these dendrites (Figures 4.4 and 4.11 (color plate)).

4.5 Endoplasmic reticulum and the Golgi apparatus of granule cells

In SEM photographs rER is easily identified by its characteristic appearance. The cytoplasmic surface of rER is studded with numerous ribosomes (Figures 4.2, 4.5, 4.7, 4.8, and 4.10). In granule cells the rER compartments are very small compared to those of other larger mammalian neurons. The rER assumes different shapes, small-flattened cisterns (Figures 4.2 and 4.5) and vesicular structures 200–500 nm in diameter (Figures 4.4, 4.7, and 4.8); there are no large aggregations of flattened cisterns. sER appears always as thin filaments around 25 nm in width and is well developed in granule cells (Figures 4.3, 4.4, 4.7–4.10). The sER forms a combined network structure together with the rER, because the sER and the rER are, of course, one continuous organelle. Thus, the sER does not appear alone without the rER. The intimate connection between mitochondria and the ER is observed almost everywhere. But, the intimate relationship between mitochondria and ER inside granule cells is assumed mainly by the abundant sER; the rER rarely participates. Thus, what should be mentioned especially on relationships between mitochondria and ER is that the mitochondria in dendrites are always surrounded and enclosed by a filamentous ER network formed mainly by sER (Figure 4.10).

The Golgi apparatus of granule cells can also be easily distinguished from other cell organelles by its specialized form of the so-called Golgi stack that consists of several cisternae piled up in a parallel array (Figures 4.4 and 4.7). It is located just above the nucleus in the vicinity of the dendritic base (Figures 4.1, 4.4, and 4.10). As shown in Figures 4.3–4.5 and 4.8, the Golgi apparatus in granule cells is poorly developed and relatively simple as compared to many other cell types where the Golgi apparatus is generally much more complicated (Tanaka *et al.*, 1986; Tanaka, 1987; Naguro and Iino, 1989a; 1989b; Naguro and Iwashita, 1992; Naguro *et al.*, 1994; Naguro *et al.*, 1997; Koga and Ushiki, 2006). Despite a poor morphological appearance, the Golgi apparatus inside granule cells maintains a particular close relationship with the mitochondrial network.

4.6 Analysis and discussion of the existence of a continuous mitochondrial network

This chapter provides *in vivo* evidence that a well-developed mitochondrial network exists in a specialized neuron, granule cells of the rat olfactory bulb, that have differentiated in the tissue and have permanently withdrawn from the cell cycle; however, it is not certain whether such granule cells always maintain one continuous mitochondrial network, because all granule cells of an olfactory bulb were not investigated in this preliminary experiment (see Section 4.13).

Using the serial thin section method to analyze whole mitochondrial shapes, the mitochondrial network or ramified giant mitochondrion has been observed in a variety of organisms, particularly in protists, algae, and fungi (see reviews Bereiter-Hahn, 1990;

Bereiter-Hahn and Vöth, 1994; Kawano *et al.*, 1995). Because of the preponderance of evidence that the mitochondrial network appears during a specific period of the cell cycle and/or the cell life cycle in many of these organisms, it was initially expected that, apart from some exceptions, the mitochondrial network formation is a phenomenon related to the cell cycle. However, ample experimental evidence from studies in a number of living cells in culture has proved that mitochondrial network formation is not exclusively dependent on the cell cycle but is also present during certain physiological and/or functional states of cells including abnormal conditions. It is now quite obvious that mitochondrial-network formation is a highly dynamic phenomenon and that mitochondria are organelles that constantly undergo remodeling through fusion and fission (Bereiter-Hahn and Vöth, 1994; Karbowski and Youle, 2003; Chan, 2006; Hoppins *et al.*, 2007; Plecitá-Hlavatá *et al.*, 2008; Bereiter-Hahn and Jendrach, 2010; Westermann, 2010). Thus, it has become apparent that mitochondria of eukaryotic cells form a network structure in most cell types. The definitive proof for the existence of mitochondrial networks was established mainly using confocal-light microscopy in living cells in culture. However, to our knowledge, heretofore only few reports have shown a single continuous mitochondrial network present in individual mammalian cells *in vivo* (Bereiter-Hahn, 1990). An intriguing remaining question is whether all mitochondria within a cell in the *in vivo* state are present as one continuous network, even temporarily or only during a restricted period.

Mitochondria of spermatozoa are often cited in the literature as one of the more prominent examples of one large mitochondrion per cell in differentiated cells. Unlike mitochondria of spermatozoa in *Drosophila* (Tates, 1971; Hales and Fuller, 1997; Alridge *et al.*, 2007), those in the mouse are very closely placed to each other as if they have been fused into one large mitochondrion but, actually, they never appear as one large interconnected mitochondrion throughout spermiogenesis (Otani *et al.*, 1988; Ho and Wey, 2007), the latter being true in a pseudo scorpion (Bawa and Werner, 1988). Skeletal muscles in mammals are also a well-known example where differentiated cells possess a well-developed mitochondrial network or mitochondrial reticulum; the spatial arrangement of a mitochondrial network in a variety of skeletal muscles in several mammalian species was investigated using a number of different methods, namely, serial section reconstruction (Brandt *et al.*, 1974; Bakeeva *et al.*, 1978; Kayar *et al.*, 1988) and high-voltage TEM (Kirkwood *et al.*, 1987), and SEM (Ogata and Yamasaki, 1985). Those studies provided excellent three-dimensional views of mitochondrial networks *in vivo*, but evidence of a mitochondrial network completely interlinking all mitochondrial material within a whole muscle fiber (per cell) was lacking (Kayar *et al.*, 1988). In short, absolute proof for the presence of just one mitochondrion per cell in mammals *in vivo* has yet to be confirmed (Dlasková *et al.*, 2010).

Thus, despite intensive investigations, even for cultured cells, the existence of one single mitochondrion per cell is still a matter of debate (see review Bereiter-Hahn and Jendrach, 2010). With the aim of gaining new insight into the nature of mitochondrial structure, Collins *et al.* (2002, 2003), using morphological and functional investigations of several cultured cells, addressed whether mitochondria represent morphologically continuous and functionally homogenous entities within single intact cells, and whether this is critical for their role in processes such as apoptosis and calcium homeostasis. A crucial question was: do mitochondria exist as discrete organelles within the cell or is the mitochondrion a single entity more akin to the endoplasmic reticulum? Since the mitochondrial structure is in dynamic balance between fission and fusion, perhaps this question should be rephrased as "do mitochondria predominantly undergo fusion to form a network or fission to form discrete organelles?" (Collins *et al.*, 2002). The experimental data of Collins *et al.* showed that the mitochondria within individual cells were morphologically heterogeneous and unconnected. This means that there were no cells having only one mitochondrion. A recent review (Bereiter-Hahn and Jendrach, 2010) emphasized the nature of single mitochondria in relation to the high-mitochondrial dynamics of continuously fusing and dividing and, by this, the single mitochondrion represents only a transient manifestation of a reticulum termed chondriome. Apart from the essential nature of mitochondrial structure, investigating what kind of cell *in vivo* has a well-developed mitochondrial network is probably more important than to ascertain the role of mitochondrial networks.

4.7 Analysis and discussion of the size of the mitochondrial network

A most notable finding of the authors' HR-FE-SEM study is the existence of filamentous portions of minimally 45 nm across in the Type-3 (Figures 4.6 and 4.7) and Type-4 (Figure 4.8) mitochondrial networks. This size, 45 nm, is well below the resolving power of the 4Pi-confocal light-microscope (Egner *et al.*, 2002), to say nothing of conventional confocal-light microscopy. 4Pi confocal microscopes are currently used for analyzing three-dimensional images of subcellular structures like mitochondria at high resolution (100 nm axial and 250 nm lateral) (Plecitá-Hlavatá *et al.*, 2008; Ježek and Plecitá-Hlavatá, 2009; Dlasková *et al.*, 2010). However, as the mitochondrial tubules described here are smaller than 50 nm in width, these cannot be detected even by 4Pi microscopy. This means that observation by confocal microscopy involves the risk that one continuous mitochondrial network with thin filamentous portions can be misinterpreted as several mitochondria having no connections. In other words, there is the possibility that apparently separated mitochondria as seen with confocal microscopy are in reality one interconnected mitochondrion with thin filamentous portions. Therefore, for *in vitro* studies with light microscopy, careful attention is needed to decide whether mitochondria have surely divided. Thus, it cannot be determined with light microscopy whether mitochondria have indeed fused or whether they are only very close to one another (Bereiter-Hahn and Vöth, 1994).

Dlasková *et al.* (2010) reported, using 4Pi confocal-light microscopy, that mitochondria form a nearly completely interconnected tubular network with a fairly uniform tubule diameter of 270±30 nm in insulinoma INS-1E cells and 280±40 nm in hepatocellular carcinoma HEP-G2 cells. The Type-1 and Type-2 mitochondrial networks in the authors' FE-SEM study consist almost exclusively of tubules around 270 nm in diameter (Figures 4.3 and 4.4). Thus, interestingly, values of the 4Pi microscopy and of the present HR-FE-SEM study are almost identical. Such values might represent a standard for large interconnected mitochondria under normal conditions.

4.8 Analysis and discussion of the diversity of mitochondrial networks

This preliminary study provides evidence that there is a high morphological diversity of mitochondrial networks in normal rat olfactory granule cells. Given that mitochondria are dynamic organelles, constantly changing their number and shape using fusion and fission, this is not surprising. However, considering that a certain functional state of cells is reflected in the shape of their mitochondrial networks, detailed investigations on this morphological diversity might be a useful tool for obtaining more insights into the functional information of a given cell.

We roughly classified the forms of mitochondrial networks into four types. Especially intriguing is the Type-3 mitochondrial network (Figures 4.6 and 4.7). To understand the possible functional meaning of this unique form, helpful information may be available from some *in vitro* studies that showed various functional states of mitochondrial networks.

In vitro studies (Stolz and Bereiter-Hahn, 1987; Bereiter-Hahn and Jendrach, 2010) have shown similar network structures with the appearance of a beads-on-a-string structure in living *Xenopus* endothelial cells injected with Ca^{2+} using phase-contrast microscopy and TEM. Judging from morphological similarities, beads and a string in their study seem to correspond to globular parts and a filamentous part in the present study. Despite a close similarity in appearance, the internal structure of the globular parts in the authors' HR-FE-SEM study differs markedly from that in equivalent structures in TEM studies. Mitochondrial cristae are well preserved in the HR-FE-SEM study, but beads (swollen area) in the TEM photographs of these completely lacked cristae. Although it is reasonable to surmise that the great differences in preservation of cristae in globular parts represent functional differences, at present we have too few data to discuss in detail the implications of the formation of globular parts. In order to identify the functional state of granule cells, it is important to investigate when and why this structure appears.

4.9 Analysis and discussion of the formation of mitochondrial beads on a string during apoptosis

Disintegration of mitochondrial networks into multiple fragmented organelles is often observed during apoptosis in *in vitro* studies, and this phenomenon has been well documented as a fundamental event along with fission (Frank *et al.*, 2001; Gao *et al.*, 2001; Karbowski and Youle, 2003; Lee *et al.*, 2004; Skulachev *et al.*, 2004; Hom *et al.*, 2007; Scorrano, 2009; Sheridan and Martin, 2010). The structure that looks like beads on a string appears at the stage prior to fragmentation (decomposition of a mitochondrial network), and has been shown to be present using both light microscopy and TEM (Frank *et al.*, 2001; Skulachev *et al.*, 2004). These morphological changes occurring during the transition process reflect an early step of mitochondrial fragmentation toward apoptosis and mitoptosis (called "the thread-beads transition" by Skulachev *et al.* (2004)).

Most olfactory-bulb granule cells are generated post-natally and continue to be added in adulthood (Altman, 1969; Kaplan and Hinds, 1977; Bayer, 1983; Petreanu and Alvarez-Buylla, 2002; Carleton *et al.*, 2003; Whitman and Greer, 2007), and so apoptosis is observed in the layers where the newly generated cells are incorporated (Najbauer and Leon, 1995; Fiske and Brunjes, 2001). Assuming that the structure that looks like beads on a string appears only in the progression of apoptosis, it is possible to decide that the granule cell having a mitochondrial network of Type-3 displays an early stage of apoptosis. However, it remains unclear as to whether this structure always appears as a step of an irreversible progression of the fragmentation process and/or whether this structure is capable of recovering its original shape (network) by escaping fragmentation.

Although mitochondrial swellings have been described in apoptosis pathways in several instances (Frank *et al.*, 2001), it is unlikely that these swellings are an indispensable prerequisite for the progression of apoptosis. This is because: (1) in many types of apoptosis the mitochondria remain morphologically normal and in many cell types the release of cytochrome c from mitochondria occurs before or in the absence of mitochondrial swelling (Martinou *et al.*, 1999); (2) the debate is still open as to whether and how mitochondrial fission participates in the decisional phase of mitochondrial apoptosis, i.e. in the release of cytochrome c (Scorrano 2009); (3) many groups have observed dramatically fragmented mitochondria in healthy cells, indicating that mitochondrial fission alone does not necessarily result in cell death (see review Sheridan and Martin, 2010).

4.10 Analysis and discussion of network formation in large cells *in vivo*

As described earlier, tissues with high and variable rates of aerobic metabolism such as skeletal muscle or heart show a prominent development of the mitochondrial network. As a possible explanation for the role of the well-developed mitochondrial network it was proposed by Skulachev (2001) that extended mitochondria represent a united electrical system that can facilitate energy delivery from the cell periphery to the cell core. In other words, it was speculated that extended mitochondria act as cables to distribute proton gradients throughout large cells.

Neurons are the most extended cell types in animals. They possess long neurites besides a cell body. Therefore, the energy produced by mitochondria has to be carried from cell bodies to places of great distance – synapses of axons and dendrites – at great energy expenditure. Is it possible to apply the cable theory to the mitochondria of neurons too?

In contrast to muscle cells, a large number of *in vitro* studies on neurons have provided evidence that mitochondria move in both anterograde and retrograde directions within axons and dendrites and travel over extremely long distances along the tracks provided by the cytoskeleton and associated proteins (Overly *et al.*, 1996; Chang and Reynolds, 2006; Zinsmaier *et al.*, 2009; Pathak *et al.*, 2010). It is widely accepted that mitochondria in neurons need to be transported and distributed in axons and dendrites in order to ensure an adequate energy supply and to provide sufficient Ca^{2+} buffering in each portion of these highly extended cells (Hollenbeck and Saxton, 2005; Wang and Schwarz, 2009).

In the present *in vivo* study the authors showed the existence of very long filamentous mitochondria within dendritic shafts of rat olfactory-bulb granular cells (Figure 4.10). Earlier, they also found long mitochondria within dendrites of olfactory receptor cells in rats (Naguro and Iwashita, 1992). Popov *et al.* (2005) similarly reported long filamentous mitochondria (of at least 36 μm) in hippocampal dendrites in rats and ground squirrels using three-dimensional reconstructions of serial-thin sections. Aside from the issue of whether such long mitochondria can move within dendrites, a fundamental question is whether they actually need to travel within dendrites. Although a large number of *in vitro* studies with cultured cells have demonstrated mitochondrial movement in terms of the transport within neuronal processes (see review by Chang and Reynolds, 2006), in the case of dendrites *in vivo* it is unclear whether such transport exists for mitochondria, but this needs to be considered in the context of the existence of long mitochondria.

The present data also indicate the structural continuity between long filamentous mitochondria in the dendrite and a mitochondrial network inside the cell body (Figures 4.4 and 4.11). The continuity of mitochondria between the dendrite and the cell body has been observed in hippocampal neurons (Popov *et al.*, 2005). A key issue, yet to be resolved, is whether such continuity is a general feature in all types of neurons *in vivo*.

Granule cells in the mammalian olfactory bulb possess one large dendrite and a few short neurites but no axons; their output is mediated by bidirectional dendrodendritic synapses located in spines (Jahr and Nicoll, 1982; Woolf *et al.*, 1991). Considering that neurons without axons, such as granule cells, do not have to produce multiple discrete mitochondria for axonal trafficking, this peculiar feature may be why a well-developed mitochondrial network exists in granule cells. However, because we do not yet know whether a well-developed mitochondrial network is present in other types of neurons *in vivo*, it is impossible at present to determine the reason for a well-developed mitochondrial network of the granule cells.

4.11 Analysis and discussion of network formation of *in vitro* cells

Recent *in vitro* studies have provided evidence for the existence of highly interconnected mitochondria in mouse embryonic fibroblasts exposed to selective stresses, which inhibit cytosolic protein synthesis. This network formation is caused by strongly increased fusion, so-called "hyperfusion" that precedes mitochondrial fission when triggered by apoptotic stimuli such as UV irradiation or actinomycin D (Tondera *et al.*, 2009). The formation of an extensive mitochondrial network caused by hyperfusion is considered to act as a survival mechanism helping to protect cells against stress-induced apoptosis (Tondera *et al.*, 2009).

Many senescent cells in culture also display an extensive elongation and/or network formation of mitochondria (Yoon *et al.*, 2006; Zottini *et al.*, 2006; Unterluggauer *et al.*, 2007; Navratil *et al.*, 2008; Mai *et al.*, 2010). Thus, it may be that this distinct morphotype has a functional role in senescent cells (Bereiter-Hahn and Jendrach, 2010). A possible role for an extensive mitochondrial network may be a protective mechanism against damaging molecules such as reactive oxygen species (Bereiter-Hahn and Jendrach, 2010; Mai *et al.*, 2010). An additional role for long and interconnected mitochondria in senescent cells may be that their presence diminishes a need for the energy-consuming processes of mitochondrial dynamics, while still allowing fast distribution and exchange of molecules (Mai *et al.*, 2010). Summarizing, these results obtained by *in vitro* studies indicate that formation of an extensive mitochondrial network plays an important role in cell survival and resistance to apoptosis.

As described earlier, granule cells in the rat olfactory bulb undergo neurogenesis from precursor cells throughout adult life. Therefore, in an olfactory bulb, while most granule cells are thought to be in the normal state, there is always a small fraction of granule cells in the stage of apoptosis or of senescence. This assumption allows a prediction that at least some of the granule cells with a well-developed mitochondrial network might represent the same functional state as cells *in vitro* in the stage of apoptosis or senescence. However, given that most granule cells are in a normal state, there must also be a specific, apoptosis-independent role of a well-developed mitochondrial network of granule cells in relation to neuronal function. As this preliminary study does not provide precise insights into life stages of granule cells, further studies are needed to assess the precise role of a well-developed mitochondrial network of granule cells.

4.12 Methodology considerations

The ODO method devised by Tanaka and Naguro (1981) was the method of choice for obtaining three-dimensional information on intracellular structures by HR-FE-SEM. The original ODO method was modified to the aldehyde-prefix ODO method (AODO method) that allowed obtaining an even better preservation of ultrastructure, especially in neuronal cells (Tanaka and Mitsushima, 1984). This AODO method was applied to investigate the morphology of mitochondria in the present study (see Section 4.14). An important property of this method is that soluble proteins and cytoskeletons are removed from the cell during specimen preparation by prolonged immersion in dilute osmium solution, a step called maceration. Through this process it became possible to observe very fine details of membranous-subcellular organelles such as mitochondria, ER, and the Golgi apparatus three-dimensionally using HR-FE-SEM.

4.13 Advantages and disadvantages of the method

Results obtained in this preliminary study indicate the existence of a continuous mitochondrial network within a cell *in vivo*. However, it is important to note that we can observe the cells cracked nearly into halves by the AODO method. Unfortunately, we cannot see the mitochondrial network of the whole cell as the other half is lost. We also cannot exclude the possibility that small mitochondria that existed individually within cells might have been removed from samples during specimen preparation. Thus, further investigations are necessary to clarify the existence of one exclusive continuous mitochondrial network in *in vivo* cells.

This method can be applied only to cells that have scant cytoplasm, because in large cells with abundant cytoplasm, well-developed ER and the Golgi apparatus often prevent observation of the spatial arrangement of mitochondria. As described above, cytoskeletal materials such as microtubules and actin filaments are removed from cells by the AODO method. Therefore, an investigation into the structural and the functional relationships between the cytoskeleton and mitochondria is impossible. However, this disadvantage is favorable for investigations on the relationship between mitochondria and ER and/or the Golgi apparatus, since the disappearance of cytoskeletal materials make it possible to observe these structures unobstructed (Figure 4.11).

In contrast to studies using electron microscopy, SEM and TEM, investigations with light microscopy can provide information on dynamic changes in the mitochondrial structure of living cells. On the other hand, even the highest resolution light microscopy modality does not deliver sufficient resolution to study mitochondrial fine structure (see Section 4.7). Combining light microscopy with electron microscopy is therefore essential to better understand mitochondrial morphology and function (Perkins *et al.*, 2009).

The present HR-FE-SEM study has some advantages relative to TEM methods. The SEM method allows rather easy exploration of a large number of cells within a sample, whereas TEM methods, reconstructions from serial thin sections or TEM tomography, are time-consuming and more restricted by the sample size. However, they do offer more complete and accurate three-dimensional images of mitochondria.

4.14 Technical aspects

Five male albino Wistar rats 16 to 20 weeks old, which had been maintained on a standard laboratory diet, were used in this study. The animals were anesthetized with an intraperitoneal injection of Nembutal (50 mg/kg of body weight). They received an intracardiac perfusion with a Ringer solution followed by 0.5% glutaraldehyde plus 0.5% paraformaldehyde in M/15 phosphate buffer solution at pH 7.4. The dissected olfactory bulbs were post-fixed in 1% osmium tetroxide in the same buffer solution for 2 hr. After rinsing with the buffer solution, they were successively immersed in 15%, 30%, and 50% demethyl sulfoxide in water for 30 min each. The specimens were frozen in liquid nitrogen and cracked with a razor blade and hammer. The cracked pieces were rinsed with the buffer solution and subsequently immersed in 0.1% osmium tetroxide in the buffer solution at 20 °C for 3 days. For conductive staining the specimens were treated with 2% tannic acid aqueous solution for 1 hr and 1% osmium tetroxide in the buffer solution for 1 hr. The specimens were dehydrated in a graded ethanol series and dried by the critical point drying method. The dried specimens were coated with 3 nm of platinum in an ion sputter coater (E-1030; Hitachi Co. Ltd, Japan) and observed with HR-FE SEM (HFS-2ST, Hitachi Co. Ltd) at 25 kV. HR-FE-SEM images allow easy identification of subcellular organelles such as mitochondria, the Golgi apparatus, and rER and sER.

Adobe Photoshop on black and white photography was used to artificially color these photographs to facilitate structure recognition for the observer.

4.15 Acknowledgments

We thank Drs Bert Ph. M. Menco and Yoshimichi Kozuka for their careful reading of the manuscript and constructive suggestions. This work was supported in part by a grant from the Ministry of Education, Culture, Sports, Science, and Technology of Japan.

4.16 Abbreviations

SEM	scanning electron microscopy
HR-FE-SEM	high-resolution field-emission scanning electron microscopy
TEM	transmission electron microscopy
ER	endoplasmic reticulum
rER	rough endoplasmic reticulum
sER	smooth endoplasmic reticulum
ODO	osmium-DMSO-osmium
AODO	aldehyde-prefix ODO

4.17 References

Aldridge, A. C., Benson, L. P., Siegenthaler, *et al.* (2007). Roles for Drp1, a dynamin-related protein, and Milton, a kinesin-associated protein, in mitochondrial segregation, unfurling and elongation during *Drosophila* spermatogenesis. *Fly*, **1**, 38–46.

Altman, J. (1969). Autoradiographic and histological studies of postnatal neurogenesis. IV. Cell proliferation and migration in the anterior forebrain, with special reference to persisting neurogenesis in the olfactory bulb. *J. Comp. Neurol.*, **137**, 433–458.

Anesti, V. and Scorrano, L. (2006). The relationship between mitochondrial shape and function and the cytoskeleton. *Biochim. Biophys. Acta*, **1757**, 692–699.

Bakeeva, L. E., Chentsov, Yu, S., and Skulachev, V. P. (1978). Mitochondrial framework (reticulum mitochondriale) in rat diaphragm muscle. *Biochim. Biophys. Acta.*, **501**, 349–369.

Bawa, S. R. and Werner, G. (1988). Mitochondrial changes in spermatogenesis of the pseudoscorpio*n, Diplotemnus* sp. *Journal of Ultrastructural Research*, **98**, 281–293.

Bayer, S. A. (1983). 3H-thymidine-radiographic studies of neurogenesis in the rat olfactory bulb. *Exp. Brain Res.*, **50**, 329–340.

Benard, G. and Karbowski, M. (2009). Mitochondrial fusion and division: Regulation and role in cell viability. *Int. J. Biochem. Cell. Biol.*, **41**, 2566–2577.

Bereiter-Hahn, J. (1990). Behavior of mitochondria in the living cell. *Int. Rev. Cytol.*, **122**, 1–63.

Bereiter-Hahn, J. and Vöth, M. (1994). Dynamics of mitochondria in living cell: shape changes, dislocations, fusion, and fission of mitochondria. *Microsc. Res. Tech.*, **27**, 198–219.

Bereiter-Hahn, J. and Jendrach, M. (2010). Mitochondrial dynamics. *Int. Rev. Cell. Mol. Biol.*, **284**, 1–65.

Brandt, J. T., Martin, A. P., Lucas, F. V., and Vorbeck, M. L. (1974). The structure of rat liver mitochondria: A reevaluation. *Biochem. Biophys. Res. Commun.*, **59**, 1097–1102.

Carleton, A., Petreanu, L. T., Lansford, R., Alvarez-Buylla, A., and Lledo, P. M. (2003). Becoming a new neuron in the adult olfactory bulb. *Nature Neurosci.*, **6**, 507–518.

Cerveny, K. L., Tamura, Y., Zhang, Z., Jensen, R. E., and Sesaki, H. (2007). Regulation of mitochondrial fusion and division. *Trends Cell Biol.*, **17**, 563–569.

Chan, D. C. (2006). Mitochondria: dynamic organelles in disease, aging, and development. *Cell*, **125**, 1241–1252.

Chang, D. T. and Reynolds, I. J. (2006). Mitochondrial trafficking and morphology in healthy and injured neurons. *Prog. Neurobiol.*, **80**, 241–268.

Chen, H., Detmer, S. A., Ewald, A. J., *et al.* (2003). Mitofusins Mfn1 and Mfn2 coordinately regulate mitochondrial fusion and are essential for embryonic development. *J. Cell Biol.*, **160**, 189–200.

Collins, T. J., Berridge, M. J., Lipp, P., and Bootman, M. D. (2002). Mitochondria are morphologically and functionally heterogeneous within cells. *EMBO J.*, **21**, 1616–1627.

Collins, T. J. and Bootman, M. D. (2003). Mitochondria are morphologically heterogeneous within cells. *J. Exp. Biol.*, **206**, 1993–2000.

Dlasková, A., Spacek, T., Santorová, J., *et al.* (2010). 4Pi microscopy reveals an impaired three-dimensional mitochondrial network of pancreatic islet beta-cells, an experimental model of type-2 diabetes. *Biochim. Biophys. Acta*, **1797**, 1327–1341.

Egner, A., Jakobs, S., and Hell, S. W. (2002). Fast 100 nm resolution three-dimensional microscope reveals structural plasticity of mitochondria in live yeast. *Proc. Nat. Acad. Sci. USA*, **99**, 3370–3375.

Fiske, B. K. and Brunjes, P. C. (2001). Cell death in the developing and sensory-deprived rat olfactory bulb. *J. Comp. Neurol.*, **431**, 311–319.

Frank, S., Gaume, B., Bergmann-Leitner, E. S., *et al.* (2001). The role of dynamin-related protein 1, a mediator of mitochondrial fission, in apoptosis. *Dev. Cell*, **1**, 515–525.

Gao, W., Pu, Y., Luo, K. Q., and Chang, D. C. (2001). Temporal relationship between cytochrome c release and mitochondrial swelling during UV-induced apoptosis in living HeLa cells. *J. Cell Sci.*, **114**, 2855–2862.

Hales, K. G. and Fuller, M. T. (1997). Developmentally regulated mitochondrial fusion mediated by a conserved, novel, predicted GTPase. *Cell*, **90**, 121–129.

Ho, H. C. and Wey, S. (2007). Three dimensional rendering of the mitochondrial sheath morphogenesis during mouse spermiogenesis. *Microsc. Res. Tech.*, **70**, 719–723.

Hoffmann, H. P. and Avers, C. J. (1973). Mitochondrion of yeast: ultrastructural evidence for one giant, branched organelle per cell. *Science*, **181**, 749–751.

Hollenbeck, P. J. and Saxton, W. M. (2005). The axonal transport of mitochondria. *J. Cell Sci.*, **118**, 5411–5419.

Hom, J. R., Gewandter, J. S., Michael, L., Sheu, S. S., and Yoon, Y. (2007). Thapsigargin induces biphasic fragmentation of mitochondria through calcium-mediated mitochondrial fission and apoptosis. *J. Cell. Physiol.*, **212**, 498–508.

Hoppins, S., Lackner, L., and Nunnari, J. (2007). The machines that divide and fuse mitochondria. *Annu. Rev. Biochem.*, **76**, 751–780.

Jahr, C. E. and Nicoll, R. A. (1982). An intracellular analysis of dendrodendritic inhibition in the turtle *in vitro* olfactory bulb. *J. Physiol.*, **326**, 213–234.

Jendrach, M., Pohl, S., Vöth, M., *et al.* (2005). Morpho-dynamic changes of mitochondria during aging of human endothelial cells. *Mech. Aging Dev.*, **126**, 813–821.

Ježek, P. and Plecitá-Hlavatá, L. (2009). Mitochondrial reticulum network dynamics in relation to oxidative stress, redox regulation, and hypoxia. *Int. J. Biochem. Cell Biol.*, **41**, 1790–1804.

Kaplan, M. S. and Hinds, J. W. (1977). Neurogenesis in the adult rat: electron microscopic analysis of light radioautographs. *Science*, **197**, 1092–1094.

Karbowski, M. and Youle, R. J. (2003) Dynamics of mitochondrial morphology in healthy cells and during apoptosis. *Cell Death Differ.*, **10**, 870–880.

Kayar, S. R., Hoppeler, H., Mermod, L., and Weibel, E. R. (1988). Mitochondrial size and shape in equine skeletal muscle: A three-dimensional reconstruction study. *Anat. Rec.*, **222**, 333–339.

Kawano, S., Takano, H., and Kuroiwa, T. (1995). Sexuality of mitochondria: fusion, recombination, and plasmids. *Int. Rev. Cytol.*, **161**, 49–110.

Kirkwood, S. P., Packer, L., and Brooks, G. A. (1987). Effects of endurance training on a mitochondrial reticulum in limb skeletal muscle. *Arch. Biochem. Biophys.*, **255**, 80–88.

Koga, D. and Ushiki, T. (2006). Three-dimensional ultrastructure of the Golgi apparatus in different cells: high-resolution scanning electron microscopy of osmium-macerated tissues. *Arch. Histol. Cytol.*, **69**, 357–374.

Lee, Y. J., Jeong, S. Y., Karbowski, M., Smith, C. L., and Youle, R. J. (2004). Roles of the mammalian mitochondrial fission and fusion mediators Fis1, Drp1 and Opa1 in apoptosis. *Mol. Biol. Cell*, **15**, 5001–5011.

Legros, F., Lombès, A., Frachon, P., and Rojo, M. (2002). Mitochondrial fusion in human cells is efficient, requires the inner membrane potential, and is mediated by mitofusin. *Mol. Biol. Cell*, **13**, 4343–4354.

Mai, S., Klinkenberg, M., Auburger, G., Bereiter-Hahn, J., and Jendrach, M. (2010). Decreased expression of Drp1 and Fis1 mediates mitochondrial elongation in senescent cells and enhances resistance to oxidative stress through PINK1. *J. cell Sci.*, **123**, 917–926.

Malka, F., Guillery, O., Cifuentes-Diaz, C., *et al.* (2005). Separate fusion of outer and inner mitochondrial membranes. *EMBO Rep.*, **6**, 853–859.

Martinou, I., Desagher, S., Eskes, R., *et al.* (1999). The release of cytochrome c from mitochondria during apoptosis of NGF-deprived sympathetic neurons is a reversible event. *J. Cell Biol.*, **144**, 883–889.

Mattenberger, Y., Di, J., and Martinou, J. C. (2003). Fusion of mitochondria in mammalian cells is dependent on the mitochondrial inner membrane potential and independent of microtubules or actin. *FEBS Lett.*, **538**, 53–59.

McBride, H. M., Neuspiel, M., and Wasiak, S. (2006). Mitochondria: more than just a powerhouse. *Curr. Biol.*, **16**, R551–560.

Naguro, T. and Iino, A. (1989a). Three-dimensional features of the pancreatic cells. In: Riva, A. and Motta, P. M. ed., *Electron Microscopy in Biology and Medicine*, Vol. 6, *Ultrastructure of the Extraparietal Glands of Alimentary Tract*. Boston, Kluwer Academic Publ., pp. 147–175.

Naguro, T. and Iino, A. (1989b). The Golgi apparatus of the pancreatic acinar cells observed by scanning electron microscopy. In: Motta, P. M. ed., *Progress in Clinical Biological Research 295, Cells and Tissues: A Three-Dimensional Approach by Modern Techniques in Microscopy.* New York, Alan R. Liss, Inc., pp. 249–256.

Naguro, T. and Iwashita, K. (1992). Olfactory epithelium in young adult and aging rats as seen with high-resolution scanning electron microscopy. *Microsc. Res. Tech.*, **23**, 62–75.

Naguro, T., Kameie, T., Breipohl, W., Iino, A., and Reutter, K. (1994). Ultrastructural aspects of the mitral cells in the rat olfactory bulb as observed by high resolution scanning electron microscopy. In: Apfelbach, R., Müller, D., Reutter, K., Weiler, E. ed., *Advances in the Biosciences*, Vol. 93, *Chemical Signals in Vertebrates VII*. Oxford, Elsevier Science Ltd, pp. 105–117.

Naguro, T., Kameie, T., and Iino, A. (1997). Intracellular structure of nerve cells in the rat olfactory bulb as seen with high-resolution scanning electron microscopy. In: Motta, P. M. ed., *Recent Advances in Microscopy of Cells Tissues and Organs*. Rome, Antonio Delfino Editore, pp. 193–198.

Najbauer, J. and Leon, M. (1995). Olfactory experience modulated apoptosis in the developing olfactory bulb. *Brain Res.*, **674**, 245–251.

Navratil, M., Terman, A. and Arriaga, E. A. (2008). Giant mitochondria do not fuse and exchange their contents with normal mitochondria. *Exp. Cell Res.*, **314**, 164–172.

Ogata, T. and Yamasaki, Y. (1985). Scanning electron-microscopic studies on the three-dimensional structure of sarcoplasmic reticulum in the mammalian red, white, and intermediate muscle fibers. *Cell Tissue Res.*, **242**, 461–467.

Okamoto, K. and Shaw J. M. (2005). Mitochondrial morphology and dynamics in yeast and multicellular eukaryotes. *Annu. Rev. Genet.*, **39**, 503–536.

Osteryoung, K. W. and Nunnari, J. (2003). The division of endosymbiotic organelles. *Science*, **302**, 1698–1704.

Otani, H., Tanaka, O., Kasai, K., and Yoshioka, T. (1988). Development of mitochondrial helical sheath in the middle piece of the mouse spermatid tail: Regular dispositions and synchronized changes. *Anat. Rec.*, **222**, 26–33.

Overly, C. C., Rieff, H. I., and Hollenbeck, P. J. (1996). Organelle motility and metabolism in axons vs. dendrites of cultured hippocampal neurons. *J. Cell Sci.*, **109**, 971–980.

Perkins, G. A., Sun, M. G., and Frey, T. G. (2009). Correlated light and electron microscopy/electron tomography of mitochondria *in situ*. *Meth. Enzymol.*, **456**, 29–52.

Pathak, D., Sepp, K. J., and Hollenbeck, P. J. (2010). Evidence that myosin activity opposes microtubule-based axonal transport of mitochondria. *J. Neurosci.*, **30**, 8984–8992.

Petreanu, L. and Alvarez-Buylla, A. (2002). Maturation and death of adult-born olfactory bulb granule neurons: role of olfaction. *J. Neurosci.*, **22**, 6106–6113.

Plecitá-Hlavatá, L., Lessard, M., Santorová, J., Bewersdorf, J., and Jezek P. (2008). Mitochondrial oxidative phosphorylation and energetic status are reflected by morphology of mitochondrial network in INS-1E and HEP-G2 cells viewed by 4Pi microscopy. *Biochim. Biophys. Acta*, **1777**, 834–846.

Popov, V., Medvedev, N. I., Davies, H. A., and Stewart, M. G. (2005). Mitochondria form a filamentous reticular network in hippocampal dendrites but are present as discrete bodies in axons: A three-dimensional ultrastructural study. *J. Comp. Neurol.*, **492**, 50–65.

Rube, D. A. and Van der Bliek, A. M. (2004). Mitochondrial morphology is dynamic and varied. *Mol. Cell. Biochem.*, **256**–257, 331–339.

Scorrano, L. (2009). Opening the doors to cytochrome c: changes in mitochondrial shape and apoptosis. *Int. J. Biochem. Cell Biol.*, **41**, 1875–1883.

Sheridan, C. and Martin, S. J. (2010). Mitochondrial fission/fusion dynamics and apoptosis. *Mitochondrion*, **10**(6), 640–648.

Stolz, B. and Bereiter-Hahn, J. (1987). Sequestration of iontophoretically injected calcium by living endothelial cells. *Cell Calcium*, **8**, 103–121.

Skulachev, V. P. (2001). Mitochondrial filaments and clusters as intracellular power-transmitting cables. *Trends Biochem. Sci.*, **26**, 23–29.

Skulachev, V. P., Bakeeva, L. E., Chernyak, B. V., *et al.* (2004). Thread-grain transition of mitochondrial reticulum as a step of mitoptosis and apoptosis. *Mol. Cell. Biochem.*, **256**, 341–358.

Tanaka, K. (1987). Eukaryotes: Scanning electron microscopy of intracellular structures. In: Bourne, G. H., ed., *Cytology and Cell Physiology, 4th Ed., Int. Rev. Cytol. Suppl. 17*. New York, Academic Press, Inc., pp. 89–120.

Tanaka, K. and Mitsushima A. (1984). A preparation method for observing intracellular structures by scanning electron microscopy. *J. Microsc.*, **133**, 213–222.

Tanaka, K., Mitsushima, A., Fukudome, H., and Kashima, Y. (1986). Three-dimensional architecture of the Golgi complex observed by high resolution scanning electron microscopy. *J. Submicrosc. Cytol.*, **18**, 1–9.

Tanaka, K. and Naguro, T. (1981). High resolution scanning electron microscopy of cell organelles by a new specimen preparation method. *Biomed. Res.*, **2** *(Suppl)*, 63–70.

Twig, G., Elorza, A., Molina, A. J., *et al.* (2008). Fission and selective fusion govern mitochondrial segregation and elimination by autophagy. *EMBO J.*, **27**, 433–446.

Tates, A. D. (1971) Cytodifferentiation during spermatogenesis in *Drosophila melanogaster*: An electron microscopes study. PhD Thesis, Leiden University, The Netherlands.

Tondera, D., Grandemange, S., Jourdain, A., *et al.* (2009). SLP-2 is required for stress-induced mitochondrial hyperfusion. *EMBO J.*, **28**, 1589–1600.

Unterluggauer, H., Hütter, E., Voglauer, R., *et al.* (2007). Identification of cultivation-independent markers of human endothelial cell senescence *in vitro*. *Biogerontology*, **8**, 383–397.

Wang, X. and Schwarz, T. L. (2009). Imaging axonal transport of mitochondria. *Meth. Enzymol.*, **457**, 319–333.

Westermann, B. (2010). Mitochondrial fusion and fission in cell life and death. *Nature Rev. Mol. Cell Biol.*, **11**(12), 872–884.

Whitman, M. C. and Greer, C. A. (2007). Synaptic integration of adult-generated olfactory bulb granule cells: basal axodendritic centrifugal input precedes apical dendrodendritic local circuits. *J. Neurosci.*, **27**, 9951–9961.

Woolf, T. B., Shepherd, G. M., and Greer, C. A. (1991). Local information processing in dendritic trees: subsets of spines in granule cells of the mammalian olfactory bulb. *J. Neurosci.*, **11**, 1837–1854.

Yoon, Y. S., Yoon, D. S., Lim, I. K., *et al.* (2006). Formation of elongated giant mitochondria in DFO-induced cellular senescence: involvement of enhanced fusion process through modulation of Fis1. *J. Cell. Physiol.*, **209**, 468–480.

Zinsmaier, K. E., Babic, M., and Russo, G. J. (2009). Mitochondrial transport dynamics in axons and dendrites. *Results Probl. Cell Differ.*, **48**, 107–139.

Zottini, M., Barizza, E., Bastianelli, F., Carimi, F., and Lo Schiavo, F., (2006). Growth and senescence of *Medicago truncatula* cultured cells are associated with characteristic mitochondrial morphology. *New Phytol.*, **172**, 239–247.

5 Is the scanning mode the future of electron microscopy in cell biology?

Paul Walther, Christopher Schmid, Michaela Sailer, and Katharina Höhn

5.1 Introduction

Scanning electron microscopy (SEM) has been treated like the red-headed stepchild of cell biological electron microscopic research in the past. This was due to the fact that only surfaces could be imaged, while most biological processes occur in the bulk of cells and tissue. In addition, the resolution of SEMs was considerably poorer than the resolution of standard transmission electron microscopes (TEM). The breakthrough has been the emerging use of field emission electron sources, introduced by Crewe, with very small energy distribution, allowing for electron probe diameters of less than a nanometer (Crewe and Wall, 1970). Since most of the specimen preparation techniques for biological samples do not conserve structural details below 1 nm anyway, modern SEM instruments are no longer limited because of resolution, and resolution-wise there is no major difference, whether such a sample is imaged with an SEM or with a standard bright field TEM. Now, the advantages of the scanning mode can be fully used!

These advantages are:

- No limitation of sample thickness. Using the secondary electron or the backscattered electron signal, bulk samples can be analyzed. In the scanning transmission (STEM) mode, thicker samples can be investigated than in the regular bright field TEM mode, since inelastic scattered electrons do not cause image blurring (Aoyama *et al.*, 2008).
- No limitation of image resolution. In regular TEM, images are usually recorded with a CCD camera with a limited amount of pixels. Current standards are 2k × 2k pixels; cameras with more pixels are very expensive. When a beam is scanned over a sample, there is, however, no limitation on the amount of pixels generated. The ZEISS system ATLAS, for example, allows for taking STEM images as large as 32k × 32k. In addition, using the scanning mode facilitates rotation of the image by just changing the scan direction; the image, once focused, stays in focus, independent of magnification changes. All these factors allow for much easier automatic image acquisition by using a scanned beam.
- A number of different signals can be used to produce an image of the specimen.
- There is no limitation on the size of the specimen chamber in an SEM. An SEM can, therefore, be combined with other instruments, such as ultramicrotomes or even

Scanning Electron Microscopy for the Life Sciences, edited by H. Schatten. Published by Cambridge University Press

focused ion beam columns (FIB). These two techniques have rapidly emerged during the past years for three-dimensional reconstruction of biological samples. With both systems the SEM basically images the bulk blockface of a truncated sample. After each image, a small portion (e.g. 50 nm or less) of the sample is removed by cutting either with an ultramicrotome installed in the SEM (Denk and Horstmann, 2004) or with a focused ion beam (e.g. Matthijs de Winter *et al.*, 2008). Both methods are very powerful, since, in contrast to TEM tomography, relatively large volumes (more than one cell) can be analyzed. The limitations of the method primarily relate to the investigator's time available and the enormous amount of data production.

5.1.1 Aim of this chapter

The aim of this chapter is to show a few out of the many preparation techniques and imaging modes possible, when using a scanning electron microscope equipped with different detectors. The goal is to obtain new insights into cell biology, but also to reconsider artifact formation due to specimen preparation.

5.2 Technical aspects

5.2.1 Conventional surface SEM (Figure 5.1A)

The human macrophages shown in Figure 5.1A were prepared for the SEM by conventional methods. Human macrophages were chemically fixed with 2.5% glutaraldehyde (in PBS and 1% saccharose) for 1 hr at room temperature. After washing with PBS, the cells were post-fixed with OsO_4 (2% in PBS) for 1 hr at room temperature. The samples were then gradually dehydrated in 30%, 50%, 70%, 90%, and 100% propanol (5 min at

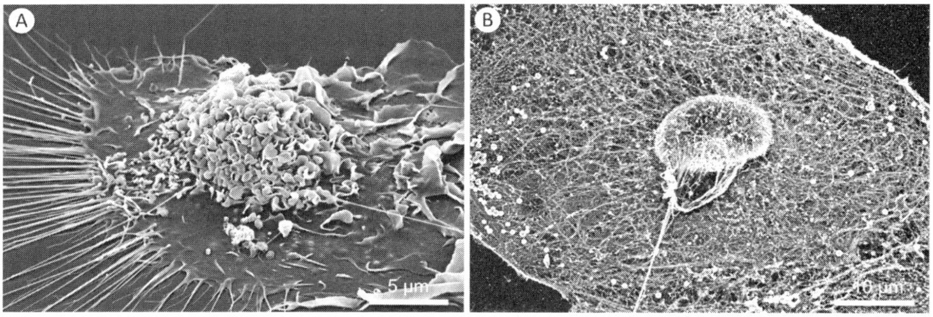

Figure 5.1 (A) A human macrophage, prepared by conventional chemical fixation and critical point drying and imaged with secondary electrons. With this method the surface of the sample, namely the surrounding plasma membrane, is imaged. The macrophages contain many cell surface protrusions such as lamelipodia, microvilli, and stress fibers. (B) A Panc 1 cell that has been extracted using triton before fixation. With this protocol the membranes and the soluble proteins are removed and cytoskeletal elements remain, especially the intermediate filaments.

each step) followed by critical-point drying using carbon dioxide as transition medium (Critical Point Dryer CPD 030, Bal-Tec, Principality of Liechtenstein). The samples were finally rotary coated in a BAF 300 freeze etching device (Bal-Tec, Principality of Liechtenstein) by electron beam evaporation with 3 nm of platinum-carbon. The thickness was controlled with a quartz crystal monitor.

The sample was then imaged in an S-5200 in-lens field emission SEM (Hitachi, Tokyo, Japan) at an accelerating voltage of 4 kV. Images were recorded using the standard secondary electron signal.

5.2.2 Extraction methods and SEM tomography (Figures 5.1B and 5.2)

The samples shown in Figure 5.1B and Figure 5.2 are Panc 1 cells prepared by a prefixation extraction method in order to visualize the intermediate filament network by high-resolution SEM (Sailer *et al.*, 2010). It is based on the protocols of Svitkina and Borisy (1998) and Svitkina (2007). After washing with phosphate-buffered saline

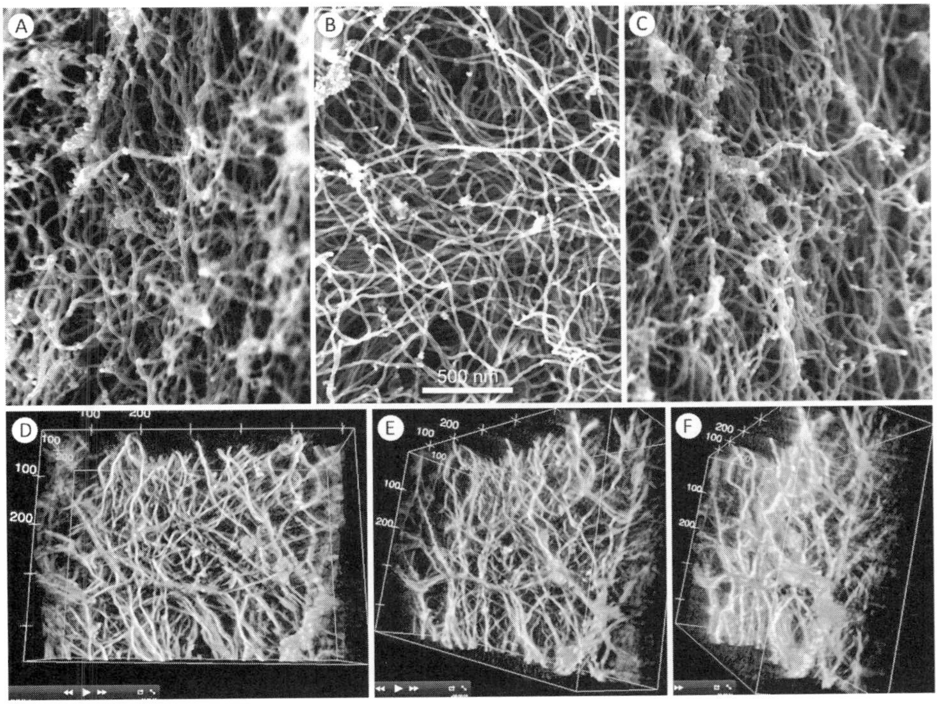

Figure 5.2 A portion of extracted Panc1 cells, where mainly the keratin intermediate filaments remained. The three-dimensional network architecture can be studied by SEM tomography (Sailer *et al.*, 2010). For this purpose the same area is imaged at different tilt angles. The upper row shows individual images at the maximum tilt of −60° (A) and 60° (C), as well as 0° (B). (D–F) For the tomogram in the lower row a tilt series of 61 images (from −60° to +60° with 2° tilt increment) was recorded and the 3-D dataset was calculated as explained in the techniques section. The 3-D dataset is visualized at different viewing angles with the 3-D viewer of the software FIJI based on ImageJ.

(PBS; pH 7.3), the cells were extracted for 25 min at around 281 K with 1% Triton X-100 (in PBS). Then, cells were washed again with PBS and chemically fixed with 2.5% glutaraldehyde (in PBS and 1% saccharose) for 1 hr at room temperature. After washing with PBS, the cells were post-fixed with OsO_4 (2% in PBS) for 1 hr at room temperature. After repeated washing with PBS, the samples were gradually dehydrated and critical-point dried as described above. Finally the samples were coated with a 5 nm layer of carbon in a freeze etching device (Baf 300, BalTec, Principality of Liechtenstein).

The samples were analyzed with an S-5200 SEM at an accelerating voltage of 5 or 10 kV. Images were recorded using the standard secondary electron signal. SEM tomography of extracted cells (Figure 5.2) was performed with the same microscope. For this purpose a holder, pre-tilted to 30°, was constructed, which allows tilting over a range from –60° to +60°, though the holder has to be turned by 180° and the sample needs to be remounted after recording half of the tilt series (Sailer *et al.*, 2010). Tomographical datasets were obtained at tilt angles from −60° to +60° at an increment of 2° with a magnification of 50 000 and an accelerating voltage of 5 kV using the secondary electron signal. This procedure resulted in 61 original images for computation of a single tomogram. Tomograms were reconstructed with the IMOD software by weighted back projection (Kremer *et al.*, 1996).

5.2.3 Cryo-SEM of high-pressure frozen and freeze fractured samples (Figure 5.3)

The cells shown in Figures 5.3, 5.4, and 5.5 were frozen using a Wohlwend HPF Compact 01 high-pressure freezer (Engineering Office M. Wohlwend GmbH, Sennwald, Switzerland; distributed in the USA by TECHNOTRADE International, Inc., 7 Perimeter Road, Manchester, NH 03103, USA).

For Figures 5.3A and 5.3B, samples were cryo-fractured as described by Walther (2008). A 300 mesh copper EM grid was dipped into a pellet of commercially available baker's yeast cells (*Saccharomyces cerevisiae*) kept in the stationary growth phase and mounted between the flat sides of two aluminum planchettes (Engineering Office M. Wohlwend GmbH, CH-9466 Sennwald, Switzerland). These sandwiches were high-pressure frozen without any cryo-protectants and without ethanol.

The sample shown in Figures 5.3C and 5.3D was high-pressure frozen as described by Höhn *et al.* (2011). A 50 μm gold spacer ring (diameter 3.05 mm, central bore 2 mm; Plano GmbH, Wetzlar, Germany) was mounted between two 170 μm thickness sapphire discs on which the human fibroblast cells were grown. These sandwiches were high-pressure frozen from the physiological state without aluminum planchettes and without the use of hexadecene.

The frozen sandwiches (either sapphire or aluminum) were mounted on a Bastacky holder (Walther, 2003). After transfer to the BAF 300 freeze-etching device (Bal-Tec, Principality of Liechtenstein), the temperature of the sample stage was raised to 155 K (vacuum about 2×10^{-7} mbar). The sandwiches were cracked with a steel knife and then double-layer coated as described by Walther *et al.* (1995) by electron beam evaporation with 3 nm of platinum-carbon from an angle of 45° and perpendicularly with about 6 nm of carbon. Thereby the additional carbon coat increases electrical conductivity (prevents charging), and, even more importantly, reduces the effects of beam damage

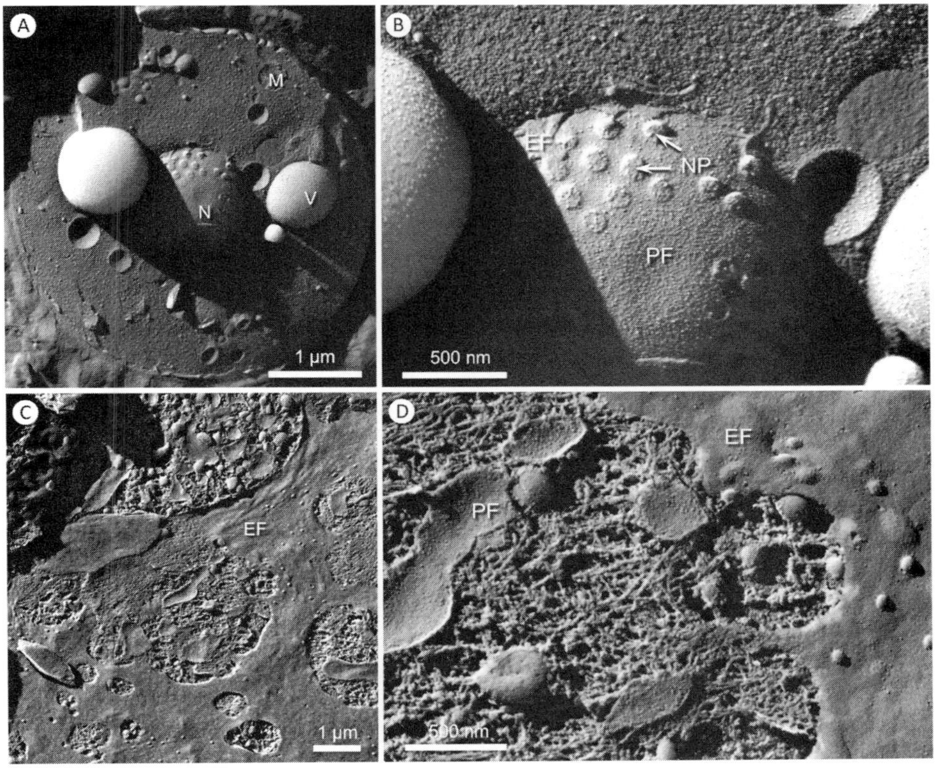

Figure 5.3 High-pressure frozen and cryofractured cells, imaged in the cryo-SEM using back-scattered electrons. (A) and (B) show a freeze fractured yeast cell at low (A) and at medium (B) magnification. A number of cellular organelles can be seen, such as the nucleus (N), nuclear membranes with the protoplasmic fracture face of the inner membrane (PF), the extraplasmic fracture face of the outer membrane (EF), nuclear pores (NP), vacuoles (V), and mitochondria (M). (C) and (D) are from a human fibroblast sample. It is assumed that the fracture plane mainly followed the basal plasma membrane that is close to the sapphire support. The so-called extraplasmic fracture face (EF) is visible, so the viewing direction is from inside the cell towards the outer half of the plasma membrane. The membrane has small cup-like infoldings that are most likely caveolae. In some areas fibrillar structures are visible. It is assumed that these are focal contacts, where the membrane is even closer to the sapphire disc and, therefore, the fracture plane passes through actin-rich regions above the basal membrane.

(Walther *et al.*, 1995). After coating, the freeze-etching device was vented with dry nitrogen and the Bastacky holder with the sample was quickly removed and immersed in liquid nitrogen, where it was mounted onto the Gatan cryo-holder 626 (Gatan, Inc., Pleasanton, CA, USA). Under liquid nitrogen (with the cryo-shield of the Gatan holder closed) the sample was transferred to the SEM and quickly inserted. After about 5 min the cryo-shield was opened. Specimens were investigated at a temperature of 153 K in the Hitachi S-5200 at a primary accelerating voltage (V_o) of 10 kV using the back-scattered electron signal. The contrast is thereby mainly formed at the underlying platinum layer, which stays in contact with the biological structure of interest (Walther *et al.*, 1995).

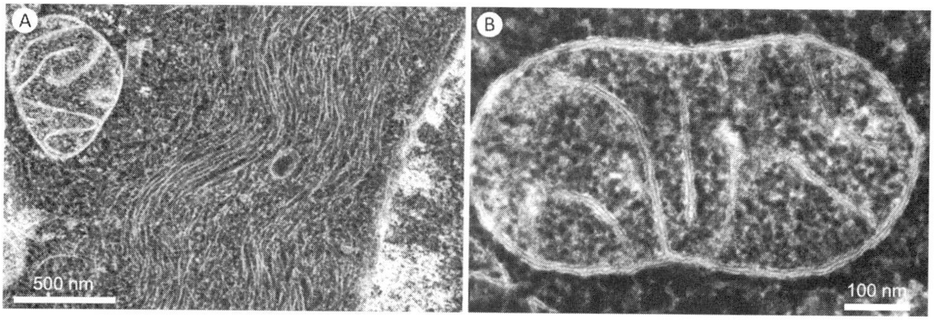

Figure 5.4 Low-voltage STEM images of high-pressure frozen and freeze substituted Panc 1 cells. The post-stained ultra-thin sections are imaged with excellent contrast, due to the enhanced scattering at 30 kV compared to 300 kV in Figure 5.5. (A) An area rich in intermediate filaments close to the nucleus at the right hand side, and a mitochondrion at the left hand side. (B) A mitochondrion at higher magnification. The double membrane is resolved with high contrast. Inner and outer membranes are in very close apposition with no visible intermembrane space.

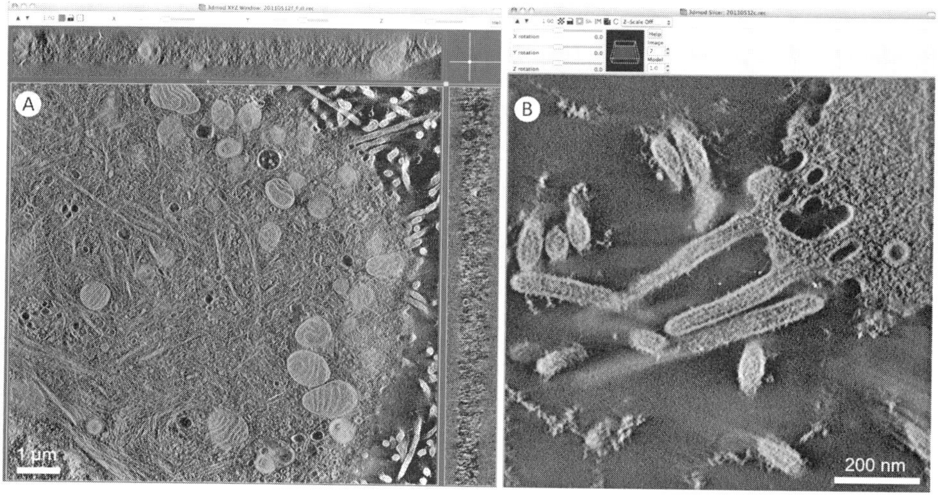

Figure 5.5 A high-voltage (300 kV) STEM tomogram of a 570 nm thick section (as measured in the electron microscope) of a high-pressure frozen Panc1 cell. The section plane consists of many intermediate filament bundles and mitochondria (A). (B) Microvilli with actin filaments inside and vesicles at the root of the microvilli that most likely are the result of endocytotic events controlled by microvilli.

5.2.4 Low-voltage STEM of high-pressure frozen and freeze substituted samples (Figure 5.4)

The Panc 1 cells shown in Figure 5.4 were high-pressure frozen as described above. During the following freeze substitution, water was replaced by the freeze substitution medium consisting of acetone, osmium tetroxide, uranyl acetate, and 5 % water (Walther and Ziegler, 2002). This procedure lasted about 16–18 hr, during which the temperature was slowly increased from 183 to 273 K. After substitution, the samples were kept at room temperature for 30 min to increase osmium fixation and then washed twice with acetone. After embedding of the samples in epon (polymerization at 333 K within 72 hr), they were cut with a microtome (Leica Ultracut UCT ultramicrotome, Leica, Vienna, Austria) using a diamond knife (Diatome, Biel, Switzerland). Ultra-thin sections with a thickness of 80 nm were mounted on copper grids for low-voltage STEM. For low-voltage STEM the sections were post-stained with lead citrate and uranyl acetate. Finally, the samples were coated with a thin layer of carbon (5 nm) on both sides.

Low-voltage STEM of 80 nm ultra-thin sections of high-pressure frozen and freeze substituted Panc 1 cells was performed with a Hitachi S-5200 equipped with a transmission detector that was used in dark field mode for best contrast. Images (1280 × 800 pixels) were recorded in STEM mode at an accelerating voltage of 30 kV.

5.2.5 High-voltage STEM tomography of high-pressure frozen and freeze substituted samples (Figure 5.5)

The Panc 1 cells shown in Figure 5.5 were prepared according to Höhn *et al.* (2011). The procedure is identical to the low-voltage STEM sample preparation described above, but the section thickness was 570 nm (instead of 80 nm). The sections were collected on bare copper grids with parallel grid bars in one direction only, to prevent the grid bars from obscuring the biological structures at high tilt angles (grids for tomography, diameter 3.05 mm, 300 bars per inch, Plano GmbH, Wetzlar, Germany). For tomography it is essential that the sections are mounted as flat as possible. It turned out to be rather difficult to attach the relatively thick sections on the grid bars. Therefore, the copper grids were coated with poly L-lysine (10% in water) before the sections were attached. The grids were then warmed on a heating table to a temperature of 60 °C to flatten the sections. Afterwards, the sections were again coated with poly-L-lysine to attach the 15 nm colloidal gold particles (Aurion, The Netherlands) on both sample sides. These gold particles served as fiducial markers for calculation of the tomograms. Finally, the mounted sections were carefully coated with 5 nm carbon from both sides by electron beam evaporation to increase electrical conductivity and mechanical stability. Improving electrical conductivity helps to reduce mass loss caused by ionization due to inelastic scattering (Walther *et al.*, 1995). Also, drift is reduced when electrical conductivity is enhanced by carbon coating. Before imaging, the samples were plasma cleaned for 10 s.

The tilt series images (−72° to +72°; 2° increment) were recorded with a 300 kV field emission STEM (Titan 80–300 TEM, FEI, Eindhoven) with an annular dark field detector (Fischione, Export, PA, USA) with a camera length of 301 mm. Illumination time per 1024 × 1024 pixel image was 18 s. A total of 73 images per tomogram were recorded. The tomograms shown in Figure 5.5 were recorded in "parallel beam mode" with a very small semi-convergence angle of 0.58 mrad, as outlined in Biskupek *et al.* (2010).

Tomograms were reconstructed by weighted back projection (WBP) or by a simultaneous iterative reconstruction technique (SIRT) with 25 iterations using the standard settings of the IMOD software package (Kremer *et al.*, 1996).

5.3 Data acquisition and analysis

This chapter is written with the aim of demonstrating the wide variety of sample preparation methods and imaging modes available with electron microscopy using a focused scanned electron beam as a probe.

5.3.1 Conventional surface SEM

Figure 5.1A shows a surface image of a human macrophage, prepared by conventional chemical fixation and critical point drying. With this method the surface of the sample, namely the surrounding plasma membrane, is imaged. Visualization is achieved using secondary electrons (Seiler, 1967) that are mainly released from the outer surface of the sample. This method is well suited to investigate the surface of the cell with all protrusions; delicate structures like stress fibers are well preserved.

5.3.2 Extraction methods and SEM-tomography

While SEM with secondary electrons is restricted to the physical surface, internal structures can be visualized as well when made accessible to the electron beam. For cytoskeletal structures this can be achieved by removing the overlying membranes and the soluble proteins by exposure to a detergent. Figure 5.1B shows a Panc 1 cell that has been extracted using triton before fixation. With this protocol some cytoskeletal elements are preserved, especially the stable intermediate filaments; these are often, but not always, arranged in bundles (compare with Figure 5A).

Figure 5.2 shows a small area of an extracted Panc1 cell, with mainly the keratin intermediate filaments remaining. The three-dimensional network architecture can be studied with SEM tomography (Sailer *et al.*, 2010; Figure 5.2). This method is similar to STEM tomography, since the sample is tilted under the electron beam and as many as 60 to 100 images of the sample are obtained at different tilt angles. These images are then back-projected by weighted back projection in order to calculate the 3-D structure. In the case of SEM tomography, the images produced with secondary electrons (Sailer *et al.*, 2010) are recorded. Low-energy secondary electrons can escape from the sample only when they are released at the very surface (Seiler, 1967). Therefore, SEM tomography works only with samples that consist of a loose fibrous network, surrounded by vacuum. This is fortunately the case for the intermediate filament network of extracted and critical point dried cells. The contrast of the intermediate filaments is very high when imaged with secondary electrons, since the signal is proportional to the surface area, which is large compared to the volume of the thin filaments.

5.3.3 Cryo-SEM of high-pressure frozen and freeze fractured samples

Cryo-SEM is a direct approach for imaging frozen samples and since the preparation is purely physical, artifact formation is limited and predictable. Cryo-fracturing is an old technique, introduced by Steere (1957) and Moor and Mühlethaler (1963) and it had its boom years in the 1970s. Since the preferred fracture plane is in the hydrophobic area of the bilayer membrane, freeze fracturing (or "freeze-etching" as it was usually called) was an important tool for exploration and verification of the "fluid mosaic model" of biological membranes (Singer and Nicolson, 1972). Branton *et al.* (1975) proposed a nomenclature for the different planes that occur by fracturing and partial freeze drying ("etching") that is also used in this work. Combining freeze fracturing with cryo-SEM has the advantage that the whole frozen and fractured sample can be analyzed on a cold stage in the SEM, and there is no need for tedious replica cleaning.

Figures 5.3A and 5.3B show a freeze fractured baker's yeast cell (*Saccharomyces cerevisiae*) at low (Figure 5.3A) and medium (Figure 5.3B) magnification. Many of the cell's organelles are clearly visible, since the fracture plane follows the membranes, as explained above. Freeze fracturing of yeast cells is easy, since they can be frozen as a three-dimensional pellet and the fracture plane will always hit some cells. One of the reasons that freeze fracturing became less popular in recent years relates to the fact that it is difficult to apply to adherent cells. An adherent cell monolayer is only a few micrometers thick and it is unlikely that the fracture plane will pass exactly through this area. Therefore, freeze fracturing of a cell monolayer remains a difficult process to achieve reproducible data. Different approaches have been tried in the past to increase reproducibility (e.g. Walther and Müller, 1997). For this work, the authors elaborated a protocol with their recently introduced sapphire discs with a thickness of 160 µm. These were chosen for their correlative light and electron microscopic approaches, since they have the same optical properties as cover slips. When fracturing a sandwich of two sapphire discs containing cultured cells on each, the probability that at least in some areas of the sample the fracture passes through the cells is relatively high. It is most likely that the basal membrane is the preferred area of the fracture plane. This allows interesting insights into this area, which is difficult to visualize by other methods. In focal contacts the fracture plane most likely passes through the actin rich area close to the membrane. Figures 5.3C and 5.3D are images from a human fibroblast sample. It is assumed that the fracture plane mainly followed the basal plasma membrane that is close to the sapphire support. The so-called extraplasmic fracture face is visible, so the viewing direction is from inside the cell towards the outer half of the plasma membrane. The membrane shows small cup-like infoldings that are most likely caveolae. In some areas fibrillar structures are visible. It is assumed that these are focal contacts, where the membrane is even closer to the sapphire disc and, therefore, the fracture plane passes through the terminal web, mainly consisting of actin filaments.

5.3.4 Low-voltage STEM of high-pressure frozen and freeze substituted samples

By introducing a detector below the sample, the SEM can easily be converted into a scanning transmission electron microscope (STEM; e.g. Reichelt and Engel, 1986).

Since the authors' Hitachi S-5200 SEM is operated at accelerating voltages from 1 to 30 kV, it is also well suited to generate information about contrast formation in low-voltage TEM. Figure 5.4 represents two low-voltage STEM images of high-pressure frozen and freeze substituted Panc1 cells. The post-stained ultra-thin sections are imaged with excellent contrast, due to the enhanced scattering at 30 kV compared to 300 kV as used in Figure 5.5. Figure 5.4A shows an area rich in intermediate filaments close to the nucleus and a mitochondrion. Figure 5.4B represents a mitochondrion at higher magnification. The double membrane is resolved with high contrast. In agreement with earlier observations (Höhn *et al.*, 2011), the inner and outer membranes are in very close apposition with no visible intermembrane space. The quality of the imaging and resolution is highly comparable to conventional bright field TEM imaging of similar samples. It remains to be proved whether low-voltage STEM is a useful alternative to TEM. As mentioned in the introduction, an advantage of STEM is the unlimited amount of pixels that can be used when recording an image. This is implemented in the Zeiss *Atlas* system.

5.3.5 High-voltage STEM tomography of high-pressure frozen and freeze substituted samples

STEM allows tomographical reconstructions of sections as thick as 1 μm (Aoyama *et al.*, 2008; Hohmann-Mariott *et al.*, 2009; Höhn *et al.*, 2011), since the image is not blurred by inelastically scattered electrons that cause chromatic aberration in projective lenses, simply due to the fact that there are no active projective lenses in a STEM. Figure 5.5 shows a high-voltage (300 kV) STEM tomogram of a 570 nm thickness section of a high-pressure frozen Panc1 cell. The 73 original images have been obtained with the so-called parallel beam mode (Biskupek *et al.*, 2010). The section plane is close to the basal surface. This area consists of many intermediate filament bundles and mitochondria (Figure 5.5A). Figure 5.5B of the same sample shows microvilli; these are actin-containing cell protrusions. The actin filaments are clearly visible. The vesicles at the root of the microvilli are most likely the result of endocytosis that involves microvilli functions.

5.4 Conclusions

The advantages of the scanning mode as discussed in the introduction can be utilized well in life science electron microscopy. It provides more flexibility in sample preparation since different electron signals can be used for image formation. The work presented in this chapter used secondary electron detection (Figures 5.1 and 5.2), back-scattered electron detection (Figure 5.3) and dark field (scattered) transmitted electron detection (Figures 5.4 and 5.5). A scanned beam allows for easier automated image acquisition, and this is an important first step towards more quantitative electron microscopy in the life sciences. The most important step for the future, however, will be quantitative image interpretation based on mathematical parameters (as shown by e.g. Lück *et al.*, 2010). This will move electron microscopy of biological samples from a more descriptive phase into more quantitative science applications.

5.5 Acknowledgments

We thank Eberhard Schmid, Reinhard Weih, and Renate Kunz for excellent technical support.

5.6 References

Aoyama, K., Takagi, T., Hirase, A., and Miyazawa, A. (2008). STEM tomography for thick biological specimens. *Ultramicroscopy*, **109**, 70–80.

Biskupek, J., Leschner, J., Walther, P., and Kaiser, U. (2010). Optimization of STEM tomography acquisition – a comparison of convergent beam and parallel beam STEM tomography. *Ultramicroscopy*, **110**, 1231–1237.

Branton, D., Bullivant, S., Gilula, N. B., *et al.* (1975). Freeze-etching nomenclature. *Science*, **190**, 54–56.

Crewe, A. V. and Wall, J. S. (1970). A scanning microscope with 5 Å resolution. *Journal of Molecular Biology*, **48**, 375–393.

Denk, W. and Horstmann, H. (2004). Serial block-face scanning electron microscopy to reconstruct three-dimensional tissue nanostructure. *PLoS Biology*, **2**(11), e329.

Lück, S., Sailer, M., Schmidt, V., and Walther, P. (2010). Three-dimensional analysis of intermediate filament networks using SEM tomography. *Journal of Microscopy*, **239**, 1–16.

Hohmann-Marriott, M. F., Sousa, A. A., Azari, A. A., *et al.* (2009). Nanoscale 3D cellular imaging by axial scanning transmission electron tomography. *Nature Methods*, **6**, 729–731.

Höhn, K., Sailer, M., Wang, L., *et al.* (2011). Preparation of cryofixed cells for improved 3D ultrastructure with scanning transmission electron tomography. *Histochemistry and Cell Biology*, **135**, 1–9.

Kremer, J. R., Mastronarde, D. N., and McIntosh, J. R. (1996). Computer visualization of three-dimensional image data using IMOD. *Journal of Structural Biology*, **116**, 71–76.

Matthijs De Winter, D. A., Schneijdenberg, C. T., Lebbink, M. N., *et al.* (2008). Tomography of insulating biological and geological materials using focused ion beam (FIB) sectioning and low-kV BSE imaging. *Journal of Microscopy*, **233**, 372–383.

Moor, H. and Mühlethaler, K. (1963). Fine structure in frozen-etched yeast cells. *Journal of Cell Biology*, **17**, 609–628.

Reichelt, R. and Engel, A. (1986). Contrast and resolution of scanning transmission electron microscope imaging modes. *Ultramicroscopy*, **19**, 43–56.

Sailer, M., Höhn, K., Lück, S., *et al.* (2010). Novel electron tomographic methods to study the morphology of keratin filament networks. *Microscopy and Microanalysis*, **16**, 462–471.

Seiler, H. (1967). Einige aktuelle Probleme der Sekundärelektronenemission. *Zeitschrift der angewandten Physik*, **22**, 249–263.

Singer, S. J. and Nicolson, G. L. (1972). The fluid mosaic model of the structure of cell membranes. *Science*, **175**, 720–731.

Steere, R. L. (1957). Electron microscopy of structural detail in frozen biological specimens. *Journal of Biophys. Biochem. Cytology*, **3**, 45–60.

Svitkina, T. M. and Borisy, G. G. (1998). Correlative light and electron microscopy of the cytoskeleton of cultured cells. *Methods in Enzymology*, **298**, 570–592.

Svitkina, T. M. (2007). Electron microscopic analysis of the leading edge in migrating cells. *Methods in Cell Biology*, **79**, 295–319.

Walther, P. and Ziegler, A. (2002). Freeze substitution of high-pressure frozen samples: the visibility of biological membranes is improved when the substitution medium contains water. *Journal of Microscopy*, **208**, 3–10.

Walther, P., Wehrli, E., Hermann, R., and Müller, M. (1995). Double layer coating for high resolution low temperature SEM. *Journal of Microscopy*, **179**, 229–237.

Walther, P. and Müller, M. (1997). Double layer coating for field emission cryo SEM – present state and applications. *Scanning*, **19**, 343–348.

Walther, P. (2003). Cryo-fracturing and cryo-planning for in-lens cryo-SEM, using a newly designed diamond knife. *Microscopy and Microanalysis*, **9**, 279–285.

Walther, P. (2008). High resolution cryoscanning electron microscopy of biological samples. In: Schatten, H. and Pawley, J. B. (ed.), *Biological Low-Voltage Scanning Electron Microscopy*. New York, Springer, pp. 245–261.

6 High-resolution labeling for correlative microscopy

Ralph M. Albrecht, Daryl A. Meyer, and O. E. Olorundare

6.1 Introduction

Correlative microscopy involves imaging the same sample by multiple imaging modalities. If labeling is involved, either the label has to be compatible with the different imaging modes or a combination of labels has to be employed. Typical correlative approaches often entail an initial examination of living or fixed, labeled cells by one or more light microscopy (LM) modes. These include wide field microscopy, laser scanning or spinning disc confocal, phase, DIC, AIC, and other interference-based imaging technology (Kachar, 1985; Olorundare *et al.*, 1987; Simmons *et al.*, 1990; Pawley, 2006; Skala *et al.*, 2007). Specialized high-resolution microscopy or "super" resolution microscopy such as STED, PALM, or STORM can increase the spatial resolution of fluorescent based labels but still in the context of LM images of structure (Hell and Wichmann, 1994; Rittweger *et al.*, 2009; Aaron *et al.*, 2011). LM is often followed by examination using one or more electron-based imaging modes. This may include use of SEM, TEM, or STEM instrumentation for structure and/or compositional analysis (Albrecht *et al.*, 2011). Various other imaging modes including one or more atomic force microscopy (AFM) approaches, X-ray microscopy, helium ion microscopy, and other imaging instrumentation can be used depending on the information desired (Russell and Butchelor, 2001). In all instances labels that can be unambiguously identified in each imaging mode must be used.

In biological systems, identification of the various molecular species participating in cellular function vis-à-vis their physical/spatial relationship to one another and to the biological structure and ultrastructure is essential. At the level of the light microscope, simultaneous multiple labeling is a powerful tool for localizing and co-localizing different antigens and epitopes in cells and tissues at spatial resolution consistent with photon based imaging. Commonly used labels for this purpose include fluorescent dyes having different excitation and emission spectra and/or various opaque markers, often gold (Au) spheres that can be detected in LM by their inflated diffraction image or intensified such that their size is increased by the chemical deposition of silver or gold on the particles following labeling (Scopsi, 1989; Simmons and Albrecht, 1989; Albrecht and Meyer, 2008).

Simultaneous detection/localization of different molecules at molecular and submolecular levels of resolution by electron microscopic imaging is somewhat less well developed. It requires small labels that can be differentiated from one another. A common

Scanning Electron Microscopy for the Life Sciences, edited by H. Schatten. Published by Cambridge University Press

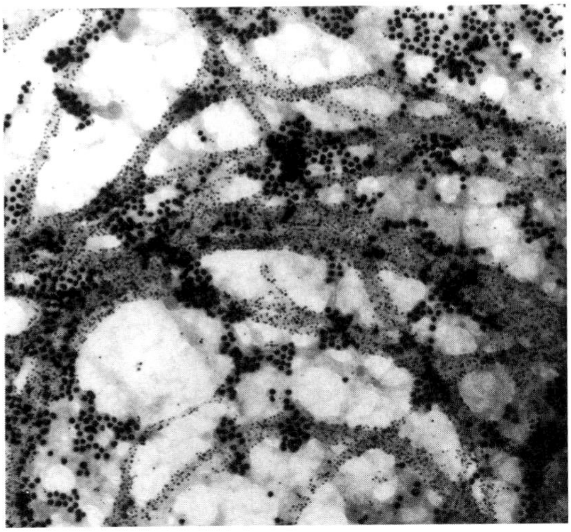

Figure 6.1 Typical labeling with two sizes, 5 nm and 15 nm, of colloidal gold particle. Size differences have to be substantial so that size overlap is not a problem. Use of the two particle sizes can be a problem where high spatial or quantitative resolution of both epitopes is desirable. Larger particles can block smaller particles and numbers of epitopes labeled per particle can vary considerably with larger particles. Here, larger particles label cell surface integrin epitopes while smaller particles label internal cytoskeletal elements.

approach has been to use colloidal metal particles of different sizes (Albrecht and Meyer, 2008) (Figure 6.1). This is not without some drawbacks. Where it is important to remain in the "molecular" range of resolution for epitope localization/co-localization the particles plus conjugated antibody or ligand have to be in the 3 to 10 nm range or even smaller. Due to variation in particle size, generally only two, possibly three, different sizes in this range can be used without substantial overlap (Park *et al.*, 1989). Particles can be synthesized with smaller variation and further size selected to increase the number of unique sizes in the sub-10 nm range but subsequent analysis of labels requires exact measuring of each particle to see to which category they belong (Albrecht and Meyer, 2002; 2003). This is usually impractical. Instrumentation to identify the very small particles accurately is also not available in all facilities. Closely located antigenic sites/epitopes can be masked by the larger colloidal markers, especially where epitope density is high and particles are in the larger size ranges. Semi-quantitative measurements or comparisons of epitope density are always very difficult, and basically impossible where particles of different size and hence different binding valence and spatial resolution are used.

The authors have investigated two approaches for simultaneous identification and localization (co-localization) of two or more epitopes. The first of these employs equally sized colloidal nanoparticles but synthesized of different heavy metal elemental composition. Gold, palladium, platinum, silver, rhodium, ruthenium, and other elements or combinations of elements can be made and conjugated to antibodies or ligands or to specific active antibody fragments or ligand fragments The particles are differentiated using energy filtering (EELS) imaging (Koeck *et al.*, 1996; Bleher *et al.*, 2008). Electron spectroscopic

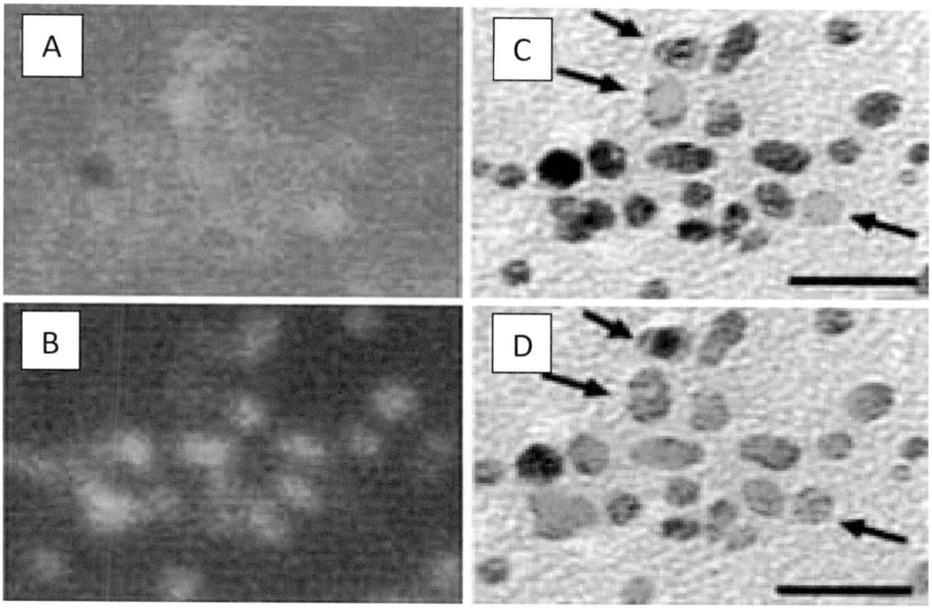

Figure 6.2 Localization of anti-α-actinin-cAu6 (A–C) and anti-actin-cPd6 (B–D) conjugates by ESI on a cluster of labels. Most of the spheres consist of Pd, only a few of the markers (arrows in C and D) show Au-specific signal. For detection of Au (bright particles in A), two images were taken before the O4,5 ionization edge at 40 eV and 50 eV for calculation of the background signal and one image above the Au O4,5-edge at 75 eV (A). (C) Elemental distribution of Au in an overlay of the Au-specific signal over a zero-loss image. For detection of Pd, background images were taken at energy losses of 280 eV and 324 eV, the maximum Pd-specific energy loss image at the M4,5-edge at 430 eV is shown in (B). Light particles correspond to Pd. The elemental distribution of Pd in an overlay of Pd-specific signal over the zero-loss image is shown in (D). Arrows indicate gold particles, the remainder are Pd particles. Bars represent 20 nm. (See plate section for color version.)

imaging (ESI) allows a number of particles to be imaged and identified simultaneously (Meyer and Albrecht, 2000; 2001; Bleher *et al.*, 2005) (Figures 6.2A–D (and color plate)). The primary downside to ESI imaging is that very thin specimens have to be used to avoid multiple loss events for beam electrons (Bleher *et al.*, 2008). It is not possible to use this approach with standard SEM imaging. Hence correlative SEM and EFTEM must be performed sequentially on the same specimen. The requirement for thin specimens limits the utility of SEM for structural information but BSE data may be useful. The recent development of aberration corrected TEM/STEM instrumentation has enabled production of very small, sub-angstrom, beam diameters with sufficient energy to permit EDX analysis on individual particles as small as 2 to 3 nm. Correlative SEM (secondary or back-scattered) or STEM imaging of biological structure can be combined with STEM-EDX analysis of particle composition and location (Figure 6.3). Use of EDX rather than EELS permits identification of particles in whole mounts and thick sections. This approach is compatible with electron tomographic imaging and reconstruction. In either case, EDX or

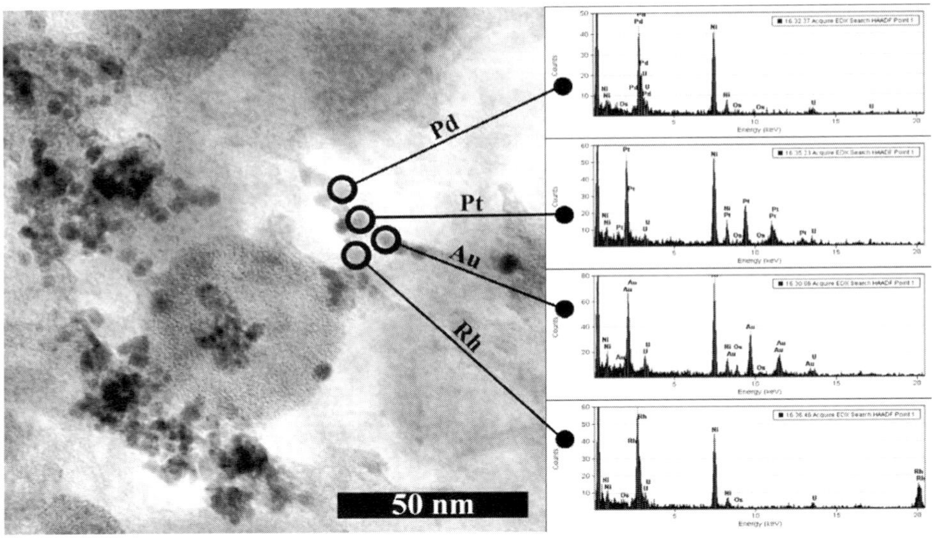

Figure 6.3 Quadruple labeled platelet. Pt 6 nm conjugated to human prothrombin; Pd 6 nm conjugated to anti alpha tubulin; Rh 3 nm conjugated to fibrinogen; and Au 8 nm conjugated to human CD42b (GPIb). Carbon peaks are from the support film, Ni peaks are from the grid, and Os peaks are from post-fixative, secondary staining.

EELS, core-shell particles can be labeled simultaneously. Core-shell particles composed of two or more elements can be distinguished from single element particles. This further increases the number of epitopes that can be labeled simultaneously.

A second approach to simultaneous multiple imaging at high spatial resolution is the use of particles of different shapes. Gold, palladium, platinum, and other heavy metal elements useful for EM labeling can be synthesized in different shapes including spheres, geometric shapes such as pyramids and cubes, and "popcorn" shaped particles (Meyer *et al.*, 2010) (Figure 6.4). While spheres can be synthesized as small as 2 nm, to date the minimum diameter for most of the shaped particles is in the 8 to 9 nm range and larger. Suitable high-resolution instrumentation such as TEM or STEM is required to visualize the smallest of the differently shaped particles clearly (Figure 6.5). High-resolution FESEM can also be used to identify particles of differing shapes (Figure 6.6). Higher accelerating voltages, 5 kV and above, in the secondary mode, or high-resolution back-scattered imaging, can be used to detect heavy metal particles such as gold, palladium, platinum, silver, ruthenium, rhodium, and others. Particles in the 3 nm range and smaller can be visualized. However, for detection of particles of different shape, particles of larger sizes, in the 5 nm to 20 nm range, are required. It should also be noted that the overall probe sizes will be larger due to the attached identifier such as Fc fragments of immunoglobulins, ligand fragments, ScFv fragments, whole antibody, or active ligand fragments. Any of these will add several nanometers to the overall probe size.

Until recently, use of EDX for detection and differentiation of smaller particles, 10 to 15 nm and smaller, of different elemental compositions, has been difficult, even with

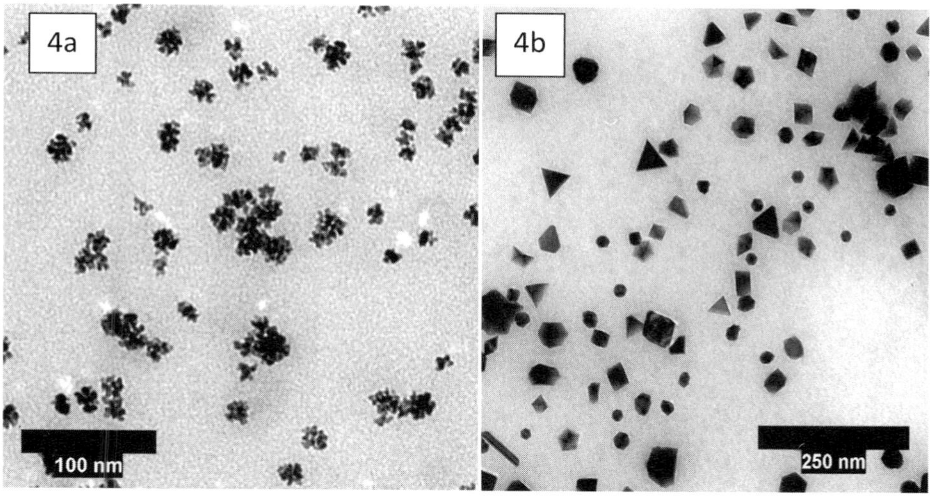

Figure 6.4 Pd nanoparticles. "Popcorn" or lobate particles (4a) and various "geometric" particles (4b).

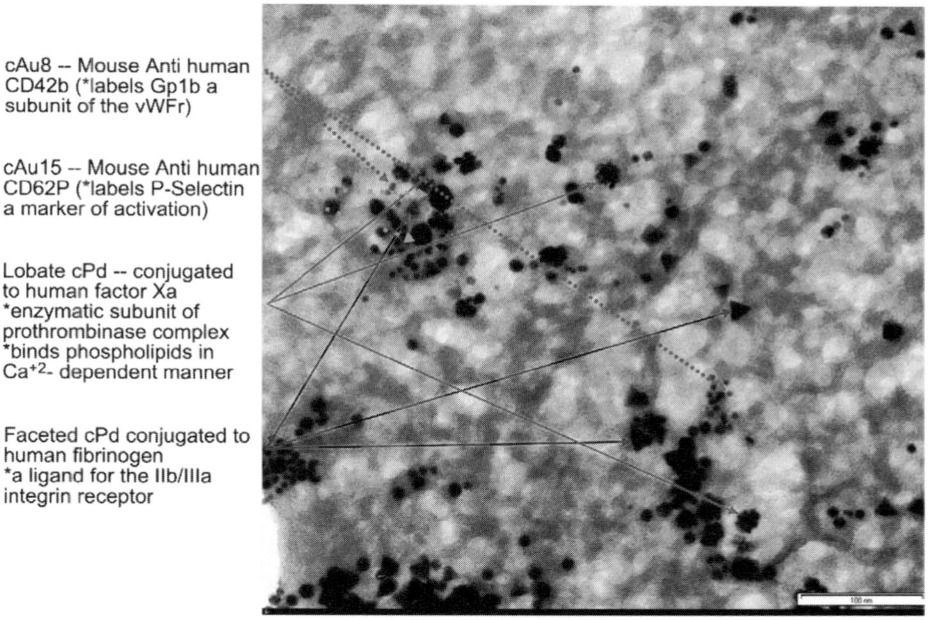

Figure 6.5 Platelet surface glycoproteins, antibody and ligand labeling.

high-resolution FESEM. This is due to beam diameters, beam energy, and sensitivity of EDX detectors. Here we show that aberration corrected STEM at 200 kV accelerating voltage can be used to differentiate particles readily in the 2 nm to 3 nm range which are directly adjacent to one another in the context of a biological specimen (Figure 6.3). This includes thick sections and whole mounts. The spherical correction of the objective lens

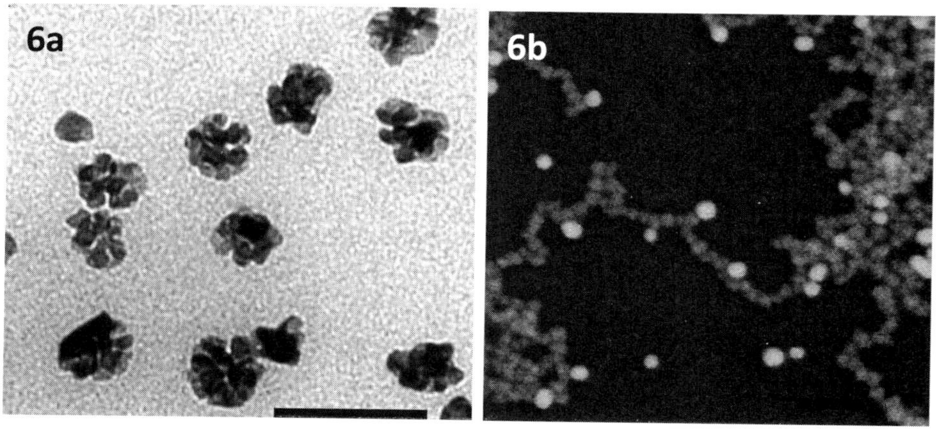

Figure 6.6 (6a) TEM of unconjugated "popcorn" palladium particles. Line = 50 nm. (6b) A LVSEM, BSE image, 5 kV accelerating voltage, of a mixture of unconjugated gold (round) and "popcorn" palladium (in chains). Line = 50 nm.

provides sub-angstrom beam diameters and sub-angstrom resolution along with EDX resolution of 136 eV or better with peak to background of 18 000/1. For the future, new generations of SEMs or possibly high-resolution proton or helium ion beam scanning instruments may substantially increase the sensitivity so that particles can be identified and their composition determined with the typical SEM type configurations and accelerating voltages (D. Joy, personal communication).

In all cases, whether particles of different composition or different shape are used, fluorescent second antibody can be used for the correlative light applications. The primary antibody or ligand, or active antibody fragment or ligand fragment, is conjugated directly to the particle so the spatial resolution for EM imaging is maximized. A second, fluorescent conjugated, antibody or antibody fragment specific for the primary antibody or primary antibody idiotype is employed. This provides sufficient fluorescent signal and avoids substantial, 99%+, fluorescence quenching, which occurs when the fluorescent labeled antibody or ligand is attached directly to the nanoparticle label, and it provides spatial resolution consistent with LM imaging modes (Kandela and Albrecht, 2007).

Another approach to simultaneous multiple labeling is the use of quantum dots otherwise known as nanocrystals. These are semiconductor crystals composed of periodic groups of II–VI, III–V, or IV–VI materials. The quantum cores range in size from 5–10 nm in diameter. At these small sizes, the materials behave differently, giving the dots high tuneability with respect to fluorescence emission and remarkable sensitivity, which allows their use in the development of multicolor fluorescence assays (Deerinck *et al.*, 2005; Sen *et al.*, 2008; Duke *et al.*, 2010). They also exhibit minimal photobleaching and a wide range of excitation wavelengths. On the downside, UV excitation can destroy the quantum cores and release toxic selenium. However, for biological applications the dots are coated with zinc sulfide plus an amphiphilic polymer and PEG to stabilize the particles and provide a surface to which active molecules can be conjugated. Although the quantum "core" particle may be of different diameter the size of the final label, regardless of original core size, can be in the 20 nm to 30 nm range or larger. This can substantially reduce their effectiveness for high-resolution labeling applications. At present, work is directed toward producing labels of smaller final size. Currently, the most common quantum core composition is cadmium–selenium (Nisman *et al.*, 2004). Other element combinations such as tellurium–selenium can be used, which may provide a basis for differentiation of particles by EELS or high-resolution EDX analysis. This would facilitate synthesis of particles having differing LM emission wavelengths without the need for fluorescent labeled second antibodies. However, in practice the lower density and small amount of metal has made the smaller quantum particles difficult to visualize and identify clearly by EM analytic procedures in biological specimens. The fluorescence yield of the particles is also lower when compared to multiple fluorescent dye molecules on a second antibody. Other possibilities for particles include yttrium or vanadium oxides doped with different rare earth elements such as europium (Patra *et al.*, 2007; Hilderbrand *et al.*, 2009; R. J. Hamers, personal communication). These particles exhibit strong fluorescence, are small, visible, and can be differentiated in EM, and are nontoxic. Such labels may be effective for correlative imaging via LM followed by helium ion scanning microscopy with luminescence and particle imaging done in the HeI microscope, although this precludes use of fluorescence on living or hydrated specimens.

Another method for live cell correlative LM and TEM labeling is the use of bioarsenical reagents for site-specific protein labeling, which was first described by Roger Tsien and colleagues in 1998. The labeling reagents FIAsH-EDT2 and ReAsH-EDT2 become fluorescent in the green and red, respectively, when they bind to recombinant proteins containing the tetracysteine motif, with the displacement of the EDT. The recombinant protein-tetracysteine is analogous to the well-known GFP type labels used for fluorescence LM studies. Fluorescence imaging is followed by TEM of the same cell after photoconversion of diaminobenzimidine and osmium staining (Gaietta *et al.*, 2002). In general, use of this methodology with SEM imaging and for simultaneous labeling of multiple targets remains to be evaluated.

6.2 Technical aspects

6.2.1 Particle synthesis and conjugation

Below methodology is described for the production of particles having different shapes and for the production of particles of different elemental composition. Methods for conjugation of typical protein identifier molecules or molecule fragments are also described.

Ovoid to spherical colloidal gold (cAu) particles may be obtained by reducing dilute solutions of tetrachloroauric acid with any of a number of agents, each of which will yield a specific particle size or range of sizes (Handley, 1989). Reduction with white phosphorus produces 3 nm or 5 nm cAu particles, while sodium citrate is commonly used to synthesize particles with diameters ranging from 16 to several hundred nm. Particles of intermediate size can be obtained by reduction with a combination of sodium citrate and tannic acid.

6.2.1.1 3 nm cAu

Some of the gold sols that Faraday made using white phosphorus and that he famously characterized in 1857 are still extant, and are located at the Faraday Museum of the Royal Society, London (Faraday, 1857). His method was subsequently modified by Turkevich *et al.* (1951) and later by Roth (1982).

To generate these small particles, the following procedure can be used: the pH of a 100 ml solution of 0.01% tetrachloroauric acid is adjusted to be slightly alkaline of neutrality, then brought to a boil. Using a magnetic stir bar to ensure rapid mixing, 0.5 ml of a saturated solution of white phosphorus in diethyl ether is rapidly injected. The color changes immediately to an orange-red, indicative of the sol formation. The sol is boiled an additional 10 to 30 min to ensure that the reaction has come to completion.

The saturated solution of white phosphorus in diethyl ether is prepared by adding finely diced shavings of phosphorus to a few milliliters of ether, followed by purging of the container with nitrogen or argon, and stirring with a magnetic stir bar until a cloudy suspension develops. The suspended particles are allowed to settle overnight before the saturated solution is used in the reaction. A saturated solution of copper sulfate is kept on hand to neutralize excess phosphorus and clean up any spills.

CAUTION: White phosphorus must always be stored under water and will spontaneously combust if left in air. Particle preparation is best carried out in a chemical hood and, as for all particle synthesis, gloves, eye protection, and appropriate laboratory clothing should be used.

6.2.1.2 3–17 nm cAu

Slot and Geuze (1985) modified a technique introduced earlier by Mühlpfordt (1982) in which a mixture of tannic acid and sodium citrate was used as the reducing agent. By varying the concentration of the tannic acid relative to the citrate concentration, one could produce monodisperse cAu sols having particles of discrete diameters ranging from 3 to 17 nm.

To make 100 ml of the sol, 1 ml of a 1% solution of tetrachloroauric acid is added to 79 ml ddH_2O and heated to 60 °C. The reducing solution is prepared by adding together 4 ml 1% trisodium citrate, 0 to 5 ml 1% tannic acid, a volume of 25 mM potassium bicarbonate equal to that of the tannic acid used, and sufficient ddH_2O to bring the total volume to 20 ml. The reducing solution is also warmed to 60 °C, after which it is added quickly to the gold solution with rapid mixing provided by a magnetic stir bar. Upon development of a reddish color, the timing of which varies from instantaneous to nearly an hour depending upon the concentration of the tannic acid, the sol is brought to a boil, then left to cool. The particle size is inversely related to the tannic acid concentration.

6.2.1.3 16–150 nm cAu

Suspensions of cAu particles larger than 16 nm are prepared as described by Frens (1973). The procedure is analogous to the tannic acid method previously described in that the strength of reducing agent used determines the particle size. However, in this case, tannic acid is not added and sodium citrate acts as the sole reducing agent. Bring 200 ml of a 0.01% solution of tetrachloroauric acid to a boil, along with a stock solution

Table 6.1 Synthesis of colloidal palladium

Sol	[Na ascorbate] (%)	Average diameter (nm)	σn/Xtnn (%)
1	0.2	6.6 ± 0.07	14.7
2	0.05	7.9 ± 0.09	17.8
3	0.01	10.1 ± 0.10	15.0
4	0.005	11.7 ± 0.17	21.5

of 1% trisodium citrate. Up to 4 ml of the citrate is then injected rapidly into the gold-containing flask with vigorous mixing. The color changes from clear to light purple to black as the auric ions are reduced to elemental gold atoms that then begin to nucleate. As additional metal condenses around the nuclei, the color again changes to various shades of red, ranging from clear, bright ruby for the smaller particle sizes, to violet for the intermediate sizes, and finally to an opaque rust with a bluish Tyndall effect for the largest particles. The sol is then refluxed an additional 20 min.

6.2.1.4 Colloids of other noble metals

The preparation of colloidal particles composed of other noble metals is similar to that of gold colloids in that a salt of the metal is chemically reduced in an aqueous medium. Sols of different mean particle diameter may be produced by using different reducing agents. However, methods analogous to Frens' citrate reduction technique or to Slot and Geuze's citrate/tannic acid technique, in which the size of cAu particles is controlled by the concentration of reducing agent used, have proven largely inadequate for synthesizing sols composed of other noble metal elements over the same range of sizes.

The authors have developed a procedure that utilizes differing concentrations of sodium ascorbate for synthesis of cPd particles having mean diameters ranging from 6.6 to 11.7 nm and having very small size variances (Meyer and Albrecht, 2003), see Table 6.1.

Apart from this particular method, a wholly different approach is usually necessary. One such approach is the Keimmethode (literally translated from the German, seed method) developed around the turn of the nineteenth century by the Nobel laureate, Richard Zsigmondy (1906). Zsigmondy's method used very small cAu particles, prepared using Faraday's white-phosphorus reduction technique, as nuclei around which additional gold could be condensed during a second step that used another reducing agent. In this way, larger particles could be produced, and the size was dependent largely upon the ratio of the number of nuclei to the total amount of gold salt. Some years later Voigt and Heumann applied Zsigmondy's work to synthesis of uniformly sized cAg particles (Voigt and Heumann, 1928). The size distribution of each preparation may be diminished by ultracentrifugation of the particles, either before or after conjugation, against a density gradient, followed by dialysis to remove the gradient material.

The authors too have found this method satisfactory for synthesis of cPt and cPd particles of graded sizes (Meyer and Albrecht, 2002), see Table 6.2.

Not only is Zsigmondy's method applicable to the synthesis of mono-metallic colloidal particles of different sizes, but it is also useful for the synthesis of multi-metallic, core-shell

Table 6.2 Synthesis of cPt using the nucleating sol procedure

Sol	Volume nucleating sol added (ml)	Average particle diameter (nm)
Nuc. sol	–	2.5 ± 0.03
1	10.0	5.5 ± 0.06
2	5.0	7.1 ± 0.06
3	2.5	9.0 ± 0.09
4	1.0	11.6 ± 0.12
5	0.5	13.4 ± 0.16

particles in which a small colloidal particle of one metal serves as the core around which a shell (or shells) of one or more other metals can be condensed (e.g. Henglein, 2000; Hodak *et al.*, 2001). The authors have used the multi-metallic, core-shell approach to make gold-coated platinum particles, which are easier to conjugate to antibody ligands than bare platinum particles (Bleher *et al.*, 2005), as well as gold-coated magnetite particles, which can be used to label cells. It is interesting to note that, once attached to a target cell, the magnetite can be inductively heated and used to produce membrane pores and specifically kill the labeled cells without damaging adjacent unlabeled cells. Neither the particles nor the oscillating magnetic field are harmful by themselves (Kaiser *et al.*, 2007).

6.2.2 Particle shapes (specific protocols useful for labeling)

6.2.2.1 6 nm faceted cPt

Particles of 6 nm cPt are prepared by reducing a dilute solution of platinum chloride with hydrogen gas. 500 ml ddH_2O, 1 ml 20% sodium citrate, and 3.25 ml 4% H_2PtCl_6 are combined in a round bottom flask and the pH is adjusted slightly above neutrality with small volumes of 1 M NaOH. The small amount of sodium citrate in this solution does not enable reduction of the platinum salt as is the case with gold salts; however, it ensures that the cPt particles, once developed, remain stable in suspension (Weiser, 1933). The flask is then heated in a boiling water bath with rapid mixing via a magnetic stir bar, flushed briefly with nitrogen gas, then flushed with hydrogen gas for five to ten minutes. A slight darkening of the initially clear solution is noticeable within one minute upon addition of hydrogen, which progresses gradually to a dark chocolate brown within 5 min. The flask is then sealed under a hydrogen atmosphere to ensure that the reaction has come to completion, and allowed to cool slowly over several hours to ambient temperature. The particles typically consist of a mixture of cubes and tetrahedrons.

6.2.2.2 6 nm faceted cPd

Particles of 6 nm cPd are prepared in a manner similar to that for cPt with the exception that 8.2 ml of a freshly prepared solution of 1% K_2PdCl_4 is used in place of the platinum salt. Darkening of the solution is immediate upon addition of hydrogen gas and progresses very rapidly to a dark chocolate brown within one minute. Cubes and tetrahedrons similarly are the shapes of the cPd particles.

6.2.2.3 18 nm faceted cPd

To make 100 ml of the sol, 200 µl of 20% trisodium citrate and 820 µl 1% dipotassium tetrachloropalladate are combined in 94 ml ddH_2O. Once the contents of the flask have reached a rapid boil, 5 ml of 6.4% ascorbic acid is added with vigorous mixing. The colloid forms instantaneously and the flask is refluxed for 20 min.

6.2.2.4 18 nm popcorn cPd

Solutions of 10% L-ascorbic acid and 1% dipotassium tetrachloropalladate are chilled in an ice-water bath. To make 500 ml cPd sol, 4.1 ml of the Pd salt solution is added to the ascorbic acid with vigorous agitation. The colloid begins to develop within a minute, the flask is returned to ice, and the reaction is allowed to come to completion overnight. The mean particle diameter is roughly 18 nm, see Figure 6.6.

6.2.3 Conjugation of colloidal gold, silver, platinum, and palladium markers (ligands, ligand fragments, antibodies, or antibody fragments)

Biologically active macromolecular protein ligands and their controls are attached to colloidal metal particles by hydrophobic interactions. These occur most stably when the pHs of both ligand and particles are adjusted to be slightly basic to the isoelectric point of the particular ligand, at which point the ligand is maximally hydrophobic. The slight net negative charge on the ligand and the negative surface charge of the particles ensure that less stable and undesirable charge-based interactions are minimized.

It is often necessary to dialyze the ligand against water or a dilute buffer prior to conjugation in order to remove excess salts that destabilize the colloidal metal particles. It is also desirable to conjugate the particles to the ligand at the lowest ligand concentration possible that will still keep the particles stable in buffered medium. Conjugation at higher ligand concentrations than necessary often results in less stable ligand packing conformations or even multiple layers, which are much more likely to dissociate from the particle and increase the concentration of unbound ligand in the medium. Free ligand subsequently competes with bound ligand for antigenic target sites and may consequently decrease both the efficiency and density of the labeling observed.

The minimum amount of ligand necessary to coat the particles in a stable monolayer is calculated empirically by means of concentration isotherms. In general, incrementally increasing amounts of ligand in microgram quantities are added to a series of test tubes followed by an equal volume of colloid, generally a milliliter or less, in order to produce a range of ligand concentrations. Following gentle agitation, one-tenth the volume of a saturated solution of sodium chloride is then added. The rapid influx of cations compresses an integral shell of negative charge surrounding each particle, whether they are conjugated to ligand or not. It is this negative surface charge that ordinarily maintains unconjugated particles in stable aqueous suspension via mutual charge repulsion. However, when the electrical charge is compressed sufficiently close to the particle surface, unconjugated particles may approach each other in near enough proximity to irreversibly flocculate as a consequence of the action of van der Waals and London dispersion forces. When stably conjugated, however, the ligand serves as a physical barrier that prevents particles from approaching one another with such close proximity to aggregate by atomic and molecular forces of attraction.

Any growth of unbound or otherwise unstable particles into larger particle clusters is accompanied by a wavelength shift of scattered light. For gold colloids, this is from red to blue, for platinum and palladium colloids, brown to grey, and for silver, yellow to red to blue-grey. Such color shifts are generally distinctive enough to be apparent to the naked eye, but may also be unambiguously measured spectrophotometrically, especially when tiny volumes are used.

Given enough time, all unstably conjugated particles will settle out of suspension, leaving behind a clear supernatant. The rate of particle flocculation is an exponential function of the valence of the cation of the particular salt used. Therefore, flocculation is more rapid when trivalent cations like Al^{+3} are used, than with divalent or monovalent cations such as Ca^{+2} or Na^{+1}, respectively. If there is sufficient ligand present in solution to stably coat each particle, then no color shift occurs following the addition of salt. In this way, the minimum amount of ligand needed to form stable particle conjugates may be readily determined as the first tube not to undergo this characteristic color shift.

If the isoelectric point of the ligand is unknown, it can be approximated in a similar manner. In this case, a constant volume of colloid at various pHs is added to tubes, each containing the same amount of ligand, and the stability of the conjugates is assessed following the addition of salt. This step may also be necessary when working with IgG antibodies, as these typically bind stably over a broader range of pH, from 7 to 10. Values published in the literature generally indicate lower values for polyclonals, around 7.2 to 7.4, but 9.0 to 9.5 for monoclonals.

Once the ligand concentration and optimal pH have been determined, ligands and particles may be conjugated in bulk. Often a stabilizing agent such as a small amount of PEG (a dimer of 20 000 MW) or bovine serum albumin (BSA), equivalent to 0.01 to 1% (w/v), may also be added to the conjugates of the larger nanoparticles (10 nm diameter and larger) to enhance stability. It is thought that the smaller molecules both cover any surface of the nanoparticle not attached to the ligand or antibody and also stabilize the ligand or antibody protein.

The conjugates are then lightly pelleted by ultracentrifugation, not only to concentrate them for storage, but also to remove any excess ligand and/or stabilizer left behind in the supernatant. This step also helps to remove aggregates, which tend to form hard pellets that adhere to the walls of the tube and cannot be resuspended. The supernatant is removed and discarded, and the pellet resuspended in medium or a buffer compatible with the tissue to be labeled. The relative particle concentration may be determined spectrophotometrically, and will, along with other variables such as ambient temperature, particle size, viscosity and the ionic strength of the medium, probe-target affinity, and steric hindrance, affect the amount of time necessary for labeling (Park *et al.*, 1988, 1989).

6.2.4 Synthesis of cRh/Fgn conjugates

The preparation of cRh and cRh/ligand conjugates is somewhat different from that for the other metal colloids in that, in the case of cRh, it is a one-step process as opposed to the two-step process for the others. Many metals, including some belonging to the group of noble metals, like Rh, do not form stable or completely stable colloidal suspensions in aqueous media unless protected from oxidation by any of a number of natural or synthetic secondary stabilizers. The authors have found it difficult to prepare stable, unprotected rhodium colloids reliably in the laboratory or to prepare stable cRh conjugates, even if the cRh itself appears stable.

To overcome this obstacle, stable cRh conjugates are made by synthesis of the cRh particles from rhodium salts in the presence of the ligand it is wished to conjugate. For labeling of IIbIIIa receptors on cell surfaces, particles are conjugated to fibrinogen, fgn, the ligand for the receptor. In this case human fgn is depleted of plasminogen, von Willebrand factor, and fibronectin and stored at –80 °C at a concentration of 15 mg/ml. Working dilutions of 2 mg/ml are prepared in ddH_2O and centrifuged at 14 000 rpm for 10 min to remove any aggregates. It should be further noted that dilutions should be made in tubes of polypropylene as opposed to polystyrene, as the latter results in irreversible protein aggregation.

cRh conjugates are prepared in a round bottom flask to which are added 48 ml ddH_2O, 645 µl 1% RhC13, 100 µl 20% sodium citrate, and 375 µl freshly prepared 2 mg/ml fgn for a final ligand concentration of 15 µg per ml of colloid. As with the other colloidal metal particle conjugates, fgn-conjugated cRh particles are stabilized with the addition of 1% (w/v) BSA, then pelleted by ultracentrifugation at approximately 12 000 *g* for 45 minutes at 4 °C. The supernatant is withdrawn and the pellet resuspended in one-tenth the original volume of HEPES-Tyrodes buffer supplemented with 2 mM $CaCl_2$. As a control for the labeling, cRh conjugated to 100 µg/mL BSA is prepared similarly by replacing the fgn in the reaction mixture with 50 µl 10% BSA. The size of the cRh metal particle portion of the markers alone, prepared in this fashion using either fgn or BSA, is approximately 3 nm. Once conjugated the labels can be used at physiological pH in standard buffer conditions.

6.3 Summary

This is a brief overview of current and developing methodology to provide correlative and high-resolution labeling of multiple epitopes simultaneously. The ability to label multiple epitopes on a single specimen, simultaneously, for both LM and high-resolution EM is important in understanding molecular interactions in a macromolecular structural/functional environment.

6.4 References

Aaron, J. S., Carson, B., and Timlin, J. A. (2011). Imaging innate immune responses using dual color stochastic reconstruction optical microscopy (STORM). *Microscopy and Microanalysis*, **17** *(Suppl. 2)*, 18–19.

Albrecht, R. M. and Meyer, D. A. (2002). All that glitters is not gold: Approaches to labeling for EM. *Microscopy and Microanalysis*, **8** *(Suppl. 2)*, 194–195.

Albrecht, R. M. and Meyer, D. A. (2008). Molecular labeling for correlative microscopy: LV, LVSEM, TEM, EFTEM and HVEM. In: *Biological Low-Voltage Scanning Electron Microscopy*. Schatten, H. and Pawley, J. (ed.). NY, Springer.

Albrecht, R. M., Olorundare, O., Oliver. J. A., and Meyer, D. A. (2011). Nanoparticle labels for co-localization and correlative imaging at high spatial resolution. *Microscopy and Microanalysis*, **17** (*Suppl. 2*), 358–359.

Bleher, R., Meyer, D. A., and Albrecht, R. M. (2005). High resolution multiple labeling for immuno-EM applying metal colloids and energy filtering transmission electron microscopy (EFTEM). *Microscopy and Microanalysis*, **11** (*Suppl. 2*), 1100–1101.

Bleher, R., Kandela, I., Meyer, D. A., and Albrecht, R. M. (2008). Immuno-EM using colloidal metal nanoparticles and electron spectroscopic imaging for co-localization at high spatial resolution. *Journal of Microscopy*, **230**, 388–395.

Deerinck, T. J., Giepmans, B. N. G., and Ellisman, M. H.(2005). Quantum dots as cellular probes for light and electron microscopy. *Microscopy and Microanalysis*, **11** (*Suppl. 2*), 914–915.

Duke, M. J., Peckys, D. B., and de Jonge, N. (2010). Liquid scanning transmission electron microscopy of individual quantum dots labeling epidermal growth factor receptors in whole cells. *Microscopy and Microanalysis*, **16** (*Suppl. 2*), 336–337.

Faraday, M. (1857). Experimental relations of gold (and other metals) to light. *Philosophical Transactions of the Royal Society of London*, **147**, 145–181.

Frens, G. (1973). Controlled nucleation for the regulation of the particle size in monodisperse gold suspensions. *Nature Physical Science*, **241**, 20–22.

Gaietta, G., Deerinck, T. J., Adams, S. R., *et al.* (2002). Multicolor and electron microscopic imaging of connexin trafficking. *Science*, **296**, 503–507.

Handley, D. A. (1989). The development and application of colloidal gold as a microscopic probe. In: *Colloidal Gold: Principles, Methods, and Applications*. Vol. **1**, Hayat, M. A. (ed.) San Diego, Academic Press, Inc., pp. 13–32.

Hell, S. W. and Wichmann, J. (1994). Breaking the diffraction resolution limit by stimulated emission: stimulation-emission-depletion fluorescence microscopy. *Optics Letters*, **19**, 780–782.

Henglein, A. (2000). Colloidal palladium nanoparticles: reduction of Pd (II) by H_2; $Pd_{core}Au_{shell}Ag_{shell}$ particles. *The Journal of Physical Chemistry B*, **104** (*29*), 6683–6685.

Hilderbrand, S. A., Shao, F., Salthouse, C., Mahmood, U., and Weissleder, R. (2009). Upconverting luminescent nanomaterials: application to *in vivo* bioimaging. *Chem. Commun.*, **28**, 4188–4190.

Hodak, J. H., Henglein, A., and Harland, G. V. (2001). Tuning the spectral and temporal response in PtAu core-shell nanoparticles. *Journal of Chemical Physics*, **114** (*6*), 2760–2765.

Kachar, B. (1985). Asymmetric illumination contrast: a method of image formation for video light microscopy. *Science*, **227**, 766–768.

Kaiser, M., Heintz, J., Kandela, I., and Albrecht, R. M. (2007). Tumor cell death induced by membrane melting via inductively heated core/shell nanoparticles. *Microscopy and Microanalysis*, **13** (*Suppl. 2*), 18–19.

Kandela, I. K. and Albrecht, R. M. (2007). Fluorescence quenching by colloidal heavy metal nanoparticles: implication for correlative fluorescence and electron microscopy studies. *Scanning*, **29**, 152–161.

Koeck, P. J. B., Schroder, R. R., Haider, M., and Leonard, K. R. (1996). Unconventional immune double labeling by energy filtered transmission electron microscopy. *Ultramicroscopy*, **62**, 65–78.

Meyer, D. A. and Albrecht, R. M. (2000). Identification of multiple colloidal labels of various metallic compositions by means of electron energy loss spectroscopy. *Microscopy and Microanalysis*, **6** (*Suppl. 2*), 322–323.

Meyer, D. A. and Albrecht, R. M. (2001). The feasibility of high resolution multiple labeling using colloidal particles of similar size but different shapes and elemental compositions. *Microscopy and Microanalysis*, **7** (*Suppl. 2*), 1032–1033.

Meyer, D. A. and Albrecht, R. M. (2002). Size selective synthesis of colloidal platinum nanoparticles for use as high resolution EM labels. *Microscopy and Microanalysis*, **8** (*Suppl. 2*), 124–125.

Meyer, D. A. and Albrecht, R. M. (2003). Sodium ascorbate method for the synthesis of colloidal palladium particles of different sizes. *Microscopy and Microanalysis*, **9** (*Suppl. 2*), 1190–1191.

Meyer, D. A., Oliver, J. A., and Albrecht, R. M. (2010). Colloidal palladium particles of different shapes for electron microscopy labeling. *Microscopy and Microanalysis*, **16**, 33–42.

Mühlpfordt, H. (1982). The preparation of colloidal gold particles using tannic acid as an additional reducing agent. *Experientia*, **38**, 1127–1128.

Nisman, R., Dellaire, G., Ren, Y., Li, R., and Bazett-Jones, D. P. (2004). Application of quantum dots as probes for correlative fluorescence, conventional, and energy-filtered transmission electron microscopy. *Journal of Histochemistry and Cytochemistry*, **52** (*1*), 13–18.

Olorundare, O. E., Goodman, S. L., and Albrecht, R. M. (1987). Trifluperazine inhibition of fibrinogen receptor redistribution in surface activated platelets. Correlative VDIC, HVEM and SEM studies. *Scanning Microscopy*, **1** (*2*), 735–743.

Park, K., Simmons, S. R., and Albrecht, R. M. (1988). Surface characterization of biomaterials by immunogold staining – Quantitative analysis. In: *Biotechnology and Bioapplications of Colloidal Gold*, Albrecht, R. M. and Hodges, G. M. (ed.). Chicago, IL, Scanning Microscopy International, pp. 41–52.

Park, K., Park, H., and Albrecht, R. M. (1989). Factors affecting the staining with colloidal gold. In: *Colloidal Gold: Principles, Methods, and Applications*. Vol. **1**. Hayat, M. A. (ed.). San Diego, Academic Press, Inc., pp. 489–518.

Patra, C. R., Bhattacharya, R., Patra, S., *et al.* (2007). Lanthanide phosphate nanorods as inorganic fluorescent labels in cell biology research. *Clin. Chem.*, **53**, 2029–2031.

Pawley, J. B. (2006). *Handbook of Biological Confocal Microscopy*, 3rd ed. NY, Springer.

Rittweger, E., Han, K. Y., Irvine, S. E., Eggeling, C. and Hell, S. W. (2009). STED microscopy reveals crystal colour centres with nanometric resolution. *Nature Photonics*, **3**, 144–147.

Roth, J. (1982). The preparation of protein A-gold complexes with 3 nm and 15 nm gold particles and their use in labelling multiple antigens on ultra-thin sections. *Histochemical Journal*, **14** (*5*), 791–801.

Russell, P. and Butchelor, D. (2001). SEM and AFM: Complementary techniques for surface investigations. *Microscopy and Analysis*, **15** (*4*), 9–12.

Scopsi, L. (1989). Silver enhanced colloidal-gold method. In: *Colloidal-Gold Principles, Methods and Applications*. Vol. **1**. Hayat, M. A. (ed.). San Diego, Academic Press, Inc., pp. 251–295.

Sen, D., Deerinck, T. J., Ellisman, M. H., Parker, I., and Cahalan, M. D. (2008). Quantum dots for tracking dendritic cells and priming an immune response *in vivo* and *in vitro*. *PLoS ONE*, **3** (*9*), e 3290.

Simmons, S. R. and Albrecht, R. M. (1989). Probe size and bound label conformation in colloidal gold-ligand labels and gold-immunolabels. *Scanning Microscopy*, **3** (*Suppl. 3*), 27–34.

Simmons, S. R., Pawley, J. B., and Albrecht, J. B. (1990). Optimizing parameters for correlative immunogold localization by video-enhanced light microscopy, high-voltage transmission electron microscopy, and field emission scanning electron microscopy. *Journal of Histochemistry and Cytochemistry*, **38**, 1781–1785.

Skala, M. C., Riching, K. M., Gendron-Fitzpatrick, A., *et al.* (2007). *In vivo* multiphoton microscopy of NADH and FAD redox states, fluorescence lifetimes, and cellular morphology in precancerous epithelia. *Proc. National Academy of Sciences* USA, **49**, 19494–19499.

Slot, J. W. and Geuze, H. J. (1985). A new method of preparing gold probes for multiple-labeling cytochemistry. *European Journal of Cell Biology*, **38** (*1*), 87–93.

Turkevich, J., Stevenson, P. C., and Hillier, J. (1951). A study of the nucleation and growth processes in the synthesis of colloidal gold. *Discussions of the Faraday Society*, **11**, 55–75.

Voigt, J. and Heumann, J. (1928). Die herstellung schutzkolloidfreier, gleichteiliger silberhydrosole. *Zeitschrift für Anorganische Chemie*, **169** (*1–3*), 140–150.

Weiser, H. B. (1933). The colloidal elements. In: *Inorganic Colloid Chemistry*, Vol. **1**. Weiser, H. B. (ed.). NY, John Wiley & Sons, Inc.

Zsigmondy, R. (1906). *Zeitschrift für Physikalische Chemie*, **56**, 65.

7 The use of SEM to explore viral structure and trafficking

Jens M. Holl and Elizabeth R. Wright

7.1 Introduction

The viral infection cycle is a complex process; every virus has developed individual mechanisms to govern replication within the host cell. Due to their small size range, from tens of nanometers to hundreds of nanometers, electron microscopy (EM) is the only imaging method available today for examinations of viral ultrastructure during infection. There are six basic stages (see Figure 7.1 A–F (and color plate)) during the infection of a host cell by a virus that can be visualized by EM. First, the virus must find a means for attaching to the host cell (Attachment, Figure 7.1). Attachment is generally mediated by binding of a viral glycoprotein or capsid to a receptor located on the cell's plasma membrane. Once the virus has attached, it must penetrate the cell membrane either by receptor mediated endocytosis or by membrane fusion (Entry, Figure 7.1). As the virus enters the cell, it undergoes an uncoating event during which the viral capsid is degraded in order to release the viral genome (Uncoating, Figure 7.1). After the viral genome is released viral proteins are synthesized, and replication occurs during which copies of the viral genome are produced (Replication, Figure 7.1). Once replication has advanced, all the components of the virus are transported to particular cellular locations in order for individual virus self-assembly to proceed (Trafficking, Figure 7.1). Finally, the virus is released from the cell through either cell lysis or, as in the case of most enveloped eukaryotic viruses, a process known as budding (Budding, Figure 7.1).

Scanning electron microscopy (SEM) has advanced to a point where we have a great many methods available to us for probing most aspects of the viral infection cycle. These technologies include: conventional secondary electron (SE) SEM for probing the surface morphologies of viruses and cells; back-scattered electron (BSE) SEM for visualizing the exact locations of immuno-labeled macromolecules; cryo-SEM for ultimate specimen preservation and high-resolution imaging; and dual-beam approaches for viewing specimens as 3-D objects. This chapter describes several of these methods and how each of them has been employed for analysis of a segment of the viral infection cycle.

Scanning Electron Microscopy for the Life Sciences, edited by H. Schatten. Published by Cambridge University Press

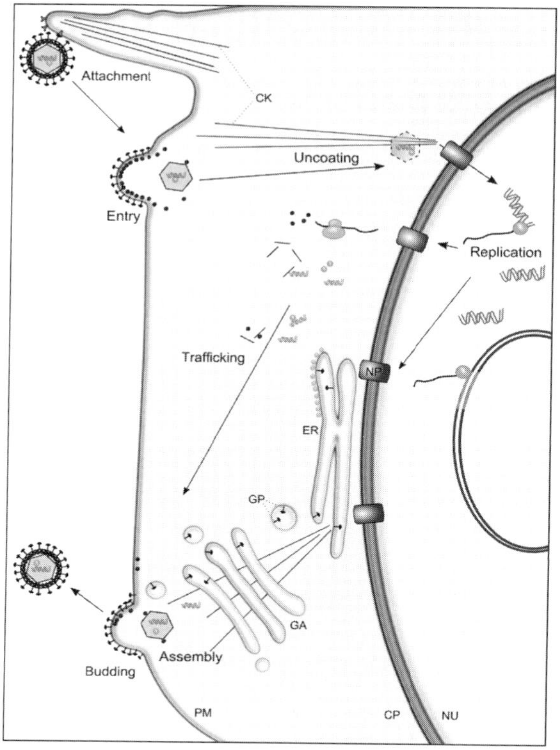

Figure 7.1 Virus assembly: A) Attachment, B) Entry, C) Uncoating, D) Replication, E) Trafficking and assembly, F) Budding. Diagram shows: cytoskeleton (CK), cytoplasm (CP), endoplasmic reticulum (ER), Golgi apparatus (GA), glycoprotein (GP), nuclear pore (NP), nucleus (NU), plasma membrane (PM). (See plate section for color version.)

7.2 Virus attachment

Most enveloped viruses use specific mechanisms for cell surface receptor binding in order to enter target cells. As compared to the size of the target cell, which is generally many micrometers in all dimensions, viruses are quite small. We know that the surfaces of living cells undergo significant dynamic rearrangements of their lipids, proteins, and other membrane-associated macromolecules. As a result, some viruses have found a way to "hijack" cellular protrusions like filopodia in order to travel rapidly to the cell body and to the hotspots of cellular endocytosis. Therefore, the challenge for SEM imaging is to locate the viruses arrayed along the cell membrane surface, during either the translocation process or the initial stages of adsorption to cellular receptors.

A recent study has shown using environmental SEM (ESEM) that Murine Leukemia Virus (MLV) travels along the filopodia of fibroblasts or HEK 293 cells when its receptor, mCAT-1, was expressed in the target cells (Lehmann *et al.*, 2005). This was also demonstrated for viruses which carried, or were pseudotyped with, Avian Leukosis Virus (ALV) or Vesicular Stomatitis Virus (VSV) glycoproteins. The traversal of MLV virions along the

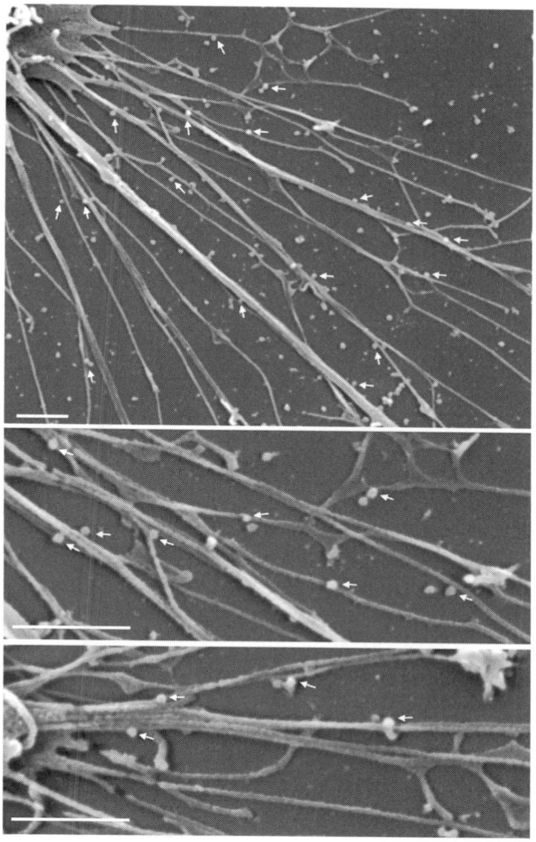

Figure 7.2 Murine Leukemia Virus (MLV) virions associated with HEK 293 cell filopodia. HEK 293 cells, which expressed the receptor, mCAT-1, were incubated with MLV and analyzed by ESEM. Arrows indicate MLV virions traversing along filopodia. Scale bars: 1 μm. © Rockefeller University Press, 2005. Originally published in *J. Cell Biol.*, **170**, 318–325.

filopodia was detectable as long as each viral glycoprotein's receptor was expressed by the target cell. Viral fusion with the cell membrane was observed predominantly at the cell body and could therefore be distinguished from receptor binding at the cell's periphery and subsequent transport along the cellular protrusions. The directed movement of viruses along filopodia could be stopped by the addition of inhibitors to actin polymerization, myosin II or certain ATPases (Lehmann *et al.*, 2005). These results indicated that actin and myosin II as well as the virus receptor may be cellular mediators of viral "surfing" along cellular protrusions. Figure 7.2 is a panel of ESEM images detailing how MLV associates with the filopodia of HEK 293 cells for transport to the cell body. Using ESEM, it was shown that MLV was attached to the filopodia of polarized MDCK cells after 30 min of incubation on ice or treatment with the specific myosin II inhibitor called blebbistatin. Cells that were kept at room temperature for 30 min did not have viral particles along the filopodia. The authors suggest that a myosin II and ATP-dependent process may be responsible for virus transport before entry.

Interactions between viruses and cellular protrusions have also been demonstrated for herpes simplex virus-1. SEM analysis showed virus particles attached to plasma membrane protrusions of CHO cells expressing nectin-1, one of the receptors for viral glycoprotein D (gD) (Shukla *et al.*, 1999). The same group also reported a cell type-dependent phagocytosis-like uptake in corneal fibroblasts of HSV-1 (Clement *et al.*, 2006). This process seems to involve actin rearrangement and dynamin assembly as well as RhoA GTPase and some tyrosine kinases (Clement *et al.*, 2006). However, the entry mechanism for HSV-1 has not been completely resolved. According to a recent review, HSV-1 and Kaposi's sarcoma-associated herpesvirus (KSHV) may be internalized by macropinocytosis (Mercer *et al.*, 2010).

7.3 Virus entry

Interactions between viruses and their receptors are often specific and multivalent, and may lead to activation of cellular signaling pathways. The cellular response is generally the internalization of the virus by several different endocytosis mechanisms (reviewed by Mercer *et al.*, 2010).

One example of activation of a signaling pathway is virus-induced macropinocytosis, or cell membrane ruffling, that was demonstrated in A431 cells after Adenovirus type 2 (Ad2) infection (Meier *et al.*, 2002). The virus entered the cell through clathrin-dependent endocytosis and also caused macropinocytosis as a side effect. The authors noted, by SEM, the formation of ruffles at the cell membrane within 10 min of infection with Ad2 (Figure 5A of Meier *et al.*, 2002).

To visualize the distribution of RSV particles on the surface of Hep-2 cells shortly after infection, Brown *et al.* (2004) used FESEM and immunogold labeling to identify the RSV G glycoprotein and CD55, a cellular protein associated with detergent resistant membrane (DRM) regions (or lipid-rafts) of the plasma membrane. Since this virus is thought to bud from DRMs, the authors wanted to determine if CD55 was also incorporated into the virus particles. In the inset of Figure 7.3, RSV particles were observed with diameters ranging from 200–300 nm. BSE SEM images of immunogold labeled RSV particles were heavily decorated with 5 nm gold particles, to the RSV glycoprotein G, and to a lesser extent with 20 nm gold particles to CD55. Microvilli-like protrusions, which appeared to be connected with the RSV particles, could be observed 4 hr after infection. This may represent an early step in virus internalization.

7.4 Virus uncoating

Most DNA viruses (e.g. Herpes viruses) rely on the cellular replication machinery in the nucleus. In contrast, most RNA viruses replicate in the cytoplasm. Exceptions are orthomyxoviruses (e.g. influenza) and retroviruses (e.g. HIV-1).

In contrast to other members of the retrovirus family, lentiviruses have the ability to replicate in non-dividing cells. The key difference is thought to be an active import

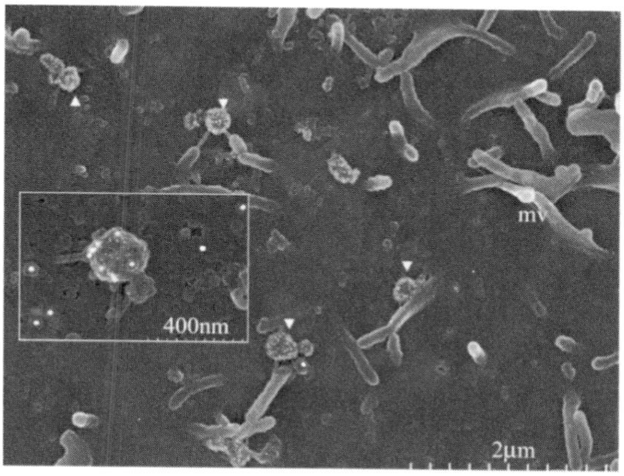

Figure 7.3 RSV virions during the initial stages of cell attachment. An overlay of SE and BSE shows RSV particles bound to cellular protrusions by filamentous tethers 4 hours post infection. Microvilli (mv) and viral particles labeled with 10 nm gold can be seen at a magnification of × 20 000. The sample in the inset shows the distribution of glycoprotein G (5 nm) and CD55 (20 nm) on the particle after immune gold labeling at a magnification of × 130 000. Copyright © 2004 Elsevier Inc. Originally published by Brown *et al.* (2004).

mechanism through the nuclear pore. Other retroviruses that cannot translocate through the pore are only able to enter the nucleus during mitosis and are therefore dependent on cell division. As seen in Figure 7.4, Arhel *et al.* (2007) demonstrated that a three-stranded DNA structure, the so-called DNA flap, has an important role in this process. This flap is formed at the end of reverse transcription by two cis-acting sequences in the HIV-1 genome, the central polypurine tract (cPPT) and the central termination sequence (CTS). Images of the nuclear surface by SE SEM showed the presence of the HIV-1 capsid protein (CA) in close proximity to the nuclear pore complex (NPC). CA could be identified by immuno-gold labeling using BSE detection (Arhel *et al.*, 2007). Decapsidation, the process of disassembling the CA matrix, seems not to occur as an immediate post-fusion event, but rather when the viral capsid is in close proximity to the NPC (Arhel *et al.*, 2007). Gold-labeled capsids could be detected sitting on NPCs when viewing overlaid SE and BSE images of cells infected or transduced with DNA flap-defective particles. Furthermore, capsids were also found to be associated with actin filaments adjacent to the nuclear membrane. However, no capsids were observed at late time points in cells transduced with WT DNA flap-containing constructs (Arhel *et al.*, 2007). Therefore the authors suggest that the presence of a DNA flap that is formed at the end of reverse transcription within the capsid acts as a viral promoting element for HIV-1 uncoating right at the nuclear pore (Arhel *et al.*, 2007).

7.5 Nuclear trafficking of viruses

Herpes viruses are large DNA viruses and consist of dsDNA enclosed by an icosahedral capsid, a protein matrix called tegument and an envelope. During infection, the newly produced capsid proteins are transported into the nucleus where they assemble into the

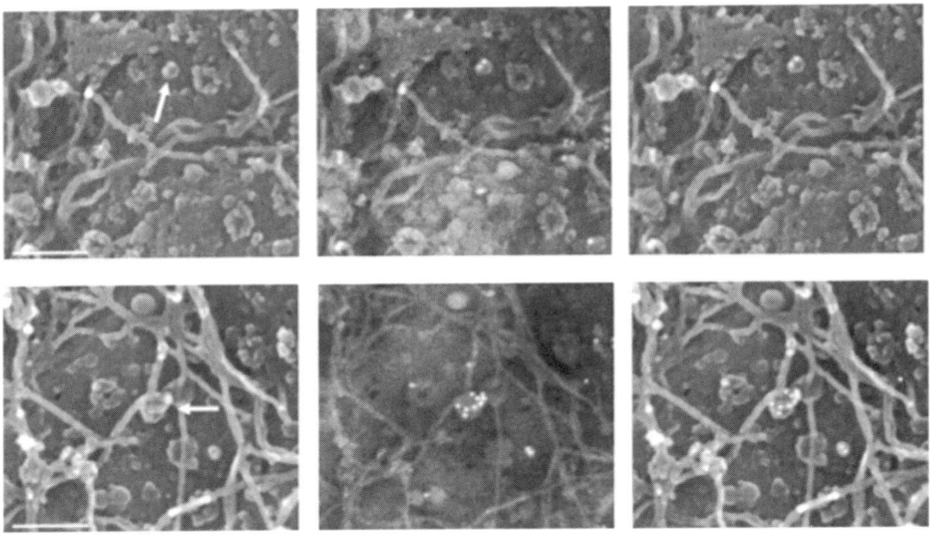

Figure 7.4 Overlay of SE and BSE of nuclear pores at late time points after transduction. SE images are shown in the left panels, BSE in the middle panels and an overlay of both on the right side. Images are taken 48 hours post transduction with HIV-1 vectors containing (*upper row*) or lacking (*lower row*) the central flap region. Arrows point to core debris in (*upper*) and to intact HIV-1 vector capsids in (*lower*). Scale bars = 200 nm. Copyright © 2007 European Molecular Biology Organization. Originally published by Arhel *et al.* (2007).

icosahedral capsid and are loaded with DNA. According to the most accepted model, the capsids which are larger than the nuclear pores bind to a primary tegument and envelope, and bud through the inner nuclear membrane into the perinuclear space. From there they are transported through the endoplasmatic reticulum and the Golgi-apparatus where they acquire their secondary tegument and envelope. A recent study that uses field emission SEM (FESEM) explored the nuclear envelope of Herpes Simplex Virus 1-infected cells (Wild *et al.*, 2009). The authors developed methods to compare subcellular structures in a freeze-dried state at ambient temperature versus cryo-frozen hydrated samples. Interestingly, 15 hr post infection they found a reduction in the number of nuclear pores in HSV-1 infected versus mock-infected cells by about a factor of 2, depending on the method used. At the same time the diameter of the nuclear pore complex (NPC) was increased with larger nuclear pores. They also observed a slight increase in the volume of the nucleus. The enlarged nuclear pores were large enough to let the HSV-1 capsid pass into the perinuclear space. A complete description of this process is given in Chapter 8.

7.6 Virus trafficking and cytoskeletal interactions

In order to direct viral proteins and genomes to the assembly and budding sites of the infected cells, viruses must be able to manipulate cellular transport mechanisms. Jeffree *et al.* (2003) were interested in the interactions between viral proteins and the cytoskeleton

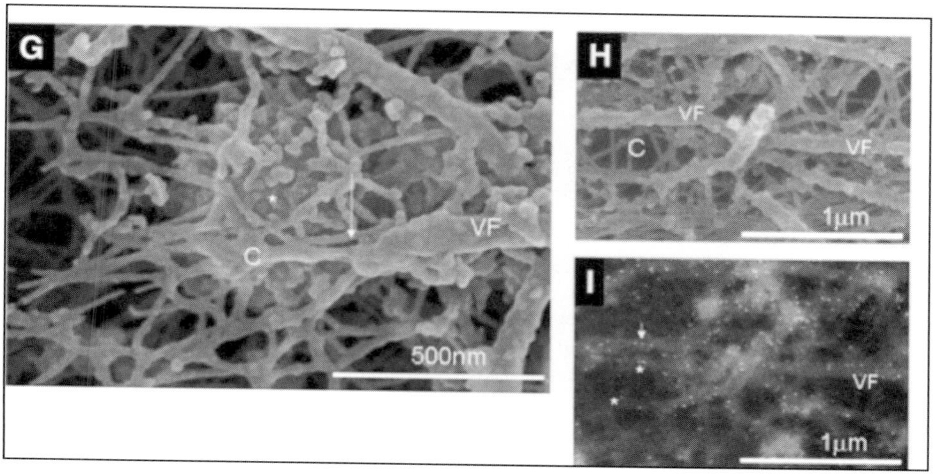

Figure 7.5 Interaction between cytoskeleton network and RSV filaments revealed by FE SEM. Virus-infected cells were extracted with Triton-X100, and cells were examined at × 90 000 (G) or × 35 000 (H, I) magnification. (G) provides ultrastructural evidence for a direct interaction of cytoskeleton network (C) and viral filaments (VF) (white arrow). Viral filaments were detected by SE (H) and BSE (I). Gold-labeled particles are visualized by white spots (indicated by white arrow). Unspecific labeling of cytoskeleton network is highlighted by a white asterisk. © 2007 Elsevier Inc. Originally published in *Virology*, **369**, (2007), 309–323.

during RSV assembly. Since SEM can only detect the surface structure of specimens, the authors used a detergent to strip the phospholipid bilayer from the cell, thereby revealing the underlying cytoskeleton. Using this method, an interaction between F-actin and RSV filaments could be observed (Figure 7.5). Further studies revealed that treatment with PI3K or Rac GTPase inhibitors prevented formation of viral filaments (Jeffree, 2003). This demonstrated that both enzymes are involved in modulation of the actin cytoskeleton, and established the role of actin as a critical cellular component in RSV replication.

7.7 Study of viral assembly and budding by SEM

7.7.1 Retroviral assembly

The Gag polyproteins of retroviruses encode for all the functions needed for assembly, budding, and release of virions. Gag polyproteins contain so-called "late domains" that are able to bind to members of the cellular endosomal sorting complex required for transport (ESCRT) complex. Within cells, ESCRT complexes are used for sorting ubiquitinated proteins into the vesicles of multivesicular bodies (MVBs). The protein HGS (also called Hrs, hepatocyte growth factor-regulated tyrosine kinase substrate) is thought to recruit other proteins to endosomal membranes. HGS also interacts with members of the ESCRT transport system e.g. Tsg101, HCRP1/Vps37A. Because Gag and HGS both interact with Tsg101 through their PT/SAP late domain, it is thought that HIV-1 Gag mimics the HGS

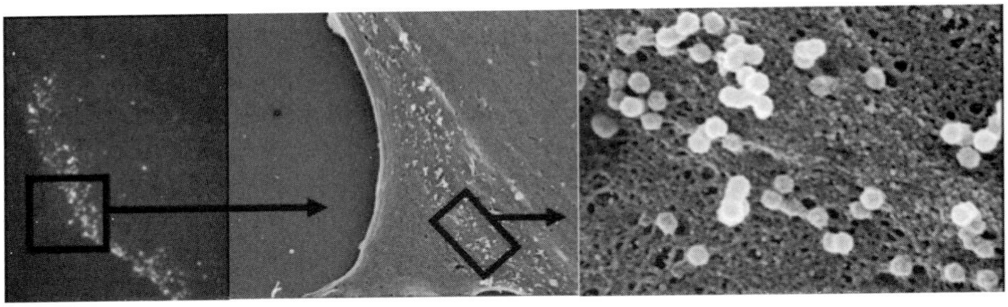

Figure 7.6 HGS late domain mutant (ASAA) blocks HIV-1 Gag particle release. HeLa cells were transfected with plasmids expressing HIV Gag-Pol and HIV Gag-GFP at a 2:1 ratio. HIV particle release could be eliminated by coexpressing a late-domain defective mutant (ASAA) of HGS. The left panel displays HeLa cells expressing high levels of Gag-GFP. The same areas (highlighted by rectangles in the left and middle panels) indicate the presence of lump-like structures (particles) when examined at higher magnification by SEM. The mutant seems to interfere with particle pinch-off and viruses remain attached to the cell surface. © by Marc C Johnson, unpublished data.

protein in order to recruit the ESCRT machinery to facilitate HIV-1 particle release. After coexpression of plasmids carrying HIV-1-Gag-Pol and HIV-1-GFP together with wild type HGS, HIV-1 virions could be observed as punctate features on the cells by confocal microscopy. Correlative SEM analysis showed many virus-like particles on the cell surface. Over-expression of an HGS variant with a late domain mutation (PSAP to ASAA) leads to large, perinuclear fluorescent structures (Bouamr *et al.*, 2007). When the same cells were examined by SEM, "lump-like structures" were observed on the cell surface (Figure 7.6). The authors argue that this is due to a late domain effect and a defect in particle release. After infection, cells displayed a more rounded shape compared with non-infected cells.

7.7.2 Glycoprotein recruitment in retroviral assembly and budding

At the end of the virus replication cycle, all the viral components must assemble together at specified cellular locations in order for new viral particles to begin the process of viral assembly. Like other enveloped viruses, retroviruses possess surface glycoproteins that are needed to infect target cells. The driving force of retroviral assembly is the Gag polyprotein. In most cases it consists of three domains, matrix (MA), capsid (CA), and nucleocapsid (NC) that can be separated by one or two spacer peptide sequences (SP). The nucleocapsid domain interacts with the RNA genome and the matrix with the cell membrane. Both functions can be substituted by protein domains that have the same function (Wills *et al.*, 1991). The CA domain is essential for the formation of a hexameric lattice structure and membrane curvature, which are important for budding (Wright *et al.*, 2007) (Briggs *et al.*, 2009).

Recent studies show that the spacer domains between CA and NC contribute to this process. Insertions or substitutions of amino acids in SP in Rous Sarcoma Virus (RSV or ASLV) and SP1 in Human Immunodeficiency Virus type 1 (HIV-1) lead to formation of chains of spherical particles on the cell surface or to a filamentous virion shape (Keller

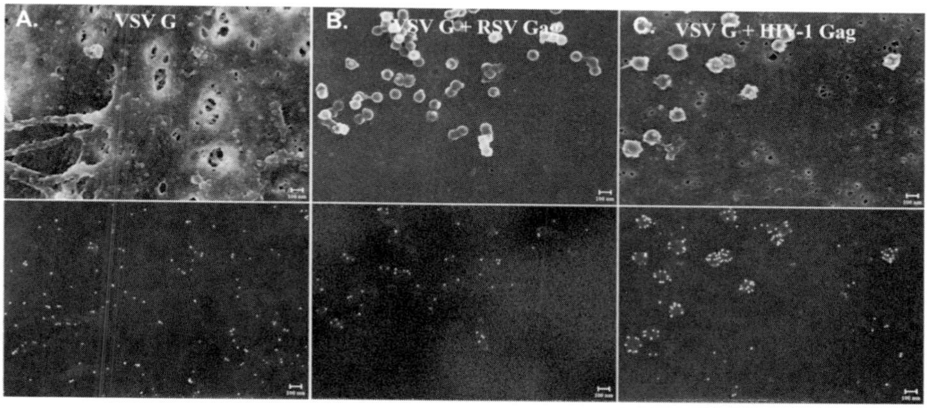

Figure 7.7 Envelope recruitment to retrovirus budding sites. VSV-G envelope protein is either expressed alone (A) or coexpressed with ASLV Gag (B) or HIV-1 Gag (C) in DF1 cells. Top images show secondary electrons and bottom images back-scattered electrons. Gold labeled VSV-G proteins are more equally distributed in (A), whereas coexpression of both ASLV Gag (B) and HIV-1 Gag (C) show local concentrations of VSV-G at both retroviral budding sites, suggesting an active recruitment by the Gag proteins. Copyright © 2009 American Society for Microbiology. Originally published in *The Journal of Virology*, **83**, (*9*), 4060–4067.

et al., 2008). An envelope glycoprotein from one retrovirus can be used to complement another retrovirus lacking its own glycoprotein. This well-established process is called "pseudotyping." It is, however, not well understood whether a direct interaction between the foreign glycoprotein and the matrix protein of the Gag polyprotein is necessary or whether the glycoproteins are randomly incorporated into the budding virus-like particles.

In their paper, Jorgenson *et al.* (2009) used SE SEM to visualize viral budding site using release-deficient viruses. In combination with the signal of BSE from gold-labeled antibodies they were able to detect the distribution of glycoproteins on the plasma membrane. Merging both signals revealed that the glycoproteins are randomly distributed over the surface when expressed alone. In the presence of ASLV or HIV-1 particles, glycoproteins from VSV or MLV were redistributed to the viral budding sites. In contrast, the ASLV glycoprotein was recruited only to its "own" budding sites. The results suggest a model where viral glycoproteins are mobile in the cytoplasmic membrane and can be recruited by Gag proteins either directly or indirectly to viral budding sites (Figure 7.7).

7.7.3 Filovirus budding

These viruses form filamentous particles that are 80 nm in diameter. One characteristic feature of Ebolavirus is the predominantly horizontal budding of viruses from the cytoplasmic membrane of infected cells (Figure 7.8B) (Noda *et al.*, 2006). Interestingly, virus-like particles that do not contain nucleocapsid (NC) seem to bud vertically (Figure 7.8A). Since matrix proteins of many negative-strand RNA viruses interact with the cytoplasmic membrane and the NC, the authors suggest a model to illustrate that

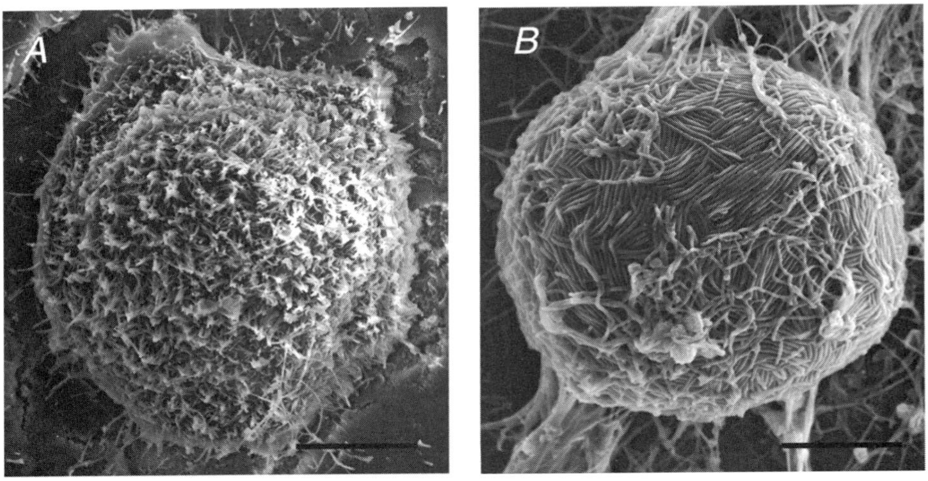

Figure 7.8 Different modes of Ebolavirus budding. Filamentous Ebola particles are budding from the cell surface vertically (A) or horizontally (B). Bars = 3 μm. © 2006 Noda, T. *et al.* (2006).

particles bud vertically when the matrix protein (VP40) has not bound enough NC protein. However, if the surface of the NC-genome complex is covered with VP40, its interactions with the cytoplasmic membrane force it into budding horizontally.

7.7.4 Cellular cofactors required for assembly and budding

A multitude of cellular factors that are used by viruses during assembly and budding processes have been identified and characterized. Continued investigations of how viruses manipulate cellular processes using cellular factors and cofactors might lead to new generations of pathway-specific antivirals. The Rab family proteins are small GTPases involved in trafficking of proteins and vesicles between the trans-Golgi network (TGN), recycling endosomes, and the cytoplasmic membrane. They have been reported to be involved in the assembly of several viruses. The Rab11 proteins are modulated by the Rab11 family interacting proteins (Rab11-FIPs) in order to direct them to certain subcellular compartments. Interestingly, Rab11 together with its protein partner FIP3 seems to be required for the formation of filamentous particles of influenza A virus (Bruce *et al.*, 2010). Looking at the surfaces of infected 293T cells by SEM revealed 10 to 20 μm long bundles of filamentous viruses or patches with individual filaments of about 0.5 μm. However, the surfaces of cells treated with siRNA against Rab11 were much smoother, lacking the bundles of filamentous virions as well as filopodia, present in the controls. In contrast, spherical structures with a diameter of 100 nm were present on the surface (Bruce *et al.*, 2010). The authors suggest that these virions are defective in budding. However, they could not detect a measurable drop in viral titer, which suggests no cessation in particle release. This could be explained by a partial reduction of Rab11 levels in siRNA treated cells. The depletion of FIP3 also blocked the formation of filamentous particles on the cell surface. In contrast, siRNA treatment against FIP2 did not reduce filament formation.

Interestingly, FIP2 has been shown to be involved in the budding of respiratory syncytial virus (RSV) from apical membrane domains. A deletion mutant of FIP2 lacking the C2 domain that binds to membrane lipids revealed the most drastic effects on particle formation (Utley *et al.*, 2008). Examination of the apical surfaces of infected MDCK cells showed filaments of 1 to 3 μm in length for the wild type FIP2 protein. Longer filaments could be found (3 to 10 μm) on infected cells transfected with the C2 deletion mutant of FIP2. The authors discuss a role of FIP in budding and pinching off from the cell membrane. Lack of the C2 domain has a negative effect on virus release and results in elongated particles connected to the cell membrane.

Another factor that influences RSV budding is the small GTPase RhoA. It is up-regulated during RSV infection. Furthermore, the inhibition of RhoA with the inhibitor *Clostridium botulinum* C3 stopped syncytia formation in HEp-2 cells but did not interfere with virus production (Gower *et al.*, 2005). Since this small GTPase has a function in actin polymerization, the authors tested whether RhoA is needed for cell-to-cell fusion. The authors found that the effector cells are very sensitive to the inhibition and showed 80% fewer fusion events, while inhibition of RhoA in the target cells had only a minor effect on cell-to-cell fusion. It seems that the infected cell relies on functional RhoA for fusion, while the uninfected target cell does not. Other studies show that depletion of RhoA leads to a decreased number and length of viral filaments (Mccurdy and Graham, 2003). In Gower *et al.*, the budding RSV is visualized by SEM (Gower *et al.*, 2005). Blocking of RhoA drastically reduced the number of budding viral filaments. Adding lysophosphaphatidic acid (LPA), a known activator of the RhoA signaling pathway, increased the number of microvilli on the cell surface of HEp-2 cells compared to noninfected cells. The same effect could also be seen a short time after infection (24 hr) with RSV, suggesting a viral effect on increased microvilli formation. Two days after infection, the surface was covered with viral filaments that could be distinguished from microvilli by their smaller diameters. The authors propose that the function of RhoA is to direct viral proteins (F) to cholesterol-rich DRMs, which have been shown earlier to have an important role in budding of other viruses like HIV-1, influenza virus or SV40 (Gower *et al.*, 2005).

The association of RSV budding and DRMs was confirmed by another study, where it was shown that the glycoprotein G partially colocalized with CD55 and CD59, which both function to deactivate the complement system and were found in Triton-X 100 insoluble membrane fractions. An SEM analysis using immuno-labeling demonstrated a concentration of CD55 and CD59 in microvilli on the surface of mock infected HEp-2 cells (Brown *et al.*, 2004). It was also demonstrated that in RSV infected cells these proteins colocalize on viral filaments with the G glycoprotein, and CD55 was detected in isolated RSV particles (Brown *et al.*, 2004). Immuno-labeling of RSV particles with anti-CD55 and anti-G antibodies established the presence of CD55 in virus membranes bound to the cell surface 4 hr after infection.

In another study, the same research group used SE SEM and BSE to examine the surfaces of RSV. They determined that the G protein was distributed as doublet clusters (Figure 7.9). The average distance between the gold particles within a doublet was about

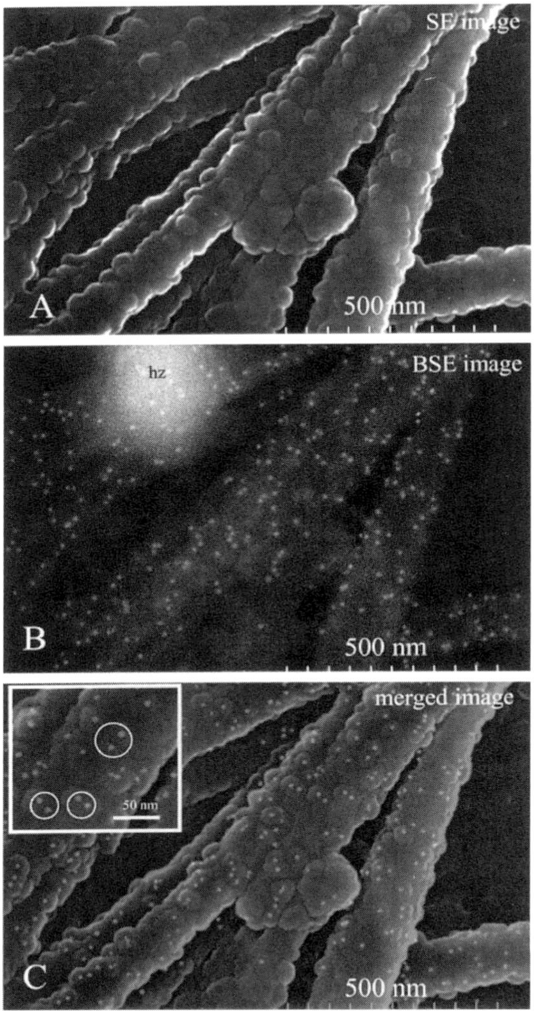

Figure 7.9 Distribution of the G protein on mature RSV filaments. Visualizations of secondary electrons (SE) and back-scattered electrons (BSE) at × 100 000 magnification are shown in (A) and (B). The image in (C) represents an overlay of (A) and (B). Immunogold labeling against glycoprotein G revealed a predominant doublet pattern of G on the virus surface. This pattern is highlighted in the enlarged inset (white circles). Copyright © 2003 Elsevier Inc. Originally published by Jeffree *et al.* (2003).

20–25 nm and the spacing between gold/G doublets was about 80 nm (Jeffree *et al.*, 2003). While the authors could not determine how many G protein spikes were accounted for within one of the clusters, it was hypothesized that the clustering of the G proteins along the virus could improve the binding of the virus to the cell membrane. It is likely that the concentration and localization of viral attachment proteins is a conserved mechanism among viruses for improving viral attachment to the cell membrane.

Tetherin (also known as CD317, HM1.24 or Bst-2) is an IFN-inducible antiviral factor that can tether viruses to the cell surface after budding (Neil *et al.*, 2008). Tetherin is able to restrict cell-to-cell spread of HIV-1. Correlative SEM and fluorescence microscopy experiments have shown that presence of the HIV-1 auxiliary protein Vpu can overcome viral retention. Jurkat cells were analyzed after 2 hr of being co-cultured with Hela cells transfected with ΔVpu HIV-1-Gag-GFP. Virions were visualized as green dots as well as in SEM by anti-env immunogold staining on the surface of Jurkat cells (Casartelli *et al.*, 2010).

In addition to its restrictive effect on HIV-1 release, in the absence of Vpu, tetherin has recently also been shown to tether Ebola virus to the cell surface. Similarly to HIV-1, Ebola also encodes for a protein, its glycoprotein (GP), which is capable of overcoming cellular restriction. Katletsky *et al.* (2009) were able to demonstrate this using SEM analysis. Hela cells expressing VP40 alone had many filamentous particles tethered to the cell surface. Coexpression of HIV-1 Vpu or Ebola GP proteins, however, could overcome the restriction and fewer filamentous particles were detected on the surface. This finding was attributed to more efficient particle budding from the cell surface.

7.8 Future prospects and technical advancements in SEM

The applicability of conventional transmission electron microscopy (TEM) is limited to samples much less than 200 nm in thickness. Mammalian cells, which can be up to 4 μm thick, must be prepared by sectioning in order to be visualized by TEM.

As a way to overcome the challenges associated with specimen thickness for TEM and SEM surface imaging limitations, groups have been developing alternative methods to prepare biological samples for imaging. Researchers have used focused ion beam (FIB) or ion abrasion (IA) technologies to "shave off" layers of a sample incrementally and then image the "block face" by SEM. By using this approach, new insights have been obtained into the structure of HIV-1 infected mammalian cells (Bennett *et al.*, 2009). Using IA-SEM the group revealed intracellular reservoirs of HIV-1 in monocyte-derived macrophages (MDM). They demonstrated the existence of channels that connect deep internal reservoirs within the cytoplasmic membrane (Figure 7.10 (and color plate)) (see also Figure 3 in Bennett *et al.*, 2009). Another paper from the same group looked into HIV-1 transmission at the virological synapse between dendritic cells and T cells. Using IA-SEM it could be demonstrated that the structures that look like filopodia in conventional TEM sections were sheets of membranes in 3-D tomograms (Felts *et al.*, 2010). Membranous sheets originating from the dendritic cell were very long and almost completely wrapped around the T cell. The authors think they identified two distinct types of contact at virological synapses. In the first group, T cells formed protrusions that directly linked to surface accessible invaginations on the dendritic cell filled with virus. The second form of synapse was formed as the cell bodies of both cell types came into close proximity with the viruses localized within the contact zone.

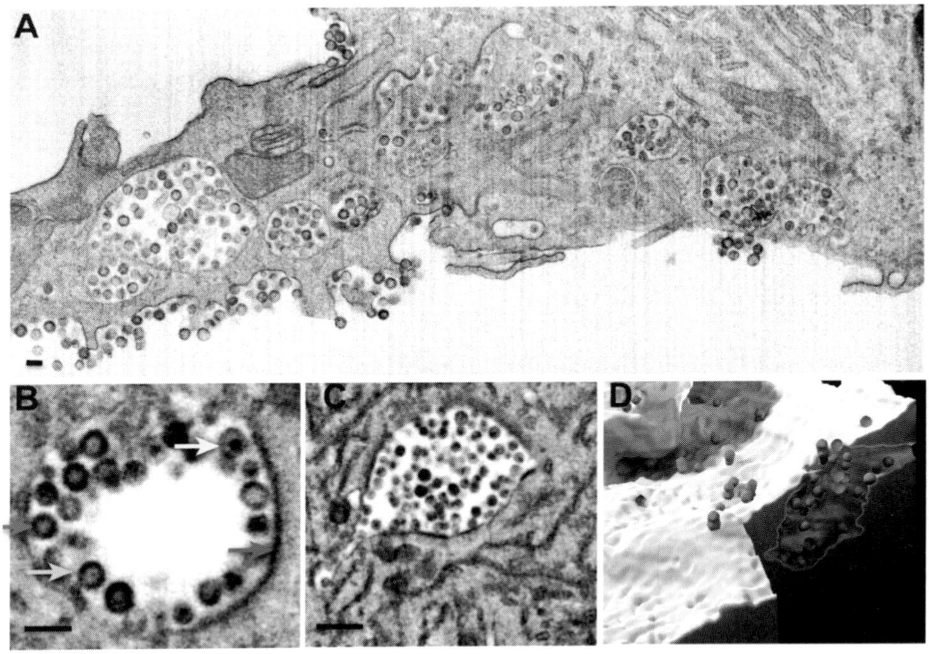

Figure 7.10 IA-SEM imaging of virion reservoirs and channels in primary HIV-1 infected MDM: (A) Single image of internal compartments, (B) Compartments magnified, with virions highlighted by arrows, (C) Compartments deep in the cell interior, (D) Segmented 3-D representation of a viral reservoir. Bars = 200 nm. © 2009 Bennett A. *et al.* (2009). (See plate section for color version.)

7.9 Conclusions

SEM and its associated methods are extremely valuable for the structural analysis of viral life cycles as they progress within the host cell. In this chapter, the authors have illustrated how use of SEM has improved understanding of the viral life cycle of a number of pathogenic viral systems. Further improvements to and advancements with SEM instrumentation and specimen preparation procedures will continue to aid in our ability to resolve many questions at the interface of cell biology and virology.

7.10 Abbreviations

ALV	Avian Leukosis Virus
ASLV	Avian sarcoma-leukosis virus, formerly known as Rous sarcoma virus
BSE	back-scattered electrons
DRM	detergent-resistant membrane
FIP	Rabll family-interacting proteins
HGS	hepatocyte growth factor-regulated tyrosine kinase substrate

HSV-1	Herpes Simplex Virus type 1, also known as Human Herpes Virus 1
KSHV	Kaposi's Sarcoma-associated Virus, Human Herpes Virus 8
mCAT-1	murine cationic amino acid transporter, ecotropic receptor for MLV
MLV	Murine Leukemia Virus
NPC	nuclear pore complex
PIK3	phosphatidylinositide-3-kinase
RSV	Respiratory Syncytial Virus
SE	secondary electrons
TMV	Tobacco Mosaic Virus
VSV	Vesicular stomatitis virus
WT	wild type

7.11 Acknowledgments

The authors wish to thank Dr. Marc C. Johnson for the use of unpublished figures and Drs. Ricardo C. Guerrero-Ferreira and Gabriella Kiss, and Mr. Grant M. Williams for helpful discussions. This work was supported through funds provided by Emory University, Children's Healthcare of Atlanta, and the Georgia Research Alliance.

7.12 References

Arhel, N. J., Souquere-Besse, S., Munier, S., *et al.* (2007). HIV-1 DNA flap formation promotes uncoating of the pre-integration complex at the nuclear pore. *The EMBO Journal*, **26**, 3025–3037.

Bennett, A. E., Narayan, K., Shi, D., *et al.* (2009). Ion-abrasion scanning electron microscopy reveals surface-connected tubular conduits in HIV-infected macrophages. *PLoS Pathogens*, **5**, e1000591.

Bouamr, F., Houck-Loomis, B. R., De Los Santos, M., *et al.* (2007). The C-terminal portion of the Hrs protein interacts with Tsg101 and interferes with human immunodeficiency virus type 1 Gag particle production. *Journal of Virology*, **81**, 2909–2922.

Briggs, J. A. G., Riches, J. D., Glass, B., *et al.* (2009). Structure and assembly of immature HIV. *Proc. National Academy of Sciences of the USA*, **106**, 11090–11095.

Brown, G., Jeffree, C. E., Mcdonald, T., *et al.* (2004). Analysis of the interaction between respiratory syncytial virus and lipid-rafts in Hep2 cells during infection. *Virology*, **327**, 175–185.

Bruce, E. A., Digard, P., and Stuart, A. D. (2010). The Rab11 pathway is required for influenza A virus budding and filament formation. *Journal of Virology*, **84**, 5848–5859.

Casartelli, N., Sourisseau, M., Feldmann, J., *et al.* (2010). Tetherin restricts productive HIV-1 cell-to-cell transmission. *PLoS Pathogens*, **6**, e1000955.

Clement, C., Tiwari, V., Scanlan, P. M., *et al.* (2006). A novel role for phagocytosis-like uptake in herpes simplex virus entry. *The Journal of Cell Biology*, **174**, 1009–1021.

Felts, R. L., Narayan, K., Estes, J. D., *et al.* (2010). 3D visualization of HIV transfer at the virological synapse between dendritic cells and T cells. *Proc. National Academy of Sciences of the USA*, **107** (30), 13336–13341.

Gower, T. L., Pastey, M. K., Peeples, M. E., *et al.* (2005). RhoA signaling is required for respiratory syncytial virus-induced syncytium formation and filamentous virion morphology. *Journal of Virology*, **79** (9), 5326–5336.

Jeffree, C. E., Rixon, H. W. M., Brown, G., Aitken, J., and Sugrue, R. J. (2003). Distribution of the attachment (G) glycoprotein and GM1 within the envelope of mature respiratory syncytial virus filaments revealed using field emission scanning electron microscopy. *Virology*, **306**, (2), 254–267.

Jorgenson, R. L., Vogt, V. M., and Johnson, M. C. (2009). Foreign glycoproteins can be actively recruited to virus assembly sites during pseudotyping. *Journal of Virology*, **83**, 4060–4067.

Kaletsky, R. L., Francica, J. R., Agrawal-Gamse, C., and Bates, P. (2009). Tetherin-mediated restriction of filovirus budding is antagonized by the Ebola glycoprotein. *Proc. National Academy of Sciences of the USA*, **106**, 2886–2891.

Keller, P. W., Johnson, M. C., and Vogt, V. M. (2008). Mutations in the spacer peptide and adjoining sequences in Rous sarcoma virus Gag lead to tubular budding. *Journal of Virology*, **82**, 6788–6797.

Lehmann, M. J., Sherer, N. M., Marks, C. B., Pypaert, M., and Mothes, W. (2005). Actin- and myosin-driven movement of viruses along filopodia precedes their entry into cells. *The Journal of Cell Biology*, **170**, 317–325.

Mccurdy, L. H. and Graham, B. S. (2003). Role of plasma membrane lipid microdomains in respiratory syncytial virus filament formation. *Journal of Virology*, **77**, 1747–1756.

Meier, O., Boucke, K., Hammer, S. V., *et al.* (2002). Adenovirus triggers macropinocytosis and endosomal leakage together with its clathrin-mediated uptake. *The Journal of Cell Biology*, **158**, 1119–1131.

Mercer, J., Schelhaas, M., and Helenius, A. (2010). Virus entry by endocytosis. *Annual Review of Biochemistry*, **79**, 803–833.

Neil, S. J. D., Zang, T., and Bieniasz, P. D. (2008). Tetherin inhibits retrovirus release and is antagonized by HIV-1 Vpu. *Nature*, **451**, 425–430.

Noda, T., Ebihara, H., Muramoto, Y., *et al.* (2006). Assembly and budding of Ebolavirus. *PLoS Pathogens*, **2**, e99.

Shukla, D., Liu, J., Blaiklock, P., *et al.* (1999). A novel role for 3-O-sulfated heparan sulfate in herpes simplex virus 1 entry. *Cell*, **99**, 13–22.

Utley, T. J., Ducharme, N. A., Varthakavi, V., *et al.* (2008). Respiratory syncytial virus uses a Vps4-independent budding mechanism controlled by Rab11-FIP2. *Proc. National Academy of Sciences of the USA*, **105**, 10209–10214.

Wild, P., Senn, C., Manera, C. L., *et al.* (2009). Exploring the nuclear envelope of herpes simplex virus 1-infected cells by high-resolution microscopy. *Journal of Virology*, **83**, 408–419.

Wills, J. W., Craven, R. C., Weldon, R. A., Nelle, T. D., and Erdie, C. R. (1991). Suppression of retroviral MA deletions by the amino-terminal membrane-binding domain of p60src. *Journal of Virology*, **65**, 3804–3812.

Wright, E. R., Schooler, J. B., Ding, H. J., *et al.* (2007). Electron cryotomography of immature HIV-1 virions reveals the structure of the CA and SP1 Gag shells. *The EMBO Journal*, **26**, 2218–2226.

8 High-resolution scanning electron microscopy of the nuclear surface in Herpes Simplex Virus 1 infected cells

Peter Wild, Andres Kaech, and Miriam S. Lucas

8.1 Introduction

Herpes virions consist of genomic double-stranded DNA enclosed by an icosahedral capsid, the tegument proteins, and the viral envelope (Roizman, 2001). During infection, newly synthesized capsid proteins are transported from the cytoplasm to the nucleus where they assemble to capsids into which DNA is packaged. This process resembles head assembly and DNA packaging in bacteriophages (Baines, 2005). Assembled capsids are transported to the nuclear periphery. There they bud at the inner nuclear membrane (INM) into the perinuclear space (PNS), acquiring tegument and the viral envelope (Granzow, 2001; Leuzinger *et al.*, 2005). Replication of herpes viruses such as herpes simplex virus 1 (HSV-1) radically alters nuclear architecture (Simpson-Holley *et al.*, 2005). In particular, the nuclear membrane becomes highly modified in the course of viral replication (Roizman, 2001). Alterations in the nuclear periphery are even more spectacular in cells infected with U_S3 deletion mutants because U_S3, a viral kinase, is involved in regulation of budding and in transportation of virions out of the PNS (Reynolds *et al.*, 2002).

The nuclear envelope, comprising the INM and the outer nuclear membrane (ONM), separates eukaryotic cells into a nuclear and a cytoplasmic compartment. The phospholipid bilayers of the INM are designated nuclear layer and luminal layer, respectively, those of the ONM luminal layer and cytoplasmic layer. Local fusions between both membranes create giant aqueous channels, the nuclear pores through which nucleocytoplasmic exchanges proceed. Elaborate protein structures of eightfold rotational symmetry, the nuclear pore complexes (NPC), are embedded into these pores (Goldberg and Allen, 1995; Lim and Fahrenkrog, 2006). Each NPC is composed of building blocks of ~30 different proteins called nucleoporins (Nup). Nucleoporins are modularly assembled to form subcomplexes within the highly dynamic NPC (Fahrenkrog *et al.*, 2004; Schwartz, 2005). The ONM continues into the rough endoplasmic reticulum (RER), which was shown to continue into membranes of the Golgi complex (Wild *et al.*, 2002).

The nuclear periphery of herpes virus infected cells has been studied intensely by conventional transmission electron microscopy (TEM) on thin sections (reviews: Mettenleiter, 2002; Mettenleiter, 2004; Mettenleiter *et al.*, 2006a). Imaging of the nuclear

Scanning Electron Microscopy for the Life Sciences, edited by H. Schatten. Published by Cambridge University Press

surface applying other techniques such as scanning electron microscopy (SEM) was rare, because the advantages of surface imaging at the subcellular level by SEM were not available to cell biologists largely due to insufficient resolution of scanning electron microscopes. This situation was considerably improved by the introduction of field emission sources. They facilitate surface imaging at much the same effective resolution for biological material as TEM (Allen and Goldberg, 1993; Hermann and Muller, 1993). An alternative methodology for surface imaging is the technique of freeze-fracture that is followed by production of replicas for imaging by TEM (Hagiwara *et al.*, 2005). Though this technology is cumbersome to apply to isolated cells it has been successfully employed to visualize the nuclear surface of HSV-1 infected BHK-21 cells (Haines and Baerwald, 1976).

The introduction of field emission scanning electron microscopy (FESEM) and the development of methodologies to image the surface of subcellular structures in the dried state (Allen *et al.*, 1997) as well as in the frozen hydrated state (Walther and Muller, 1999) led to detailed information on subcellular structures such as the nuclear surface. Recently, the authors have investigated the nuclear surface of bovine herpes virus 1 (BHV-1) (Wild *et al.*, 2005) and of HSV-1 (Wild *et al.*, 2009) infected cells by FESEM in the dried and frozen hydrated state.

This chapter presents data obtained by cryo-FESEM of Vero cells infected with wild type (wt) HSV-1 and with the well-characterized U_S3 deletion mutant R7041(ΔU_S3) (Purves *et al.*, 1987; Purves *et al.*, 1991) as well as some data obtained by FESEM of dry-fractured cells. It describes viral interactions with the INM and the ONM as well as alterations in nuclear pores in the context of capsid nucleus to cytoplasm translocation. To understand these processes better, some images of the nuclear surface taken by transmission electron microscopy from cells prepared *in situ* employing cryo-fixation and freeze-substitution (CFTEM) are included.

8.2 Technical aspects

8.2.1 Infection and preparation of cells for microscopic analysis

Vero and HeLa cells were grown in Dulbecco's modified minimal essential medium (DMEM) supplemented with penicillin (100 U/ml), streptomycin (100 g/ml) and 10% fetal bovine serum (FBS) for 2 days. Then cells were inoculated with HSV-1 strain F, the U_S3 deletion mutant R7041(ΔU_S3), and the repair mutant R2641 (kindly provided by B. Roizman, University of Chicago, IL, USA) at a multiplicity of infection (MOI) of 5 plaque forming units (PFU) per cell and incubated at 37 °C for 10 to 20 hr.

8.2.2 Cryo-field emission scanning electron microscopy (Cryo-FESEM)

HSV-1 infected (MOI 5) and mock infected cells grown in cell culture flasks were trypsinized and centrifuged at 150 g for 8 min. The pellet was resuspended in fresh medium, collected in Eppendorf tubes, and fixed with 0.25% glutaraldehyde in medium at

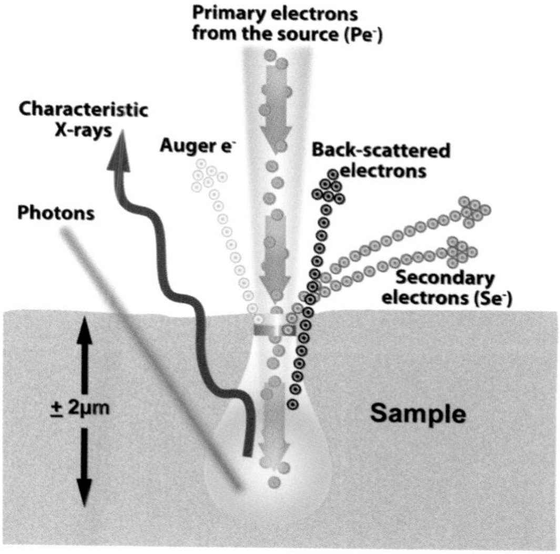

Figure 1.2 The incident electron beam interacting with the specimen produces emission of low-energy secondary electrons (SE), back-scattered electrons (BSEs), light emission (cathodoluminescence), characteristic X-ray emission, specimen current and transmitted electrons, and others as shown.

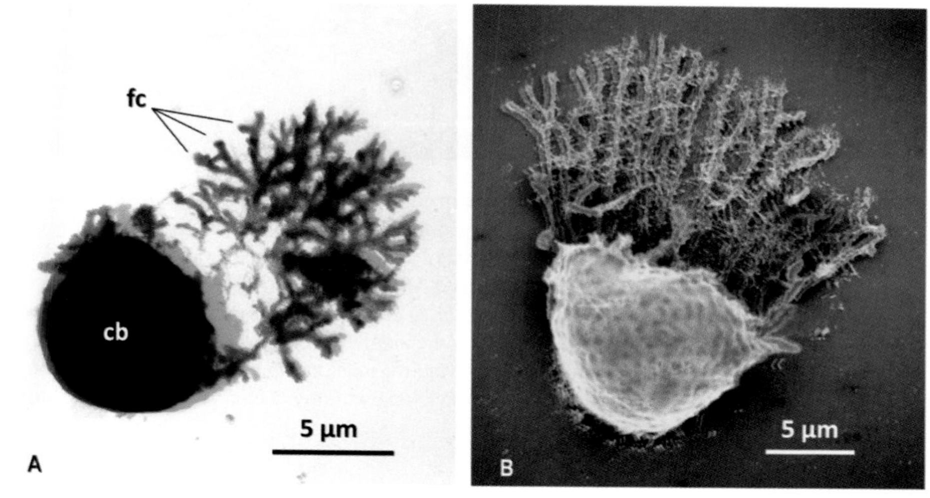

Figure 3.1 Comparison of high-voltage TEM (HVEM) with high-resolution SEM of triton-treated, fixed, crawling sperm of the nematode intestinal parasite, *Ascaris suum*. (A) Activated sperm cell crawling on coated grid, fixed and CPD. The pseudopod has been de-membranated, revealing the "bottlebrush" arrangement of major sperm protein filaments into fiber complexes (*fc*) that represent its machine for crawling. Because of thickness limitations of TEM, most of the cytoskeleton was removed, so that the full array of complexes cannot be seen. Accelerating voltage (Vacc) = 1000 kV; in stereo, left eye red. (B) This cell is from the same HVEM preparation, but Pt-coated and viewed with HRSEM. The branching and composition of the fiber complexes are clearer, as is the loss of all but the adherent branches of the cytoskeleton caused by the detergent treatment. It also demonstrates the integrity of the cell–body plasma membrane, which is resistant to detergent disruption, and some of its unique surface features. Vacc = 1.5 kV; in stereo, left eye red.

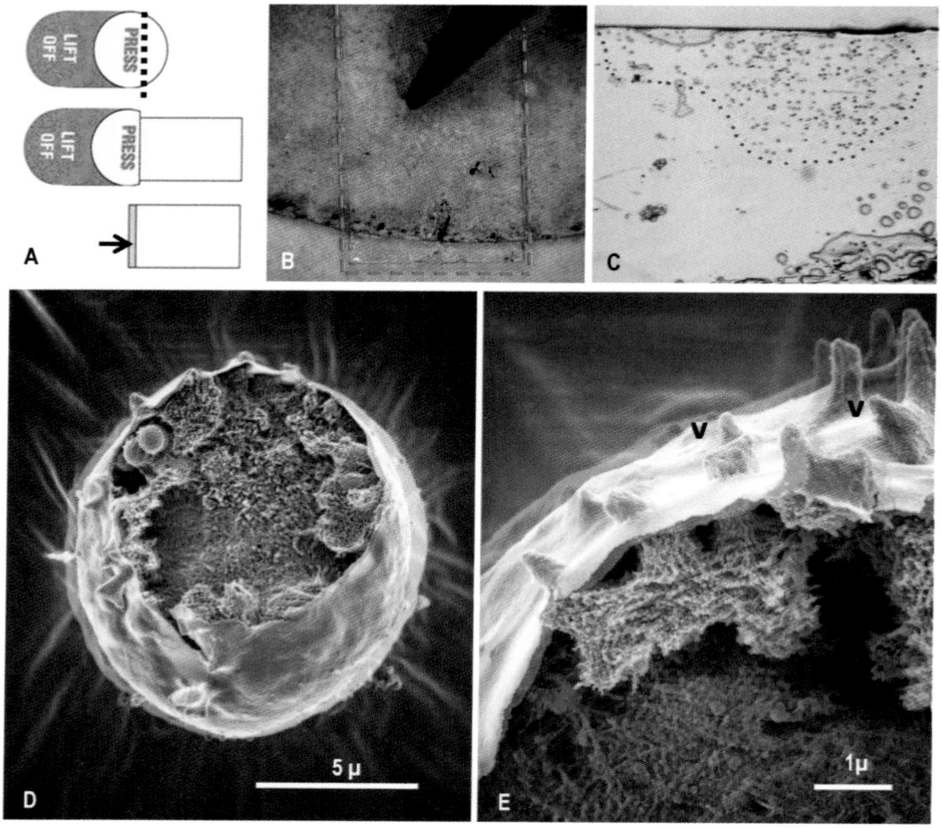

Figure 3.3 Tape-ripping. Adherent *Ascaris* sperm are fixed, CPD, then torn open with double-stick tape and both surfaces Pt coated. (A) Thin strip of adhesive from an adhesive mounting tab is applied to the end of a cut coverslip. (B) Small area of adhesive strip (dotted outline) is gently pressed against cells on glass with forceps. (C) After adhesive is pulled away, numerous cell fragments are seen on ripper strip adhesive, shown in D and E. (D) *Ascaris* sperm cell body where it joins the pseudopod. The smaller corpuscles at the interface are mitochondria, the larger spherical particles are refringent storage granules. (E) Showing the relationship between the external villipodia (*v*) which adhere to the substrate and the cytoskeleton that forms them. Vacc = 1.5 kV; in stereo, left eye red.

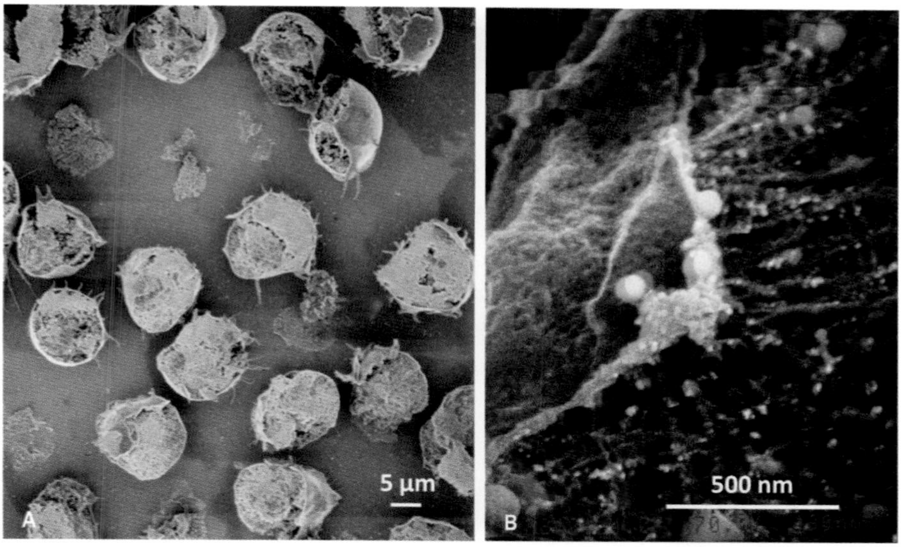

Figure 3.4 Wet-ripping and immunolabeling. *Ascaris* sperm cells attached to glass are lightly fixed, then ripped open by adhesion to an overlying coverslip treated with poly-D-lysine (PDL). The ripped cells were rinsed in blocking buffer, and incubated with primary antibody against nematode *major sperm protein*, the major constituent of the fibrous cytoskeleton of the pseudopodium. Cells were then incubated with anti-species Ig adsorbed to 10 nm colloidal gold. (A) A view of the PDL "ripper" slip showing the bases of numerous cells which have been torn away from the substrate, with many views of pseudopod cytoskeleton and cell body. Vacc = 1.5 kV. (B) Anti-MSP-gold labeling of the fiber complexes associated with pseudopod membrane, imaged with the back-scatter electron detector. Bright dots are colloidal gold. Note that the label is localized to filaments and the interior of the pseudopod membrane. Vacc = 4 kV, back-scatter mode; in stereo, left eye red.

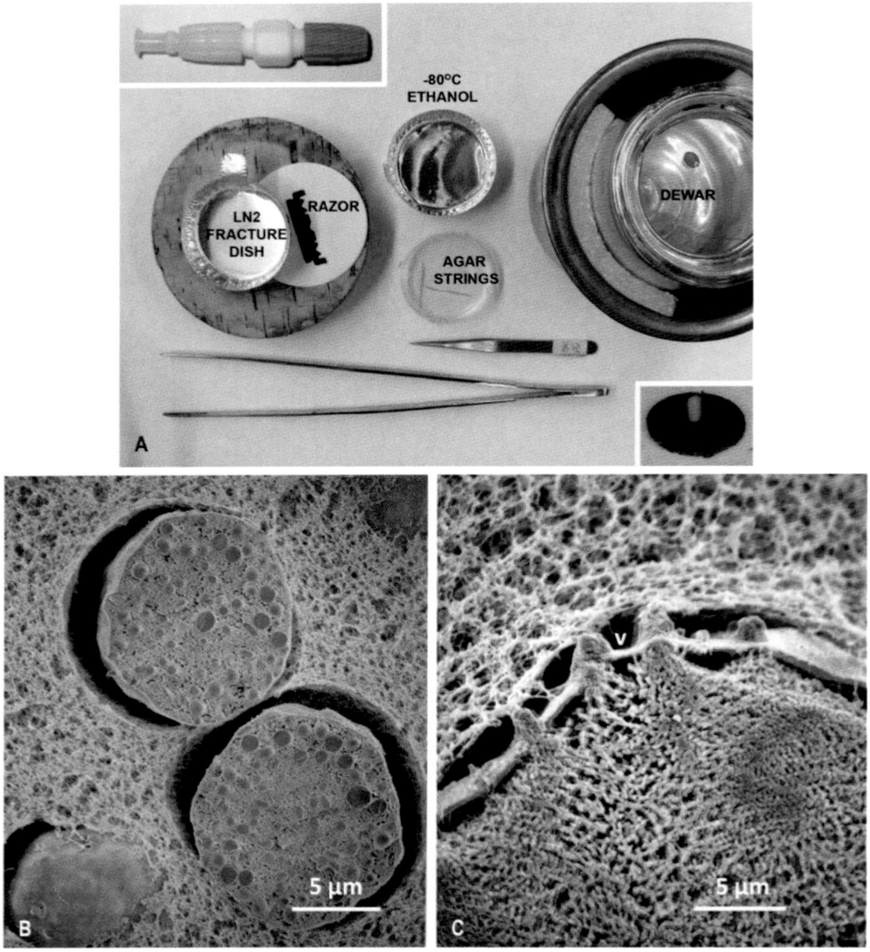

Figure 3.5 Fractured agar strings. *Ascaris* sperm are fixed, rinsed in buffer and embedded in liquefied agar, chilled, extruded into strings, dehydrated to absolute ethanol, frozen in liquid nitrogen and fractured, CPD, then mounted on conductive tape, fractured end up, presenting numerous three-dimensional sections. (A) Setup for fracturing agar strings. Upper inset: capillary tube ejector. Lower inset: a fractured stub mounted on a conductive adhesive disc for use in an in-lens HRSEM. (B) Two sections through the cell bodies of sperm cells. Agar strands retract with drying, leaving a view of the cell's exterior. Vacc = 1.5 kV; (C) A fracture section through a pseudopod showing some of the fine structure of the cytoskeleton's association with the pseudopod membrane, the villipodia (*v*). Vacc = 1.5 kV; in stereo, left eye red. Compare to tape-ripped villipodia in Figure 3.3B.

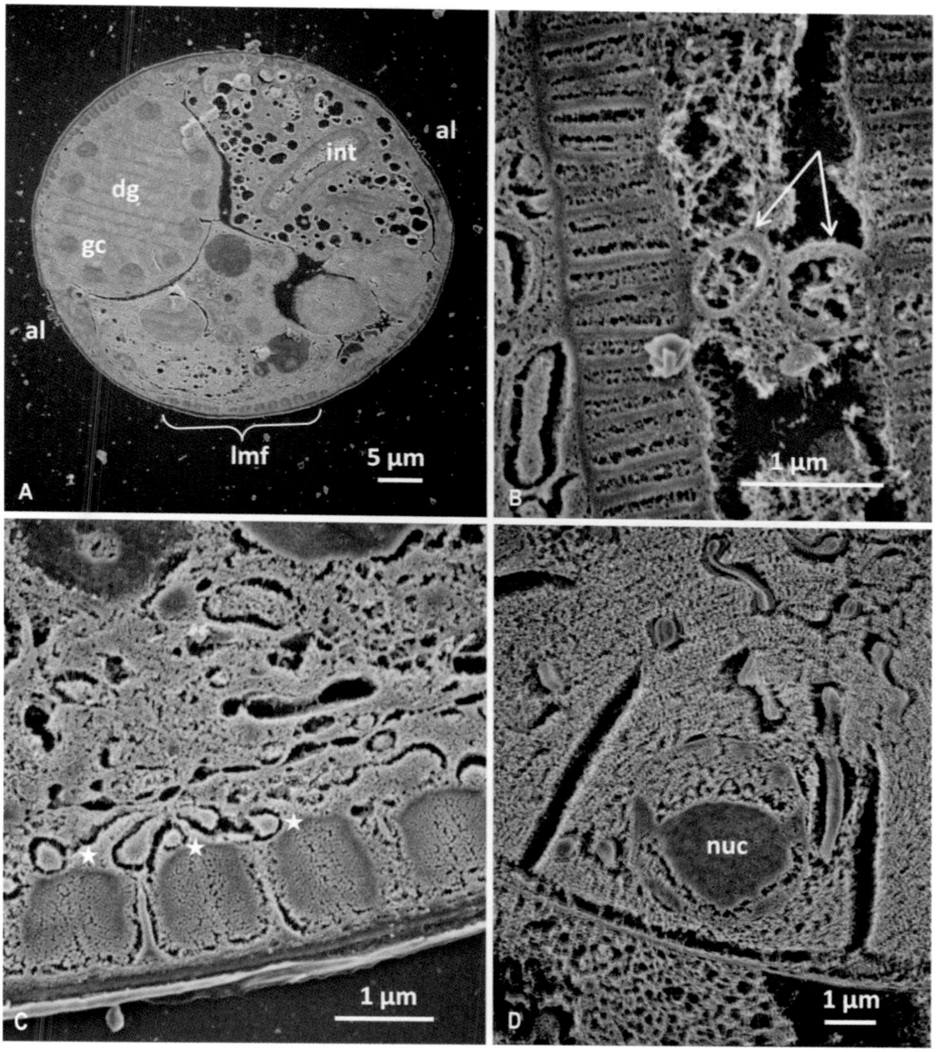

Figure 3.6 De-embedding of plastic sections. Hermaphrodite *Caenorhabditis elegans* were fixed by freeze-substitution and embedded in an Epon–Araldite mixture. Blocks were cut in 300–500 nm thick sections, which were transferred to silanized glass strips. Plastic was removed as described and the strip Pt-coated. (A) Low-power cross section through distal third of worm, showing features comparable to TEM micrographs of *C. elegans*. Major features: intestine (*int*), ventral longitudinal muscle fibers (*lmf*), alae (*al*) extensions from cuticle, distal gonad (*dg*) of syncitial germ cells (*gc*). (B) An enlarged view of the intestinal wall and lumen, showing the carcasses of two *E. coli* (arrows). (C) Enlarged view of ventral longitudinal muscle cells and muscle arms (*stars*) directed toward their synapse with the ventral nerve cord. (D) Section of a germ cell in the distal gonad. While limited by section thickness, stereo enhances some of the tubular elements that join the cells with the center of the gonad, the rachis. Nucleus, *nuc*. calibration bar = 1 µm. In stereo, left eye red. All micrographs, Vacc = 1.5 kV.

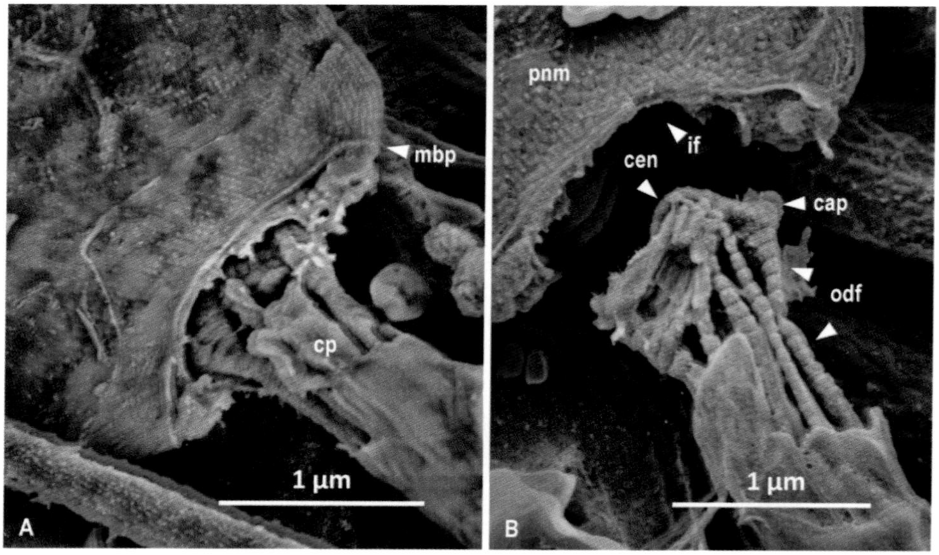

Figure 3.8 Cryo-SEM of de-membranated bull sperm. Fresh bull sperm (American Breeders Service, DeForest, WI, USA) was briefly incubated with Triton X-100 detergent to expose sperm tail elements without damage to other structures, then flash-frozen under high-pressure LN2, etched to expose cells in ice, coated while frozen with Pt, and observed frozen with an LN2-cooled specimen holder. The membrane of the neck area containing the connecting piece appears to be most vulnerable to the brief detergent treatment. (A) The intact connecting piece (*cp*) is seen in its normal relationship to the margin of the basal plate (*mbp*), with centriole and outer dense fibers just visible inside. (B) The components of the connecting piece have been pulled away from the implantation fossa (*if*), showing the proximal centriole (*cen*), at an angle to the axis of the tail, the striated outer dense fibers (*odf*) and capitulum (*cap*). The proximal centriole nucleates the first mitotic spindle in the zygote in some species. The striking embossing on the post-nuclear outer membrane (*pnm*) is not seen in chemically fixed sperm and corresponds to hexagonally packed surface glycoproteins known to reside there. Vacc = 1.5 kV; in stereo, left eye red.

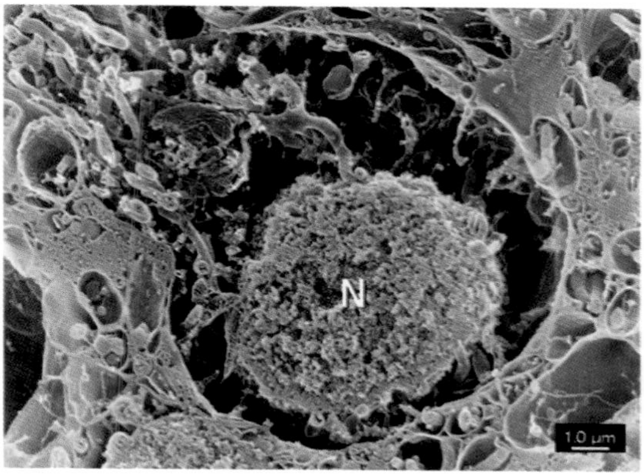

Figure 4.1 HR-FE-SEM image of a cracked granule cell in the rat's olfactory bulb. The nucleus (N), situated at the center of the small cell body, occupies a large area of the central region and conceals the spatial configuration of cell organelles from the view. This granule cell has one apical dendrite (upper left corner). ER (magenta), Golgi apparatus (green), mitochondrial outer membrane (blue), mitochondrial cristae (yellow).

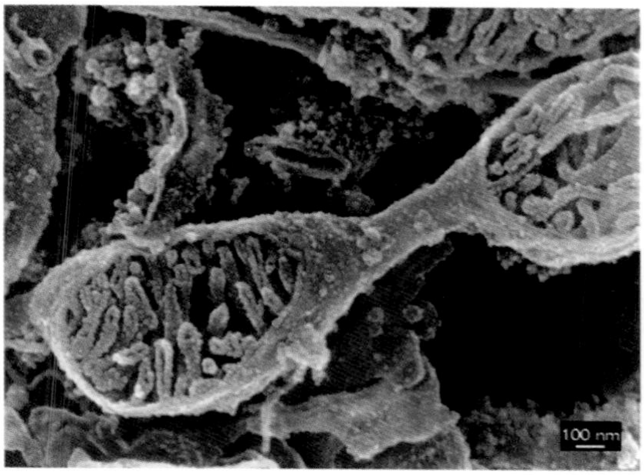

Figure 4.2 High-magnification view of mitochondrial network Type-4 in an olfactory-bulb granule-cell body. Mitochondrial cristae (yellow) are well preserved and do not show any deformation. Golgi apparatus (green), ER (magenta), mitochondrial outer membrane (blue).

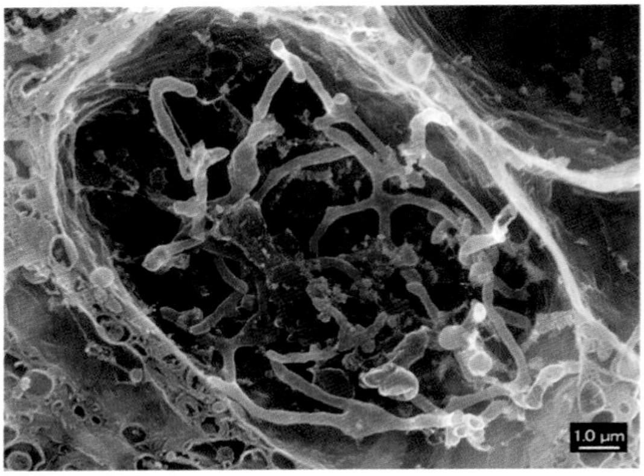

Figure 4.3 Low-magnification view of mitochondrial network Type-1. A large mitochondrion (blue) appears to be in the form of a single continuous network structure extending throughout the cell body. The widths of tubular components of a mitochondrial network are fairly uniform in size, about 250–300 nm. Golgi apparatus (green), ER (magenta), mitochondrial cristae (yellow).

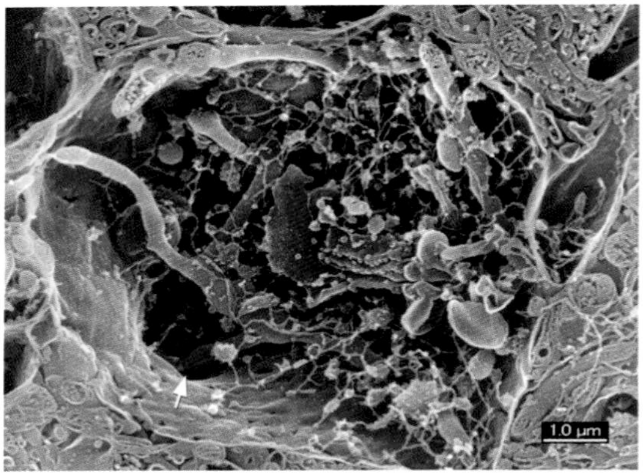

Figure 4.4 Mitochondrial network Type-1 in an olfactory-bulb granule-cell body. A large hole in the bottom left corner represents an entrance to the apical dendrite. Note an elongated part of the mitochondrial network in the cell body that enters into the dendrite (arrow). Multipartite rER forms vesicles or irregular cisterns and is always connected with filamentous sER, around 25 nm in diameter. They form an additional (additional to the mitochondrial network) network system joining together in the cell body. Golgi apparatus (green), ER (magenta), mitochondrial outer membrane (blue), mitochondrial cristae (yellow).

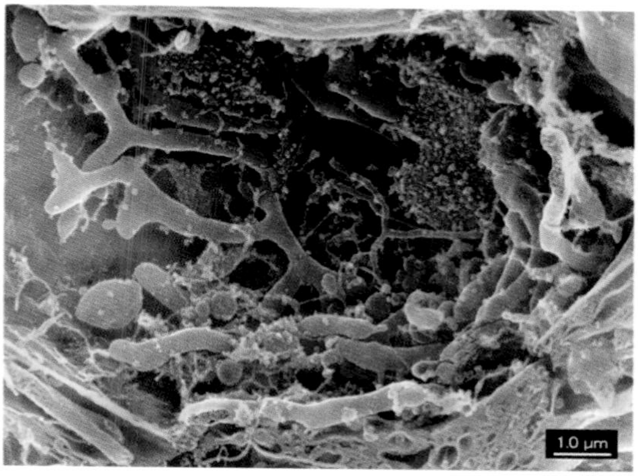

Figure 4.5 Mitochondrial network Type-2 in an olfactory-bulb granule-cell body. The network is composed of two parts different in width. Filamentous parts, around 75 nm in diameter across, originate from the tips of tubular parts of normal size about 250–300 nm. The rER shows a wide variation in shape and size, while the sER appears as a filamentous structure of about 25 nm. Golgi apparatus (green), ER (magenta), mitochondrial outer membrane (blue), mitochondrial cristae (yellow).

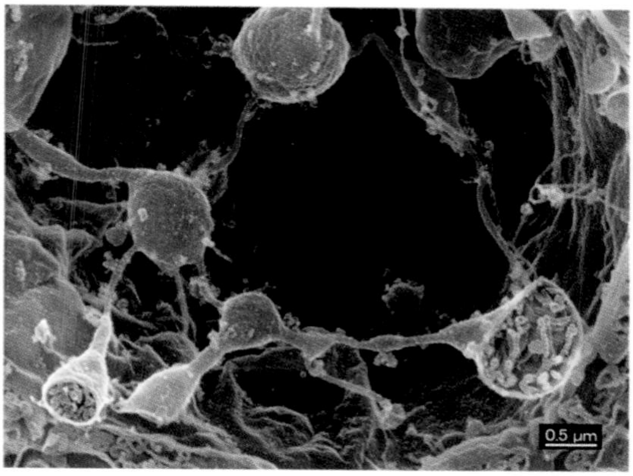

Figure 4.6 Mitochondrial network Type-3 in an olfactory-bulb granule-cell body: a necklace-like loop, consisting of globular parts and filamentous parts. Note the mitochondrial cristae (yellow) in globular parts. They are well preserved without any deformation. Mitochondrial outer membrane (blue), ER (magenta).

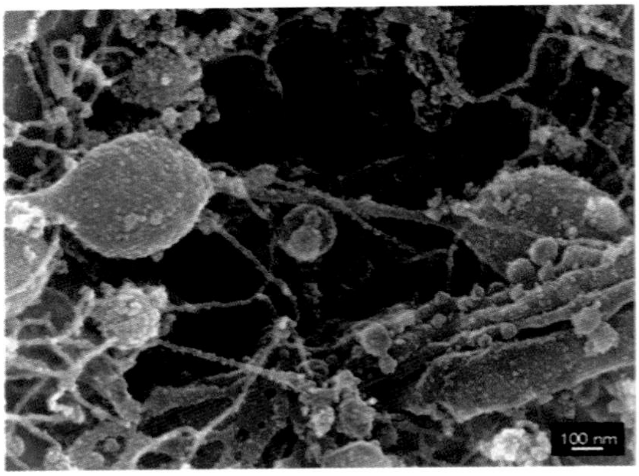

Figure 4.7 Higher magnification image of mitochondrial network Type-3 in an olfactory-bulb granule-cell body. A long filamentous mitochondrial network connecting two globular parts is 45 nm in width. This size is almost twice that of a filamentous tubule of sER that runs parallel to the long axis of the filamentous part of a mitochondrion. Golgi apparatus (green), mitochondrial outer membrane (blue), ER (magenta).

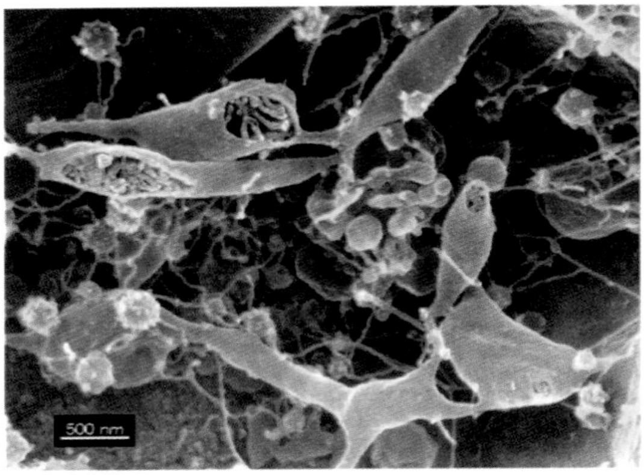

Figure 4.8 Mitochondrial network Type-4 in an olfactory-bulb granule-cell body. The HR-FE-SEM image shows the structural complexities of this mitochondrial network of this Type-4. Due to a wide variety of structural appearances, it is difficult exactly to describe the morphological features of this type of mitochondrial network. Golgi apparatus (green), mitochondrial outer membrane (blue), ER (magenta).

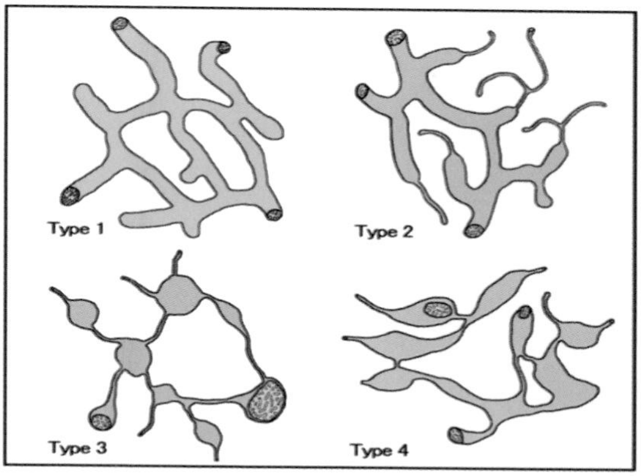

Figure 4.9 Types of mitochondrial network. The mitochondrial network of Type-1 is composed of tubular parts rather uniform in size. Type-2 is composed of two kinds of tubular part differing in diameter. Type-3 is composed of two parts, globular and filamentous. Type-4 is composed of irregularly shaped parts. Mitochondrial outer membrane (blue), mitochondrial cristae (yellow).

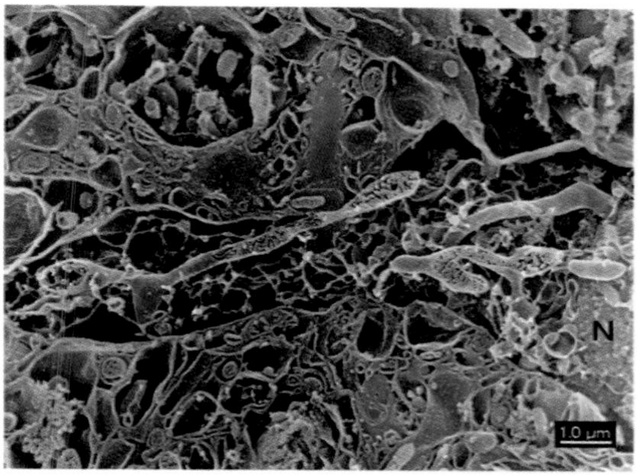

Figure 4.10 Longitudinal section of an apical dendritic trunk of an olfactory-bulb granule cell. Mitochondria in a dendritic trunk display long slender tubules. In this HR-FE-SEM photograph, the direct continuity between mitochondria in the dendrite and the mitochondrial network in a cell body is unclear (but see Figures. 4.4 and 11). Abundant filamentous sER (magenta) also forms a network structure. Nucleus (N), Golgi apparatus (green), mitochondrial outer membrane (blue), ER (magenta).

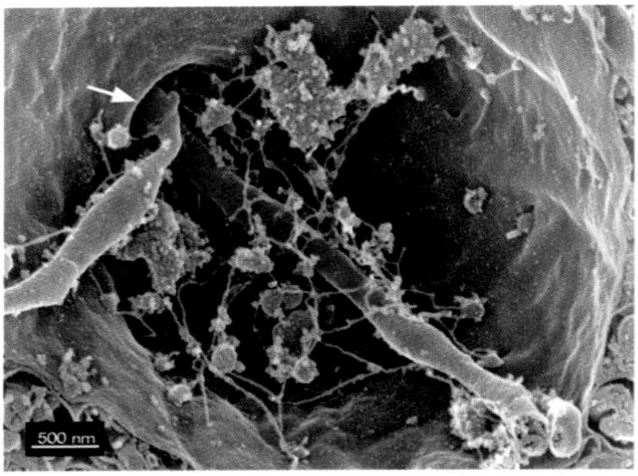

Figure 4.11 Base of an apical dendrite viewed from the cell body. Long tubular mitochondria enter into a dendrite from the cell body passing through a hole (arrow) that borders the dendrite and the cell body. Mitochondrial outer membrane (blue), mitochondrial cristae (yellow), ER (magenta).

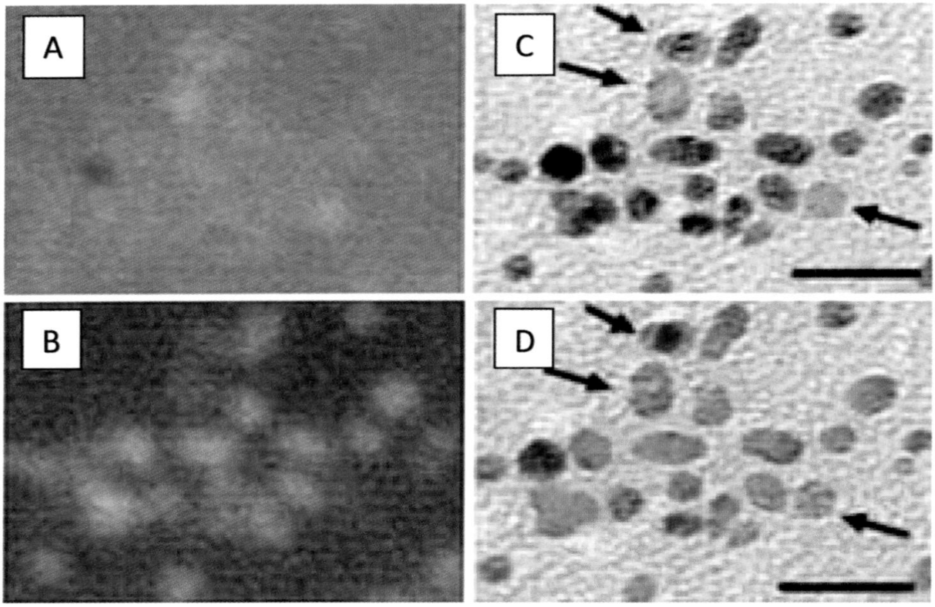

Figure 6.2 Localization of anti-α-actinin-cAu6 (A–C) and anti-actin-cPd6 (B–D) conjugates by ESI on a cluster of labels. Most of the spheres consist of Pd, only a few of the markers (arrows in C and D) show Au-specific signal. For detection of Au (bright particles in A), two images were taken before the O4,5 ionization edge at 40 eV and 50 eV for calculation of the background signal and one image above the Au O4,5-edge at 75 eV (A). (C) Elemental distribution of Au in an overlay of the Au-specific signal over a zero-loss image. For detection of Pd, background images were taken at energy losses of 280 eV and 324 eV, the maximum Pd-specific energy loss image at the M4,5-edge at 430 eV is shown in (B). Light particles correspond to Pd. The elemental distribution of Pd in an overlay of Pd-specific signal over the zero-loss image is shown in (D). Arrows indicate gold particles, the remainder are Pd particles. Bars represent 20 nm.

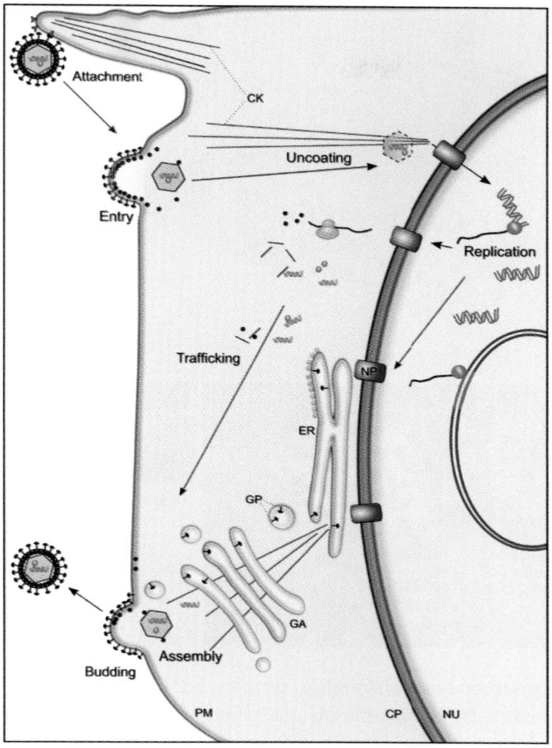

Figure 7.1 Virus assembly: A) Attachment, B) Entry, C) Uncoating, D) Replication, E) Trafficking and assembly, F) Budding. Diagram shows: cytoskeleton (CK), cytoplasm (CP), endoplasmic reticulum (ER), Golgi apparatus (GA), glycoprotein (GP), nuclear pore (NP), nucleus (NU), plasma membrane (PM).

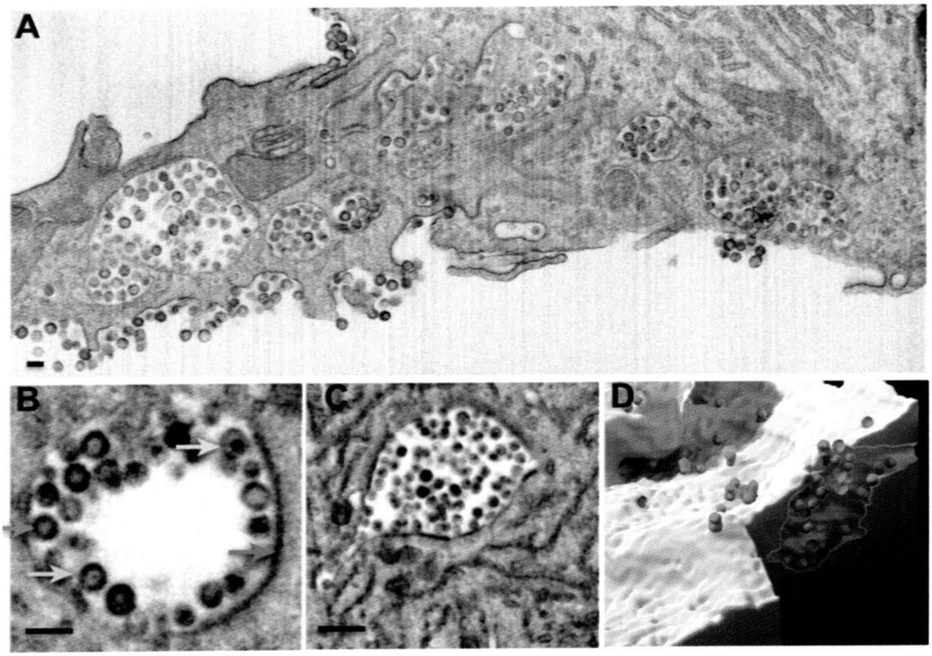

Figure 7.10 IA-SEM imaging of virion reservoirs and channels in primary HIV-1 infected MDM: (A) Single image of internal compartments, (B) Compartments magnified, with virions highlighted by arrows, (C) Compartments deep in the cell interior, (D) Segmented 3-D representation of a viral reservoir. Bars = 200 nm. © 2009 Bennett A. *et al*. (2009).

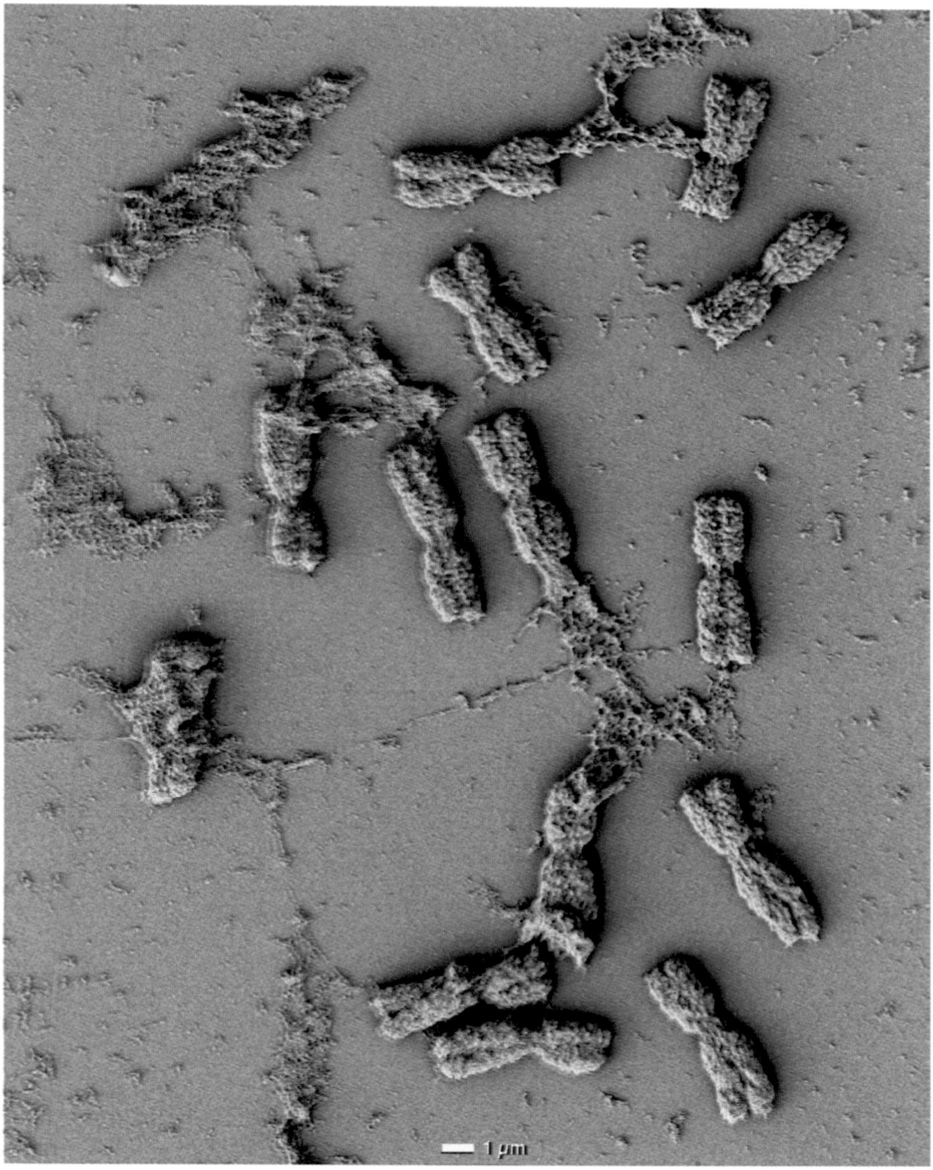

Figure 9.3 Anaglyph micrograph showing 3-D topography of a barley mitotic metaphase spread (2n = 14). Sister chromatids are distinguishable, and in some cases are separated at the telomeres. Centromeres exhibit a distinct primary constriction. (Use red-blue glasses.)

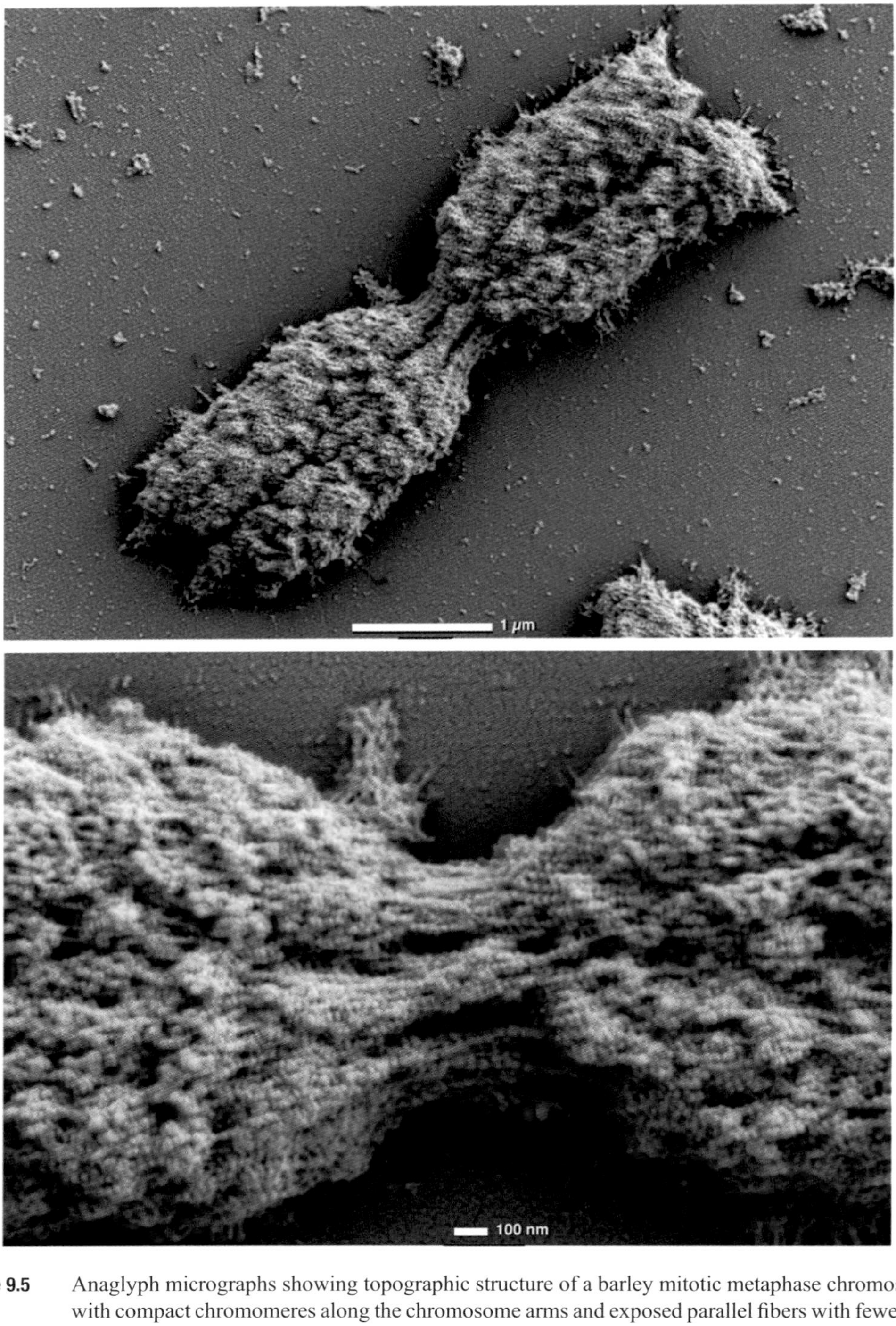

Figure 9.5 Anaglyph micrographs showing topographic structure of a barley mitotic metaphase chromosome with compact chromomeres along the chromosome arms and exposed parallel fibers with fewer and smaller chromomeres at the centromere. At higher magnification (lower image) 30 nm fibrils are visible along the entire chromosome, in particular at the centromere. (Use red-blue glasses.)

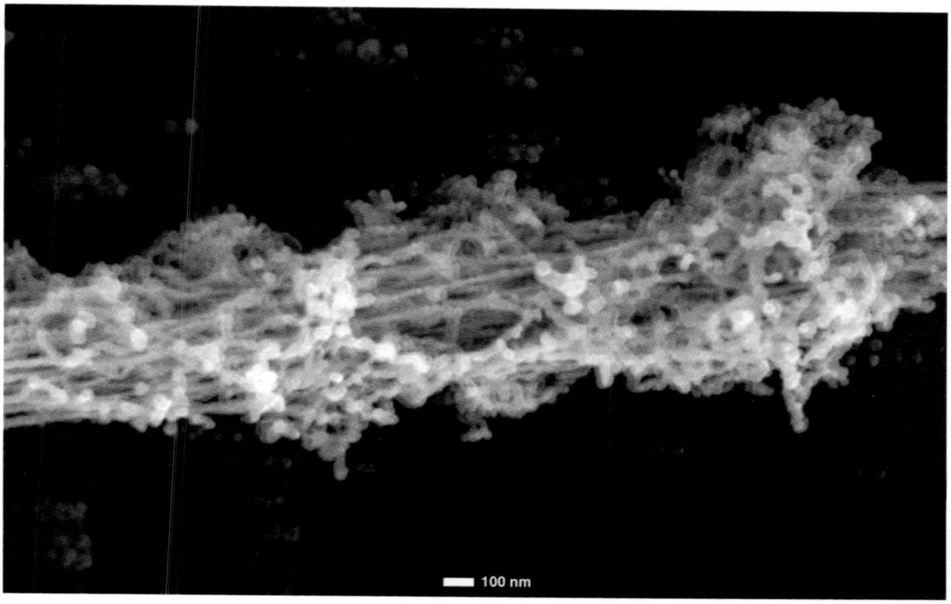

Figure 9.8 Anaglyph micrograph showing the stretched centromere of a barley chromosome after controlled decondensation. Parallel fibrils are distributed throughout the centromere interior, and chromomeres loosen to loops of fibrils in the range of 30 nm. (Reprinted from Wanner and Formanek, 2000, with permission from Elsevier.) (Use red-blue glasses.)

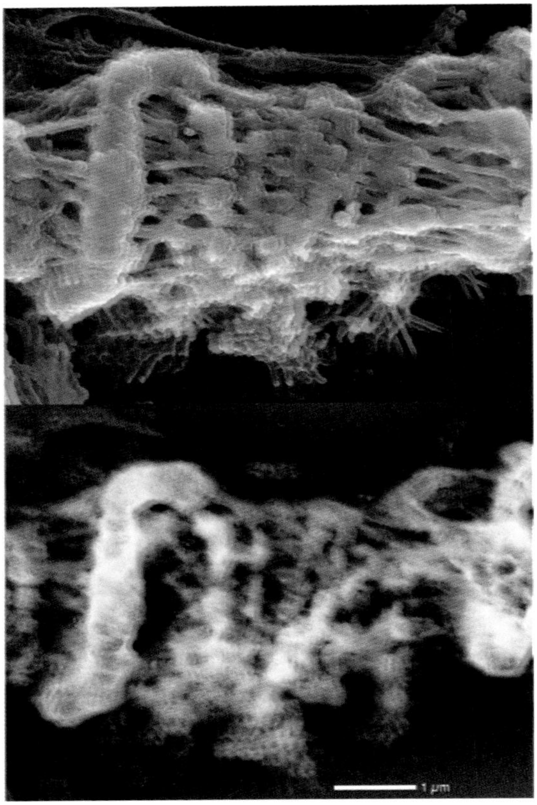

Figure 9.12 Anaglyphs of a mutant *Drosophila melanogaster* polytene chromosome (with distortion of banding) stained with platinum blue showing structural elements (compact chromatin bands and parallel fibrils; SE, upper image) as well as corresponding 3-D DNA distribution (BSE, lower image). (Use red-blue glasses.)

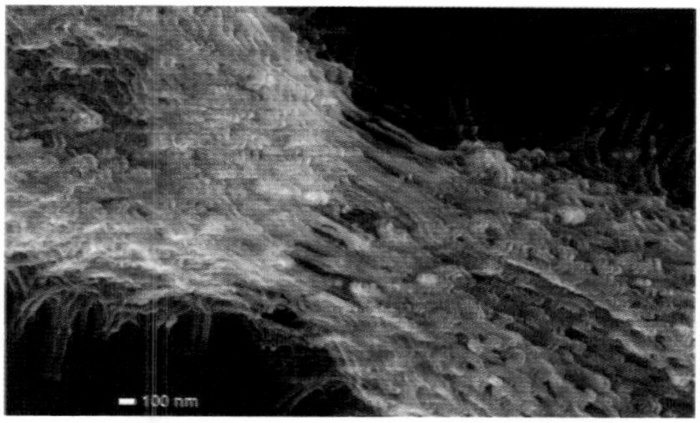

Figure 9.14 Anaglyph image of superimposed SE and BSE signals showing 3-D distribution of markers immunolabeling phosphorylated histone H3 (serine 10). The centromere is characterized by parallel fibrils, lacking chromomeres and exhibiting very few H3P markers. The majority of the markers are localized in the pericentric chromatid arms. (Reprinted from Wanner and Schroeder-Reiter 2008, with permission from Elsevier.) (Use red–blue glasses.)

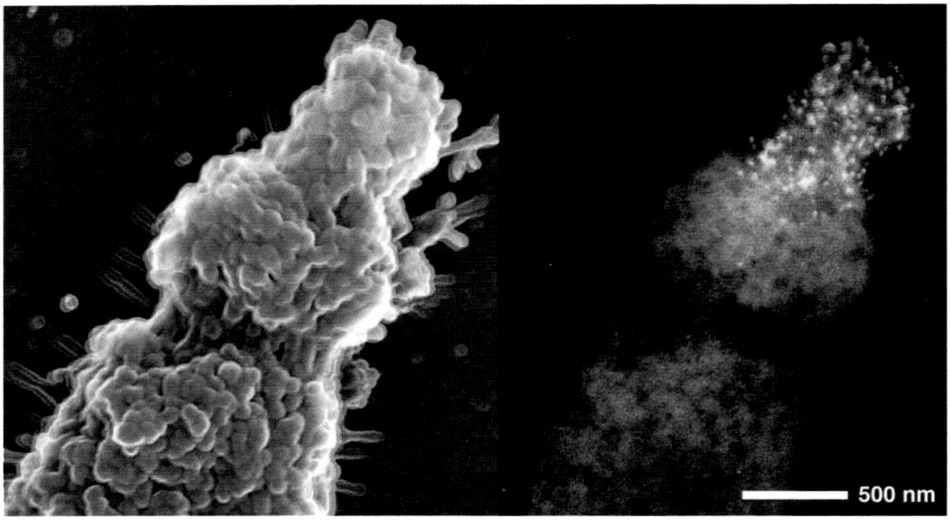

Figure 9.15 Anaglyph micrographs showing substructural elements (SE image, left) of a plant chromosome (*Ozyroë biflora*, Hyacinthaceae), as well as 3-D marker distribution labeling 45S rDNA with *in situ* hybridization (BSE image, right), identifying a peg-like terminal structure as a NOR. (Use red–blue glasses.)

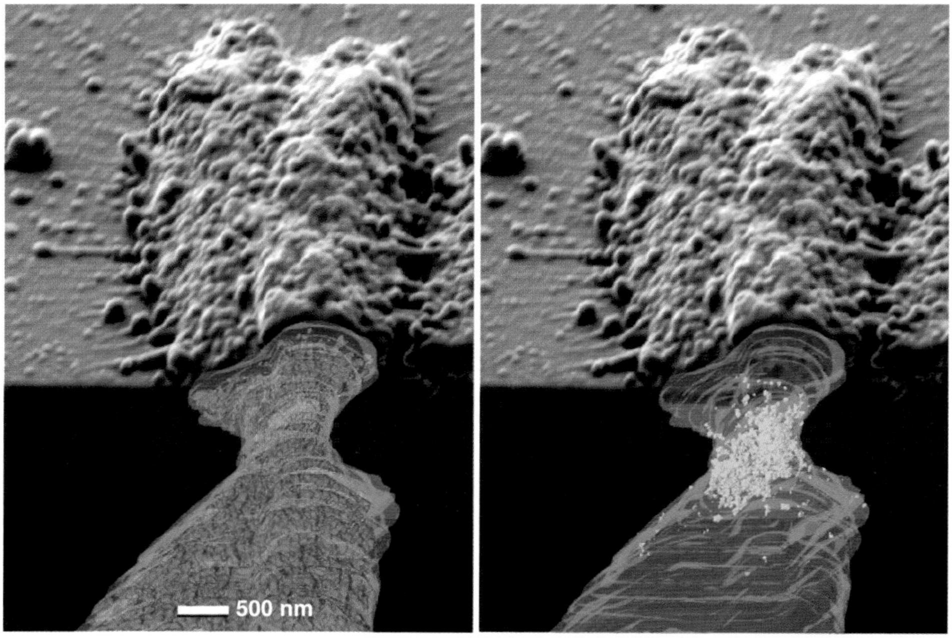

Figure 9.18 3-D reconstruction of an isolated barley chromosome immunolabeled for the centromere-specific histone H3 variant CENH3. The interior chromatin exhibits a dense network. The majority of the CENH3 signals localize to the centromere interior, indicating that it is a component of the inner kinetochore. (Reconstructed from a FIB/FESEM series of 162 images with 15 nm milling steps.)

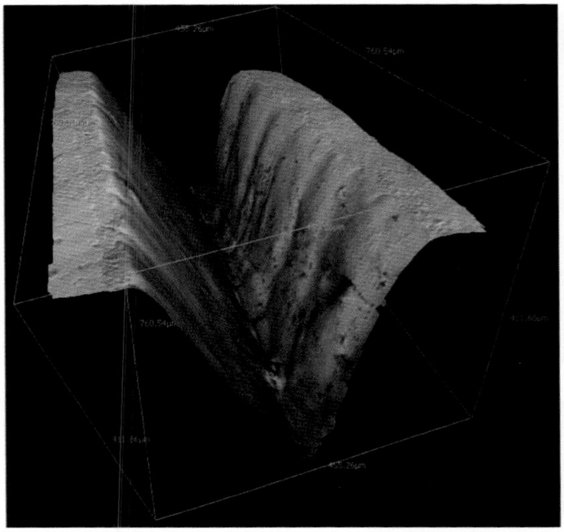

Figure 13.7 3-D model of a laser generated groove in an enamel. Courtesy of Drs C. Xu and Y. Wang, unpublished.

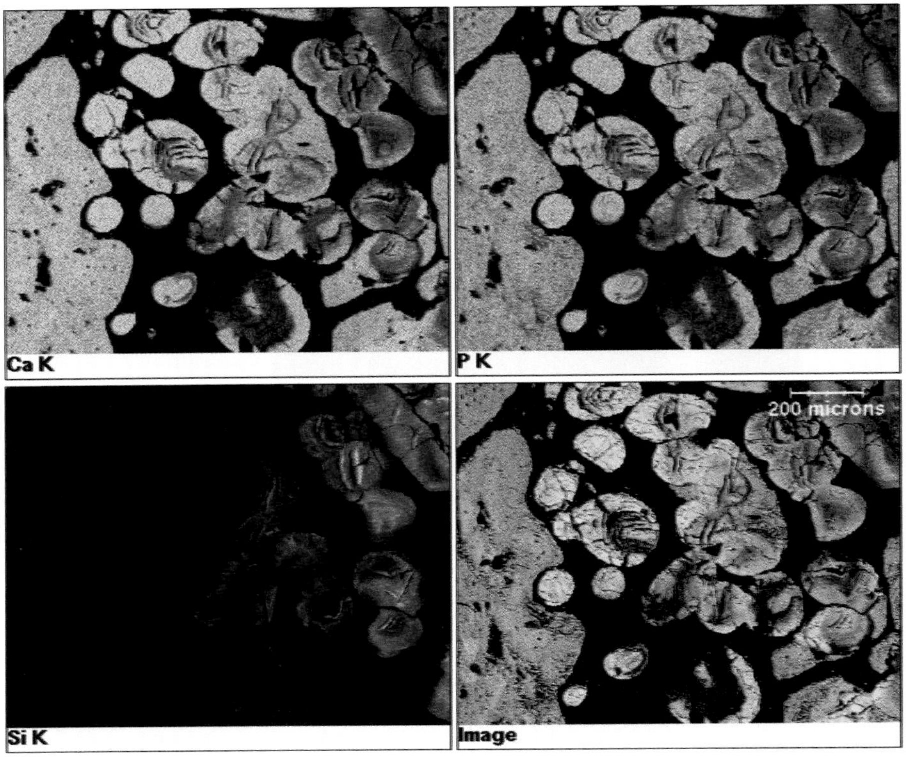

Figure 13.13 X-ray map (for Ca, P, and Si) and BSE image of bone healing with the help of scaffold of bioactive glass. Round glass particles are in different stages of conversion to hydroxyapatite (particles closest to bone are fully converted and have Ca and P in their composition, but no Si). Bar 200 μm. Courtesy of Dr. Bi, unpublished.

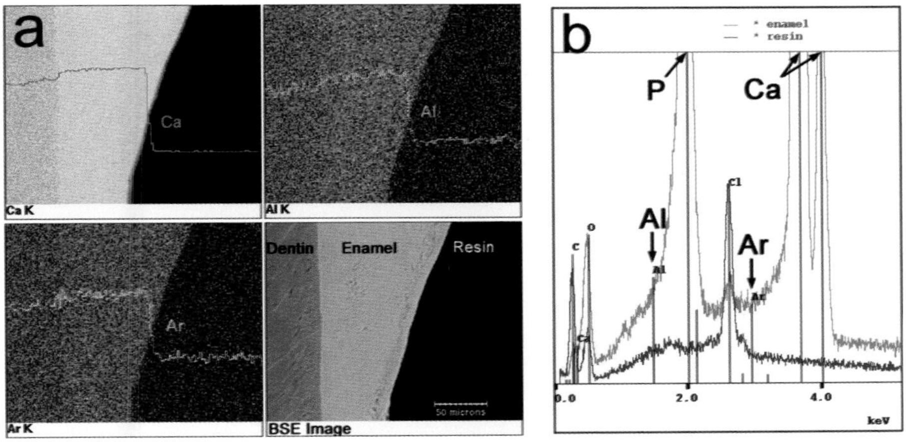

Figure 13.14 (a) Maps and scan lines of Ca, Al, and Ar obtained from polished cross-section of resin embedded tooth. There was no Al nor Ar in the specimen; registered distribution of X-ray intensities was actually a distribution of background intensities. Bar 50 μm. (b) Superimposed spectra of enamel and of embedding resin; substantial difference in background intensities recorded; no characteristic X-ray peaks for Al and Ar were detected.

4 °C for 30 min. The suspension was kept in the tubes at 4 °C until cells were sedimented. Subsequently, the supernatant was removed. A copper grid (300 mesh/inch^2, 12 μm thickness) was dipped into the sedimented cells, sandwiched between two flat aluminum specimen carriers (Type B: BAL-TEC, Leica Microsystems, Heerbrugg, Switzerland), the surfaces of which were scratched with a scalpel, and immediately frozen in a high-pressure freezing machine HPM 010 (BAL-TEC). After freezing, the sandwich was mounted under liquid nitrogen on a designated specimen holder for freeze-fracturing in the VCT 100 cryo-preparation box and transferred to a freeze-fracturing device BAF 060 (BAL-TEC) using the VCT 100 cryo-transfer system (BAL-TEC) designed according to Ritter *et al.* (1999). The specimen was fractured at −120 °C by removing the top aluminum specimen carrier with the hard metal knife in the BAF 060. The fractured surfaces were partially freeze-dried ("etched") at −105 °C for 2 min in a vacuum of about 10^{-7} mbar. Then the specimen surface was coated with 2 nm platinum/carbon by electron beam evaporation at an angle of 45°, and with 1 nm platinum/carbon tilting the electron beam gun between 0° and 90°. The specimen was retracted into the transfer shuttle of the VCT 100 system and transferred under high vacuum onto the cryo-stage in the SEM (LEO Gemini 1530, Zeiss, Oberkochen, Germany). Specimens were imaged at −115 °C (the saturation water vapor pressure of the specimen corresponding approximately to the vacuum in the chamber of 5×10^{-7} mbar) and at an acceleration voltage of 5 kV using the in-lens secondary electron detector.

8.2.3 Field emission scanning electron microscopy (FESEM)

In order to examine size and shape of NPCs a protocol for *in situ* preparation of nuclei according to Allen *et al.* (1998) was applied. Cells were grown on glass cover slips for 2 days and then infected with HSV-1 at a MOI of 5 and incubated at 37 °C for up to 15 hr. Then cells were fixed with 1% formaldehyde + 0.025% glutaraldehyde in 0.1 M Na/K-phosphate, pH 7.4, at room temperature for 1 min, and permeabilized with 0.5% Triton-X-100 for 5 min. After post-fixation with 1% osmium tetroxide at 4 °C for 30 min, cells were treated with 1% aqueous thiocarbohydrazide at room temperature for 30 min. After additional fixation with 1% osmium tetroxide at room temperature for 30 min, cells were dehydrated with a graded series of ethanol starting at 70%, and critical point dried (BALTE CPD 030).

The dry samples were fractured as follows. The glass cover slip with the dry cells on it was laid on an adhesive film on a SEM sample holder. This was firmly touched with another adhesive sample holder and pulled away without sideways movement. The surface on which the cells were grown and the adhesive surface were coated with 5 nm platinum by sputter coating in a high-vacuum sputtering device (BAL-TEC SCD500). The coated samples were examined in a FESEM (LEO Gemini 1530, Zeiss) at an acceleration voltage of 3 kV using the in-lens secondary electron detector.

8.2.4 Cryo-fixation for transmission electron microscopy (CFTEM)

Cells grown on sapphire discs, infected with HSV-1 (MOI 5), and incubated at 37 °C for 8 to 17 hr were fixed with 0.25% glutaraldehyde *in situ* for safety reasons by adding glutaraldehyde to the culture medium at 37 °C, and then frozen in a high-pressure freezing machine (HPM 010, BAL-TEC) as described by Monaghan *et al.* (2003). Frozen cells were transferred into a freeze-substitution unit (FS 7500, Boeckeler Instruments, Tucson, Arizona, USA) precooled to −88 °C for substitution with acetone and subsequent fixation with 0.25% glutaraldehyde and 0.5% osmium tetroxide at temperatures between −30 °C and +2 °C, as described in detail in (Wild *et al.*, 2001; Wild 2008), and embedded in Epon. Sections 50 to 60 nm thickness were stained with uranyl-acetate and lead-citrate and analyzed in a transmission electron microscope (CM12, FEI, Eindhoven, The Netherlands) equipped with a CCD camera (Ultrascan 1000; Gatan, Pleasanton, CA) at an acceleration voltage of 100 kV.

8.3 Discussion of data

8.3.1 The nuclear surface in freeze-fractured cells

To visualize subcellular structures in the frozen hydrated state, well-frozen cells need to be fractured. Ideally, freeze-fracture should be performed on *in situ* frozen cells. The authors were able to freeze-fracture cells grown on petriPERM discs (Wild *et al.*, 2005). However, the rate of successful preparations was very low. Therefore, suspensions of cells were used that were detached from the culture plates by trypsinization. For safety reasons, detached cells were fixed with 0.25% glutaraldehyde for 30 min prior to freezing. Fixation has also the advantage that all cellular processes are arrested so that probably few disturbances take place during the rather long preparation procedure between harvesting and freezing. However, this procedure is far from the ideal of freezing cells immediately after removal from the medium without pretreatment, as can be easily applied for TEM. Fracturing of frozen hydrated cells does not create completely arbitrary surfaces. Rather, the fracture planes run preferentially along the hydrophobic center of the lipid bilayer of cell membranes, i.e. along the center of the ONM or INM. However, many of the fracture planes run through the cytoplasm yielding cross-fractures of cells and organelles, e.g. the nucleus (Severs, 2007).

Convex fracturing through the ONM gives a view onto the luminal face of the ONM where nuclear pores appear commonly as slight depressions because the NPCs are embedded in pores formed by the ONM and INM (Figure 8.1). Convex fracturing through the INM gives a view onto the nuclear face of the INM which contains many proteins appearing as small dots. NPCs at the nuclear face stand out as button-like structures. They occasionally are removed during fracturing (Walther *et al.*, 1992). The diameter of NPCs at the INM is on average 120 nm; that of nuclear pores at the ONM varies between 95 and 140 nm. Concave fracturing of nuclear membranes gives a view onto the cytoplasmic face of the ONM and the luminal face of the INM (Severs, 2007). The images presented here have all but one (Figure 8.9) been taken from convexly fractured nuclei.

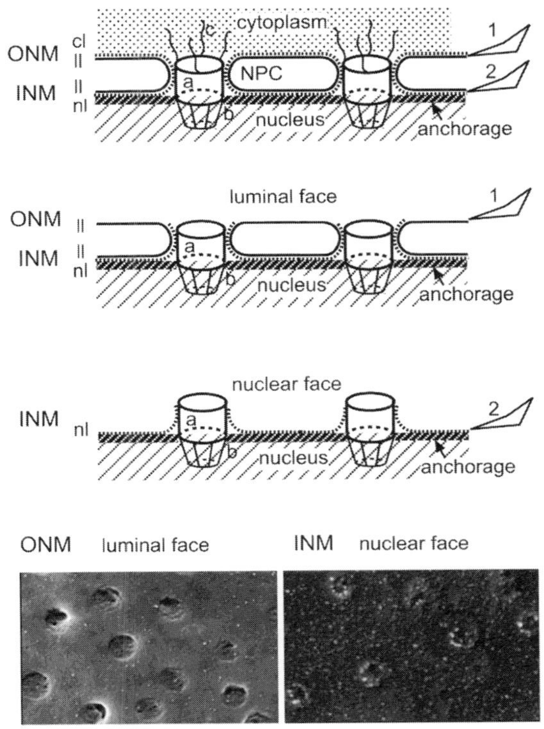

Figure 8.1 The nuclear envelope comprising the outer nuclear membrane (ONM) and inner nuclear membrane (INM) with embedded nuclear pore complexes (NPC) consisting of the central channel (a), the nuclear basket (b), and the cytoplasmic filaments (c). The phospholipid bilayer of the ONM is composed of the cytoplasmic layer (cl) and the luminal layer (ll) delineating the perinuclear space, that of the INM of the luminal layer and the nuclear layer (nl). Convex fracturing results either in splitting of the ONM (1) giving a view onto the luminal face with NPCs located within the pores, or in splitting of the INM (2) giving a view onto the nuclear face and the NPCs which are attached to the nuclear layer by NPC proteins. The nuclear layer of the INM is attached to the nuclear matrix by lamins. The luminal face is rather smooth whereas the cytoplasmic and nuclear face is rough due to the numerous proteins embedded in the cytoplasmic and nuclear layer. Concave fracturing gives a view onto the cytoplasmic face of the ONM and to the luminal face of the INM, respectively. All images but one (Figure 8.9c) have been taken from convex fracture planes, and, for simplicity, the luminal face of the ONM is indicated by "o"; the nuclear face of the INM by "i".

8.3.2 Nuclei expand during infection

Nuclei of cells grown as a monolayer are triaxial ellipsoids with long a- and b-axes but very short c-axis (Wild *et al.*, 2009). During trypsinization, cells become sphere-like. However, the nuclei remain triaxial ellipsoids (Figure 8.2) although with great variations of shape. How much these variations are due to the preparation procedure remains unclear. During HSV-1 infection, nuclear volume increases by factors of 1.5 (Wild *et al.*, 2009) to 2 (Simpson-Holley *et al.*, 2005). The nuclei are still triaxial ellipsoids but show many irregularities at the surface (Figure 8.3A). Increase of the volume of an

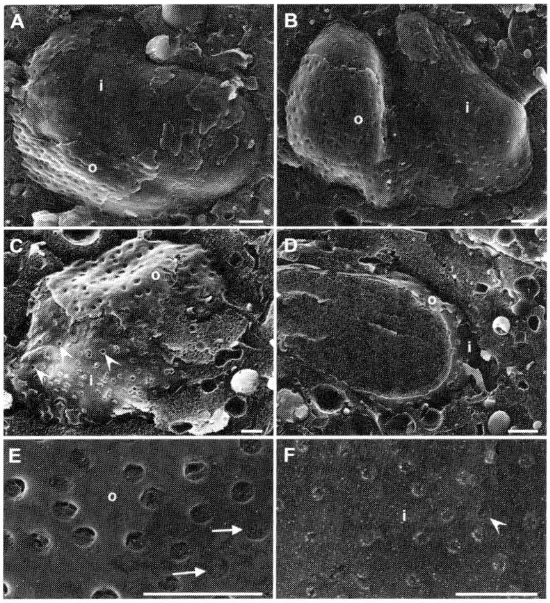

Figure 8.2 Nuclei in mock infected cells: the nuclear shape of mock infected Vero cells depicted from 3-D confocal images and TEM images is a triaxial ellipsoid. Cryo-FESEM revealed many variations; some of them may be caused or enhanced during preparation (A–C). The fracture plane is often not along the nuclear membranes but within the nucleus (D). Nuclear pores appear as distinct low depressions at the luminal face of the ONM (o) whereas the NPCs at the nuclear face of the INM (i) are visible as round button-like structures. Occasionally, NPCs protrude slightly towards the cytoplasm (E: arrows), or they have been removed during fracturing leading to depressions at the INM (C, F: arrowheads). Bars 500 nm.

ellipsoid by a factor of 1.5 results in enlargement of its surface by a factor of ~1.31. However, calculation of the nuclear surface on the basis of the axes measured on confocal images revealed only an enlargement by a factor of ~1.06 (Wild *et al.*, 2009). This underestimation is possibly due to methodological inaccuracies, and also does not consider irregularities of the nuclear surface seen in SEM and TEM images.

8.3.3 Capsids bud at the inner nuclear membrane

In HSV-1 infected cells, the main event at the nuclear surface is the budding of capsids. The result of a capsid budding at the inner INM is an enveloped virion with a diameter of about 200 nm within the PNS (Roizman, 2001). These virions can accumulate in the PNS (Campadelli-Fiume *et al.*, 1991; Skepper *et al.*, 2001). The process of budding is rather rarely captured by TEM (Leuzinger *et al.*, 2005), despite the fact that hundreds of capsids bud at the INM within a short period of time. Budding of capsids could only recently be demonstrated by SEM (Wild *et al.*, 2009). Interestingly, the number of budding capsids was 1.83 /μm^2 at 12 hr of incubation but only 0.1 /μm^2 at 15 hr, suggesting that the release of capsids via budding peaks soon after the onset of budding at about 8 hr of incubation

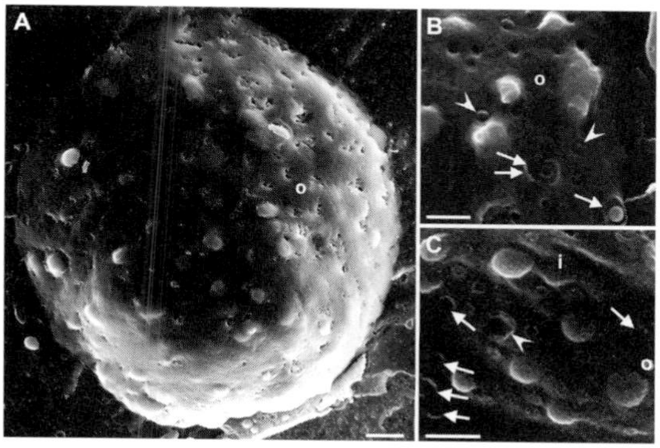

Figure 8.3 Capsids formed in the nucleus are released via budding at the INM (i) acquiring tegument and the viral envelope. The arising virus pushes the ONM (o) towards the cytoplasm forming numerous bulges. (A) Low-power image of the ONM luminal face showing numerous budding capsids as well as irregularities of the nuclear surface and of pores, some of them being damaged. (Image courtesy of *E. S. Virology*, **429**, 2012.) (B) Some of the NPCs protrude markedly into the cytoplasm (arrowheads). The ONM is focally removed giving a view onto a virus close to completion of budding (arrow), and to the INM where budding had taken place (double arrows; for details see Figure 8.4). (C) Budding capsids at the INM and many pores without NPCs (arrows). The cavity (arrowhead) may represent a location where budding had taken place or a dilated nuclear pore. Bars 500 nm.

but ceases rapidly thereafter. Budding of capsids may be delayed in mutants, e.g. in U_S3 deletion mutants and its repair mutant (Figure 8.3A) showing 20 budding capsids at least on ~25 μm^2 at 16 hr post inoculation. Though there are many distortions, such as damaged nuclear pores, budding capsids are easily recognizable even when they are covered by the ONM. At higher magnification of a well-preserved nucleus (Figure 8.3B), budding capsids are seen to push the ONM towards the cytoplasm resulting in distinct bulges. At the nuclear face of the INM, budding capsids appear as half sphere-like particles (Figure 8.3B).

Budding starts with close apposition of a capsid to the INM and concomitant deposition of a protein (most probably a viral protein that accomplishes budding) at the INM that becomes the viral envelope (Leuzinger *et al.*, 2005; Wild *et al.*, 2005). Then the capsid pushes the INM towards the PNS whilst at the periphery the INM is pulled behind the budding capsid for fission to give rise to an enveloped virion. Tegument proteins are concomitantly deposited around the capsid. The result is an indentation in the nucleus as apparent in Figure 8.3B. Early and late phases of budding capsids containing DNA imaged by TEM are displayed in Figures 8.4A and 8.4B, respectively. The deposited protein appearing as a dense layer on the viral envelope remains on virions in the PNS (Figure 8.4C). Capsids are also enveloped despite the fact that they do not contain DNA (Figure 8.4C–E). The dense layer is absent in virions within Golgi cisternae and in extracellular virions. This is one reason that led to the assumption that the envelope of perinuclear virions fuses with the ONM releasing tegument and the capsid into the cytoplasmic matrix. In this context, the cytoplasmic capsid is assumed to be re-enveloped by Golgi membranes (Skepper *et al.*, 2001). However, the process taking place at the

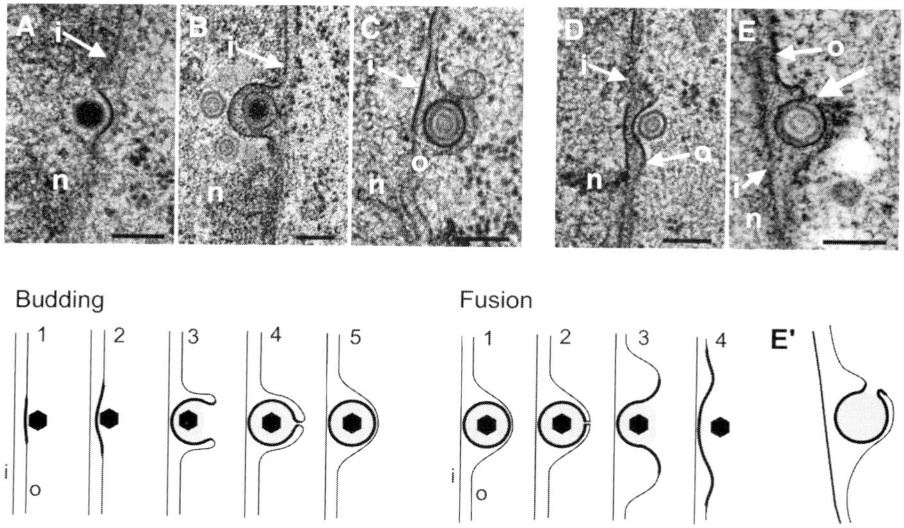

Figure 8.4 Transmission electron micrographs of HSV-1 infected Vero cells after high-pressure freezing and freeze-substitution showing an early (A) and a late phase (B) of budding capsids (containing DNA) from the nucleus (n) through the INM (i) into the PNS, a virion within the PNS (C), and an early (D) and late phase (E) of budding capsids (without DNA) at the ONM (o). Note the course of the ONM in E, which is accordingly drawn in panel E′. This is the result of pulling the ONM around the capsid for fission (thick arrow) rather than the beginning of fusion. The results of budding are fully enveloped virions in the PNS. The electron dense envelope is due to the presence of budding proteins. If the process taking place at the ONM was fusion the entire dense viral envelope would be inserted into the ONM, and the releasing capsid would not stick to the ONM as depicted in the schematic presentation of budding and fusion. Bars 200 nm.

ONM has nothing in common with fusion. If the process at the ONM is budding rather than fusion, the capsids in the cytoplasmic matrix must gain access to it other than by de-envelopment. There is strong evidence that nuclear capsids can be released directly into the cytoplasm via dilated nuclear pores (Leuzinger *et al.*, 2005; Wild *et al.*, 2005; Wild *et al.*, 2009). These capsids then bud at the ONM, RER membranes or at membranes of the Golgi complex and vacuoles of possible Golgi origin.

8.3.4 Capsids bud also at the outer nuclear membrane

The interaction of capsids with the ONM shows all characteristics of budding (Darlington and Moss, 1968; Leuzinger *et al.*, 2005; Wild *et al.*, 2005). Figures 8.4D and E also demonstrate that the process taking place at the ONM is identical to the budding process at the INM. It starts by the capsid pushing the ONM toward the nucleus and concomitant deposition of an electron-dense substance. Capsids connect to the membrane by tegument proteins (Grunewald *et al.*, 2003). They are firmly attached to the budding membrane so that they are not removed from the ONM even during freeze-fracturing as shown by cryo-FESEM (Figure 8.5F). Fusion, however, starts by formation of the fusion pore in the

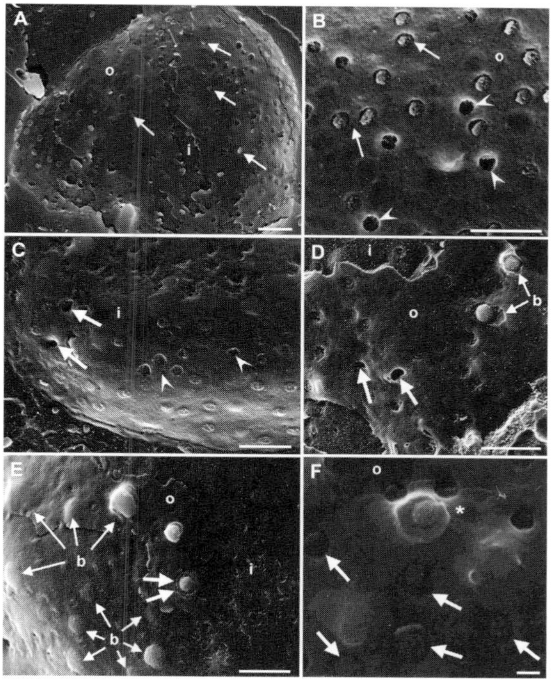

Figure 8.5 Number of nuclear pores decreases and size of pores increases during HSV-1 infection. (A–D) Many NPCs protrude markedly into the cytoplasm (arrows) and may be removed during fracturing (B–C: arrowheads). Some nuclear pores become dilated, appearing as holes with well-delineated borders (C–E: thick arrows), or are occupied by the NPC and probably nuclear material as indicated by thick arrows in panel F. In panel D, the ONM is partially removed, giving a view onto two virions (b) that obviously stick to the INM. (E) Early stages of budding capsids (b) are seen on the INM. A particle of ~130 nm is present in a circle delineated by the INM (double thick arrows). This particle with a very different surface than budding capsids most probably represents a capsid escaping the nucleus through a dilated nuclear pore. (F) A particle (asterisk) of ~130 nm size sticks to the ONM, forming a rim around the particle. This structure is considered very likely to be a capsid budding at the ONM from the cytoplasm into the PNS, because budding implies that capsid and tegument are firmly attached to the membrane so that they are not split from the membranes during fracturing. Bars A–E 500 nm; F 100 nm.

microsecond range (Haluska *et al.*, 2006). The result of fusion would be a capsid surrounded by tegument for release into the cytoplasm, and the dense viral envelope would be inserted in its full size into the ONM (Figure 8.4). Fusion pores, which are crucial for recognizing fusion, have to the authors' knowledge never been shown. To interpret the interaction of capsids with the ONM as late phases of fusion rather lacks fundamentals and is used to support the theory of the uneconomic pathway of capsid translocation from the nucleus to the cytoplasm via envelopment and de-envelopment, as discussed (Campadelli-Fiume *et al.*, 2006; Mettenleiter *et al.*, 2006b).

The mechanisms of budding are unknown. In HSV-6, clathrin is assumed to be involved in budding of capsids (Mori *et al.*, 2008). If that were true, clathrin would need to be cleaved from the viral envelope as it is in coated vesicles to uncover the fusion machinery. Demonstration of clathrin failed in HSV-1 infected cells (unpublished data). There is increasing evidence that budding of capsids is accomplished by the viral glycoprotein gK (Hutchinson *et al.*, 1995). If gK were indeed the budding protein it would be expected that gK needs to be cleaved from the envelope to uncover the fusion machinery. Fact is that virions with a dense envelope are present in RER cisternae. These virions have either been intraluminally transported from the PNS or they are there because capsids have bud from the cytoplasm into RER cisternae. The consequences are: i) facilitation of intraluminal transport involves protection from fusion, and ii) if capsids can bud at RER membranes, which was shown recently (Leuzinger *et al.*, 2005), they can also bud at the ONM since the ONM is part of the RER.

8.3.5 Budding of capsids requires membranes

Capsids escaping the nucleus by budding at the INM require the INM for envelope formation. Assuming virions are sphere-like particles of 200 nm in diameter (Grunewald *et al.*, 2003), the surface area of a single virion equals ~125 000 nm^2. The approximate mean interpore area between four neighboring pores was 206 200 nm^2 in mock infected cells (Wild *et al.*, 2009). The maximal interpore area was about 681 500 nm^2, and the minimal interpore area 11 600 nm^2. A budding capsid requires 125 000 nm^2 from the INM, i.e. it reduces the nuclear surface by 125 000 nm^2. Surprisingly, the mean interpore area measured in HSV-1 infected cells 15 hr post inoculation was close to a factor of 2 larger than in mock infected cells. The maximal interpore area was even 10 times larger than in mock infected cells, whereas the minimal interpore area was similar in mock and HSV-1 infected cells. These data suggest highly dynamic processes taking place at the nuclear envelope in the course of nuclear exit of capsids, leading to enlargement of the interpore area. On average, 1.83 capsids were shown to bud per 1 μm^2 nuclear surface at 12 hr post inoculation requiring 228 750 $nm^2/\mu m^2$ nuclear membrane or 22.8%. Expressing the number of budding capsids per the entire nuclear surface of 450 to 480 μm^2, a total number of 823 to 878 capsids will bud more or less simultaneously provided that budding takes place over the entire nuclear surface. Consequently, a total area of 103 to 109 μm^2 of the INM would be required to provide enough membranes for envelopment within a short period of time (Wild *et al.*, 2009). Budding is probably not as fast as fusion. Fusion in exocytosis is completed in the subsecond range (Knoll *et al.*, 1991; fusion of influenza virus with liposome membranes requires less than one minute Kanaseki *et al.*, 1997). Enlargement of the nuclear volume and the enormous requirement of membranes for budding is speculated to result in dilation of nuclear pores.

The fate of virions derived from budding at the INM is controversially discussed. Capsids are assumed to be de-enveloped by fusion of the viral envelope with the ONM (Skepper *et al.*, 2001). Consequently, the INM would undergo constant reduction whereas the ONM and the adjacent RER would get continually enlarged. To the authors' knowledge, such deformations have never been described. Conversely, one might assume that membrane constituents inserted into the ONM recycle back to the INM so that a continuous flow would be maintained between INM and ONM and vice versa. Also for this concept, proof does not yet exist. However, the more profound crux of this concept remains in neglecting the fact that virions can accumulate in the PNS (Darlington and Moss, 1968; Spring *et al.*, 1968; Baines *et al.*, 1991; Campadelli-Fiume *et al.*, 1991; Torrisi *et al.*, 1992; Skepper *et al.*, 2001; Leuzinger *et al.*, 2005), or can be transported into adjacent cisternae of the RER (Schwartz and Roizman, 1969; Whealy *et al.*, 1991; Gilbert *et al.*, 1994; Radsak *et al.*, 1996; Stannard *et al.*, 1996; Granzow *et al.*, 1997; Wild *et al.*, 2002). Recently, it was proposed that perinuclear virions are transported via RER cisternae directly into Golgi cisternae, of which membranes were shown to be connected to RER membranes (Wild *et al.*, 2002; Leuzinger *et al.*, 2005). In this context, the INM would remain as the final viral envelope. In HSV-1 deleted of U_S3 (Reynolds *et al.*, 2002) or of glycoprotein B (Farnsworth *et al.*, 2007) extremely large numbers of virions accumulate in the PNS. The membranes of all these virions must be provided other than by recycling. Furthermore, cryo-FESEM images suggest that budding may take place simultaneously more or less all over the nuclear surface. Simultaneous budding of 1.83 capsids/μm^2 giving a total of more than 800 capsids would require a total amount of ~100 μm^2 of membranes that must be provided prior to and/or simultaneously to budding. Interestingly, 98% of the deletion mutant R7041(ΔU_S3) accumulates in the PNS (unpublished data), implying that membranes must be supplied for formation of the viral envelope by budding at the INM. Determination of ^{3}[H]-choline incorporation into nuclear membranes clearly demonstrated that phospholipid synthesis is stimulated by HSV-1 (Sutter *et al.*, 2012), and that U_S3 is involved in its regulation (Wild *et al.*, in press). ^{3}[H]-choline is also incorporated into cytoplasmic membranes, mainly for maintenance of the Golgi complex. The Golgi complex is a crossroad in HSV-1 envelopment. Capsids bud at Golgi membranes forming an enveloped virion within a transport vacuole. This process is called wrapping (Granzow, 2001). Alternatively, it seems likely that virions are transported from the PNS via RER into Golgi cisternae. Vacuoles containing virions are then detached from Golgi membranes by fission (Wild *et al.*, 2002; Leuzinger *et al.*, 2005). Therefore, the amount of membranes needed is enormous.

8.3.6 Nuclear pores dilate during infection

NPCs form a key transport barrier between the cytoplasm and the nucleus. In non-dividing cells, nucleocytoplasmic shuttling of macromolecules is tightly controlled by their selective translocation through the 30 to 50 nm long (Stoffler *et al.*, 1999) and 9 nm wide (Pante and Aebi, 1993) NPC central channel. This channel was shown to be expandable, enabling the transport of macromolecules with diameters of up to ~39 nm (Pante and Kann, 2002). Negatively stained nuclear pores on the surface of carbon coated copper grids measure 125 nm in diameter and are occupied by the NPC (Pante and Aebi, 1996; Walther *et al.*, 2001). In HSV-1 infected cells, intact pores with the NPC measure ~120 nm at the INM of freeze-fractured nuclei. However, NPCs are often removed. The diameter of pores at the ONM varies considerably due to shrinkage and beam damage. In contrast to mock infected cells, NPCs very often protrude through the pores into the cytoplasm after HSV-1 infection. In addition, there are a few gaps larger than 120 nm at both the INM and the ONM (Figures 8.5 A–E). Nuclear pores may also be dilated and filled with material of probably nuclear origin and/or capsids (Figures 8.5E and F). Quantification revealed that about 40%

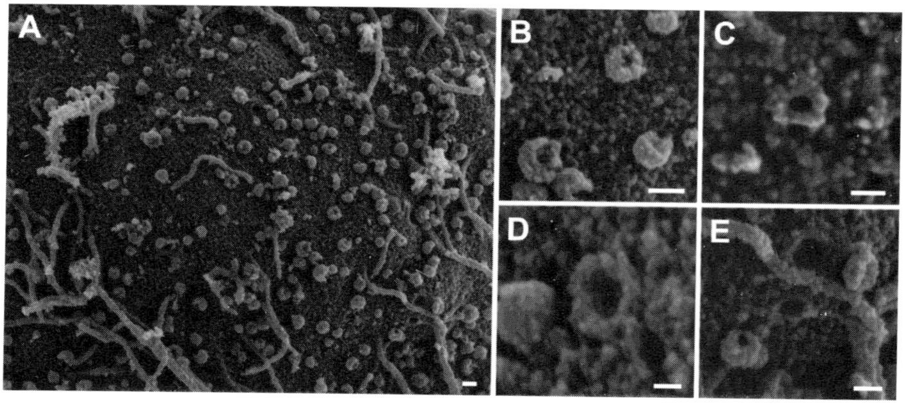

Figure 8.6 Nuclear surface of HSV-1 infected Vero cells at 17 hr post inoculation imaged by FESEM after dry-fracturing showing low numbers of NPCs (A), normal sized NPCs (B), and enlarged NPCs with dilated central channels (C–E). Bars 100 nm.

of the nuclear pores in HSV-1 were larger than 140 nm measured on both the INM and the ONM on cryo-FESEM images (Wild *et al.*, 2009). However, only a limited number of nuclear pores were severely enlarged, measuring up to 560 nm.

To verify that nuclear pores dilate in the course of HSV-1 infection, we examined NPCs by FESEM in cells after dry-fracturing *in situ*. The central channel of NPCs in dry-fractured Xenopus egg nuclei measures 35 to 45 nm (Goldberg *et al.*, 1997). The mean central channel diameter was 28.7 nm in NPCs of mock infected cells (Wild *et al.*, 2009). The mean central channel diameter in HSV-1 infected nuclei, however, was 92 nm. The overall diameter of NPCs was 135 nm in mock infected cells but 198 nm in HSV-1 infected cells at 15 hr of incubation (Figure 8.6).

The question whether or not the larger protrusions and gaps at the nuclear surface are related to nuclear pores can best be answered by thin sections perpendicularly through these protrusions or gaps. Imaging the nuclear surface on thin sections obtained after high-pressure freezing and freeze-substitution *in situ* revealed gaps distinctly bordered by nuclear membranes suggesting them to be dilated nuclear pores (Figure 8.7). Nuclear material containing capsids was found to protrude through such gaps into the cytoplasm. It never merged with the cytoplasmic matrix. Such gaps measured up to 700 nm (Wild *et al.*, 2009). The pathway for nuclear capsids to gain direct access to the cytoplasm was suggested to be via impaired nuclear pores taking place late in infection (Leuzinger *et al.*, 2005; Wild *et al.*, 2005; Wild *et al.*, 2009). Release of capsids through gaps in the nuclear envelope has been reported for another member of the herpes virus family, the simian agent 8 (Borchers and Oezel, 1993) and pseudo-rabies virus (Klupp *et al.*, 2011).

8.3.7 Nuclear pore number declines during infection

Nuclear pores may exhibit a distinctly non-random distribution over the nuclear surface with a minimum interpore distance that possibly influences their distribution (Maul *et al.*,

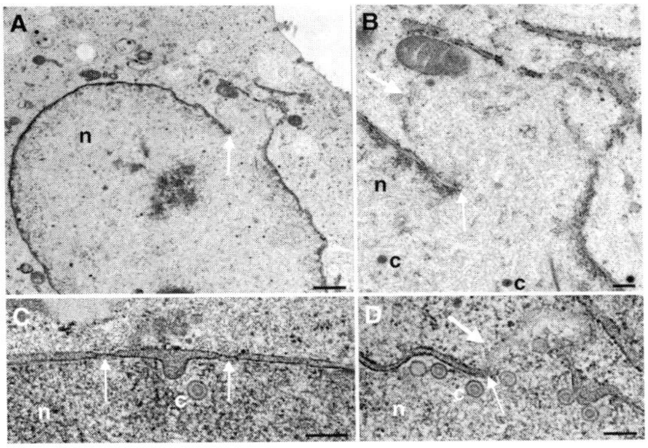

Figure 8.7 Transmission electron micrographs of nuclei (n) in HSV-1 infected Vero cells after high-pressure freezing and freeze-substitution. (A,B,D) Dilated nuclear pores distinctly delineated at one side (arrows) through which nuclear material containing capsids (c) protrudes into the cytoplasm without merging with it (thick arrow). (C) Intact nuclear pores (arrows) on each side of an indentation that most probably is the result of a budding capsid (see Figure 8.4B). Bars A: 1 μm, B–D: 200 nm.

1971). In some cells nuclear pores occur in regular geometric arrays (Franke, 1974); in others, such as 3T3 cells, no indications of regular distributions were found (Kubitscheck *et al.*, 1996). The number of pores per unit area of the nuclear envelope varies considerably among different cell types. A general value for higher eukaryotic cells is 10 to 20 pores /μm^2 (Maul, 1977) giving a total number of 3000 to 4000 pores per nucleus in somatic cells (Gorlich and Kutay, 1999). However, only 1900 pores were detected on 3T3 cell nuclei by confocal microscopy (Kubitscheck *et al.*, 1996). The number of nuclear pores in mock infected HeLa cells was 10.7 pores/μm^2 nuclear surface in freeze-fractured cells and 29.1 pores/μm^2 nuclear surface in dry-fractured cells, respectively (Wild *et al.*, 2009). The difference of close to a factor of 3 is probably due to preparation artifacts, such as shrinkage, occurring during fixation, permeabilization, dehydration, and drying. The mean number of pores in HSV-1 infected cells was 6.7 pores/μm^2 in freeze-fractured cells and 4.7 pores/μm^2 nuclear surface in dry-fractured cells which is 4.3 and 2.3 times, respectively, less than in mock infected cells. This discrepancy possibly occurred because nuclear pores were altered in HSV-1 infected cells and thus their identification in dry-fractured cells was difficult or even impossible.

The mean nuclear surface calculated on the basis of the a-, b-, and c-axes measured in confocal images was 450 μm^2 in mock infected cells and 480 μm^2 in HSV-1 infected cells (Wild *et al.*, 2009). Expressing the mean number of pores counted on cryo-FESEM images per the total nuclear surface area revealed a mean of 4800 pores/nucleus in mock infected cells but only a mean of 2250 pores/nucleus in HSV-1 infected cells, suggesting that the number of pores declined by a factor of 2.1 within 15 hr of incubation. The mean pore numbers identified by immunostaining using antibodies against nuclear pore proteins Nups 62, 90, and 152 or Nup153 were ~2000 in mock infected cells and ~1560 in HSV-1 infected cells. Estimation of pore numbers on cryo-FESEM images might result in overestimation because the spherical surface of the nucleus cannot be satisfactorily considered. On the other hand, estimation of pore numbers on fluorescent images probably results in underestimation because the resolution power is not sufficient to allow clear separation of pores located in close proximity. The minimal interpore distance was 6.7 nm in mock infected cells but 25.8 nm in HSV-1 infected cells. Nevertheless, the mean number of nuclear pores in HSV-1 infected cells was lower by a factor of 1.8 than in mock infected cells. This corresponds to the data of cryo-FESEM and thus strongly supports the idea of decrement of pore numbers in the course of HSV-1 infection. This is confirmed by the 1.75 times larger mean interpore distances in HSV-1 infected cells compared to mock infected cells, which cannot be related to the increase in nuclear surface alone (Wild *et al.*, 2009).

8.3.8 Alterations at the nuclear periphery in the absence of U_S3

A viral kinase, U_S3, enables translocation of capsids from the nucleus to the cytoplasm. Budding of capsids at the INM is assumed to be inhibited in cells infected with U_S3 deletion mutants of HSV-1 or pseudo rabies virus (Wagenaar *et al.*, 1995; Klupp, 2001; Reynolds *et al.*, 2001). HSV-1ΔU_S3 virions have been shown to accumulate in pouches formed by nuclear membranes (Reynolds *et al.*, 2002; Poon *et al.*, 2006). Cryo-FESEM of Vero cells infected with the well-characterized mutant R7041(ΔU_S3) revealed that the INM forms simple or highly complicated folds (Figures 8.8 and 8.9). The nuclear face of such folds (Figures 8.8E, F) as well as small areas of the cytoplasmic face of the ONM (Figure 8.9C) are smooth. This is probably caused by insertion of viral proteins resulting in densely stained membranes (Figure 8.10). The INM also forms pouches, which protrude deeply into the nucleus or into the cytoplasm. Large pouches contain dozens of virions. In addition, odd structures are formed which are located close to the nucleus. These structures might be related to nuclear membranes or to Golgi membranes. The Golgi complex is huge but is not involved in envelopment of R7041(ΔU_S3) capsids (Senn *et al.*, 2006). There are large areas devoid of nuclear pores. The number of nuclear pores is decreased as in wt HSV-1 infection (unpublished data). The number of dilated nuclear pores is similar to that in mock infected cells. Approximately 98% of all virus particles are in the PNS indicating that viral transportation out of the PNS is impaired. R7041(ΔU_S3) virions are infective (Senn *et al.*, 2006), implying that virions derived from budding through the INM into the PNS acquire all proteins – tegument proteins and glycoproteins – essential for infectivity. Recent demonstration of some proteins from isolated perinuclear virions (Padula *et al.*, 2009) suggests that this applies also to wt HSV-1.

8.3.9 Functionality of nuclear pores during infection

The NPC controls nuclear import and export of proteins up to 39 nm in diameter (Pante and Kann, 2002). The NPC comprises the spoke–ring complex, which encircles the central channel complex called the transporter. The spoke complex is framed by the cytoplasmic and nuclear rings that serve as attachment sites for the cytoplasmic

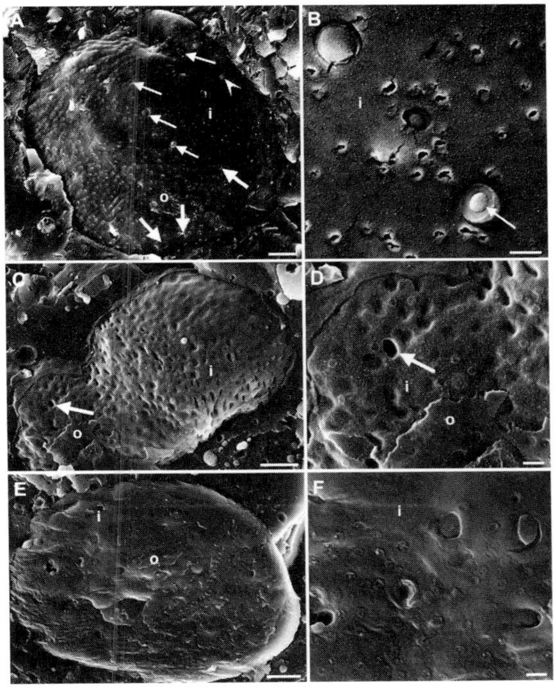

Figure 8.8 Nuclei of Vero cells 17 hr after inoculation with the U_S3 deletion mutant R7041. (A) Overview showing folds of the nuclear membrane forming button-like structures (arrows), budding capsids (arrowhead). Nuclear pores were damaged during preparation. Nevertheless, dilated pores can easily be recognized (thick arrows). (B) Details of the membranous folds, one of them with a budding capsid (arrow). (C, D) Nucleus with multiple small folds of the INM and two dilated nuclear pores (thick arrows). (E, F) Intact nuclear pores are irregularly distributed. Numerous folds form large button-like structures and excavations. Bars A,C,E: 1 μm, B,D,F: 200 nm.

filaments and the nuclear basket, respectively (Yang *et al.*, 1998; Suntharalingam and Wente, 2003). Nup62 is located at the cytoplasmic entry side of the transporter (Schwarz-Herion *et al.*, 2007). Locations of Nup90 and Nup152 are to the authors' knowledge not defined. Nup153 is part of the nuclear basket (Pante *et al.*, 1994; Pante *et al.*, 2000). Nup153 is known to be required for NPC assembly and anchoring of the NPC to the nuclear envelope (Gerace and Burke, 1988) but is also involved in nuclear import and export (Ball and Ullman, 2005). Nup153 was shown to be down regulated in HSV-1 infected cells (Ray and Enquist, 2004). Proteins, such as capsid proteins and tegument proteins, have to be imported into the nucleus for HSV-1 replication. The question arises whether nuclear import is affected in HSV-1 infected cells. In poliovirus infected HeLa cells nuclear import was drastically inhibited in accordance with degradation of NPC components (Gustin and Sarnow, 2002). The number of pores with diameters smaller than 140 nm was ~2 /μm^2 equaling a total of 960 pores per entire nuclear surface in HSV-1 infected cells at 15 hr of inoculation. NPC proteins are not lost during infection (Hofemeister and O'Hare, 2008; Wild *et al.*, 2009) and, hence, nuclear import and export

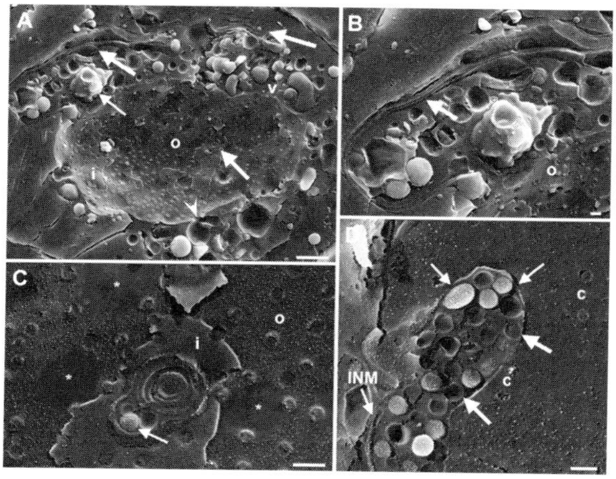

Figure 8.9 Nuclei of Vero cells 17 hr after inoculation with the U_S3 deletion mutant R7041. Concave fracturing gives a view onto the cytoplasmic face of the ONM (o) and the luminal face of the INM (i). (A) Overview showing irregular distribution of intact nuclear pores, odd structures (thin arrow), membranous structures (thick arrows) outside and inside of the nucleus, and vacuoles (v), some of them being connected to the INM (arrowhead). (B) The multiple membrane layers (thick arrow) associated with vacuoles are considered likely to be related to the Golgi complex. (C) Folds of the INM forming cylinder-like structures and excavations that contain virions (arrow). Smooth areas (asterisks) of the ONM are presumed to be due to insertion of viral proteins as in the INM. (D) Fracture plane through the nucleus showing a large pouch formed by the INM containing numerous virions mostly, but not always, of spherical shape (arrows). Many virions were removed during fracturing (thick arrows). There are a few capsids (c) within the nucleus. Bars A: 1 μm, B–D: 200 nm.

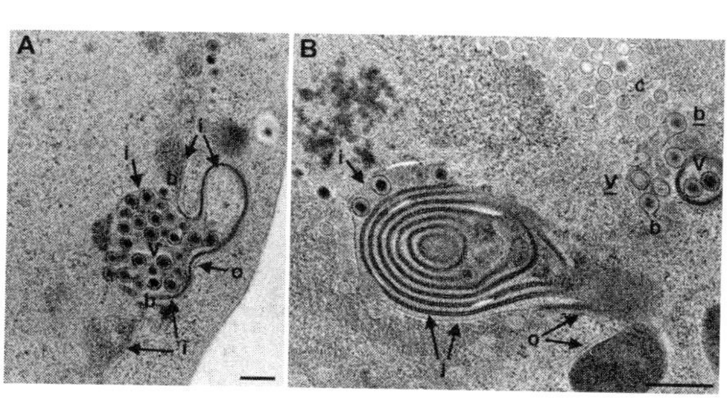

Figure 8.10 TEM images of sections through the nucleus of cells with the U_S3 deletion mutant, R7041. (A) A simple fold of the INM (i) and ONM (o) associated with a large pouch containing numerous virions (V) reaches into the cytoplasm. A few capsids (b) bud at the INM. (B) Multiple folds of the INM with two virions and a budding capsid are located at the nuclear periphery. On the right side there are capsids without DNA (c), budding capsids (b) where fission from the INM failed (*b*), membranous folds associated with DNA containing virions (V) and virions without DNA (*V*). Note that the INM is thickened at sites of the folds caused by insertion of viral proteins. Bars 200 nm.

are probably not disturbed in unaffected pores. However, if import and export are affected in dilated pores, the number of unaffected pores is considered likely to be still sufficient to fulfill the requirements for protein shuttling through NPCs. Processing of viral proteins necessary for virus replication starts early in infection (Honess and Roizman, 1974), whereas dilation of nuclear pores is a rather late event. Further, dilation of nuclear pores does not necessarily imply impairment of nuclear import and export as long as nucleoporins form NPCs. Using dextran beads, the nucleocytoplasmic barrier for molecules bigger than 70 kDa was shown to have remained intact throughout HSV-1 infection (Hofemeister and O'Hare, 2008), suggesting that selectivity of nuclear export is not disturbed for large molecules. The mechanisms for exporting dextran beads are not necessarily the same as for exporting capsids. Further, a small number of severely dilated pores may be sufficient for release of hundreds of capsids.

8.4 Conclusions

The process of budding at the INM is rarely encountered by TEM despite hundreds of capsids being transported via budding into the PNS. Cryo-FESEM showed for the first time (Wild *et al.*, 2009) that budding is a very frequent process indeed. Surprisingly, high budding activity was found at 10 to 12 hr post inoculation whereas budding activity was very low at 15 hr post inoculation with wt HSV-1, indicating that release of capsids via budding out of the nucleus peaks soon after the appearance of first virions in the cytoplasm at 8–9 hr post inoculation, but ceases soon thereafter. Capsids budding at the INM result in fully enveloped virions in the PNS. They differ from virions in the extracellular space only by the envelope that is covered by a protein most probably of viral origin. This protein is assumed to be responsible for budding and for transportation of virions from the PNS into adjacent RER and further into Golgi cisternae where it is cleaved.

Budding of capsids requires a large amount of membranes. Hence substantial amounts of membrane constituents need to be inserted into the INM for viral envelope formation during budding of capsids and for maintenance of membrane integrity of expanding nuclei. Incorporation of [^{3}H]-choline into nuclear membranes indicates that *de novo* synthesized phospholipids are used for supplying the necessary amount of membrane.

Cryo-FESEM also revealed decline of nuclear pore numbers and dilation of nuclear pores. Decline of nuclear pore numbers was also shown by confocal microscopy using antibodies against nucleoporins, and by FESEM of *in situ* dry-fractured cells. Dry-fracturing also demonstrated enlargement of NPCs and enlargement of the NPC central channel. If the enlarged NPCs consist of nucleoporins, indeed, one may speculate that nucleoporins are transported from other sites to these enlarged NPCs to maintain its integrity. Whether functional integrity of NPCs is also maintained needs to be investigated. It has to be borne in mind that dilation of nuclear pores is a late event in HSV-1 infection, that hundreds of the pores remain intact, that NPCs in slightly dilated pores do not necessarily lose functional integrity, and that only a few pores dilate severely.

TEM of high-pressure frozen and freeze-substituted cells yielded details of the budding process at both the INM and ONM. It also demonstrated dilation of nuclear pores through which capsids escaped to gain direct access to the cytoplasmic matrix. These capsids can bud at the ONM, and at membranes of the RER, the Golgi complex, and of vacuoles. The molecular mechanisms guiding nuclear export of capsids via dilated nuclear pores need to be identified.

8.5 Acknowledgments

The authors thank Bernard Roizman for the mutants R7041(ΔU_S3) and R2641, Anna Paula de Oliveira for providing virus and tissue cultures, and Elisabeth Schraner for excellent technical assistance. This study was supported by the Foundation for Scientific Research at the University of Zürich, Switzerland.

8.6 References

Allen, T. D., G. R. Bennion, S. A. Rutherford, *et al.*, 1997. Macromolecular substructure in nuclear pore complexes by in-lens field-emission scanning electron microscopy. *Scanning*, **19**, 403–10.

Allen, T. D. and M. W. Goldberg, 1993. High-resolution SEM in cell biology. *Trends Cell Biol.*, **3**, 205–8.

Allen, T. D., S. A. Rutherford, G. R. Bennion, *et al.*, 1998. Three-dimensional surface structure analysis of the nucleus. *Methods in Cell Biology*, **53**, 125–38.

Baines, J. D., P. L. Ward, G. Campadelli-Fiume, and B. Roizman, 1991. The UL20 gene of herpes simplex virus 1 encodes a function necessary for viral egress. *Journal of Virology*, **65**, 6414–24.

Baines, P. A., 2005. Herpes simplex virus type 1 infection induces activation and recruitment of protein kinase c to the nuclear membrane and increased phosphorylation of lamin B. *Journal of Virology*, **80**, 494–504.

Ball, J. R. and K. S. Ullman, 2005. Versatility at the nuclear pore complex: lessons learned from the nucleoporin Nup153. *Chromosoma*, **114**, 319–30.

Borchers, K. and M. Oezel, 1993. Simian agent 8 (SA8): Morphogenesis and ultrastructure. *Zentralblatt Bakteriologie*, **279**, 526–36.

Campadelli-Fiume, G., F. Farabegoli, S. Di Gaeta, and B. Roizman, 1991. Origin of unenveloped capsids in the cytoplasm of cells infected with herpes simplex virus 1. *Journal of Virology*, **65**, 1589–95.

Campadelli-Fiume, G., B. Roizman, P. Wild, T. C. Mettenleiter, and T. Minson, 2006. The egress of herpes viruses from cells: The unanswered questions. *Journal of Virology*, **80**, 6716–19.

Darlington, R. W. and L. H. Moss, 1968. Herpes virus envelopment. *Journal of Virology*, **2**, 48–55.

Fahrenkrog, B., J. Koser, and U. Aebi, 2004. The nuclear pore complex: a jack of all trades? *Trends in Biochemical Sciences*, **29**, 175–82.

Farnsworth, A., T. W. Wisner, M. Webb, *et al.*, 2007. Herpes simplex virus glycoproteins gB and gH function in fusion between the virion envelope and the outer nuclear membrane. *Proc. Natl. Acad. Sci. USA*, **104**, 10187–92.

Franke, W. W., 1974. Structure, biochemistry and functions of the nuclear envelope. *International Review of Cytology, Supplement*, **4**, 71–239.

Gerace, L. and B. Burke, 1988. Functional organization of the nuclear envelope. *Annual Review of Cell Biology*, **4**, 335–74.

Gilbert, R., K. Ghosh, L. Rasile, and H. P. Ghosh, 1994. Membrane anchoring domain of herpes simplex virus glycoprotein gB is sufficient for nuclear envelope localization. *Journal of Virology*, **68**, 2272–85.

Goldberg, M. W. and T. D. Allen, 1995. Structural and functional organization of the nuclear envelope. *Current Opinion in Cell Biology*, **7**, 301–9.

Goldberg, M. W., C. Wiese, T. D. Allen, and K. L. Wilson, 1997. Dimples, pores, star-rings, and thin rings on growing nuclear envelopes: evidence for structural intermediates in nuclear pore complex assembly. *Journal of Cell Science*, **110**, 409–20.

Gorlich, D. and U. Kutay, 1999. Transport between the cell nucleus and the cytoplasm. *Annual Review of Cell & Developmental Biology*, **15**, 607–60.

Granzow, H., 2001. Egress of alpha herpes viruses: comparative ultrastructural study. *Journal of Virology*, **75**, 3675–84.

Granzow, H., F. Weiland, A. Jons, *et al.*, 1997. Ultrastructural analysis of the replication cycle of pseudorabies virus in cell culture: a reassessment. *Journal of Virology*, **71**, 2072–82.

Grunewald, K., P. Desai, D. C. Winkler, *et al.*, 2003. Three-dimensional structure of herpes simplex virus from cryo-electron tomography. *Science*, **302**, 1396–8.

Gustin, K. E. and P. Sarnow, 2002. Inhibition of nuclear import and alteration of nuclear pore complex composition by rhinovirus. *Journal of Virology*, **76**, 8787–96.

Hagiwara, A., Y. Fukazawa, M. Deguchi-Tawarada, T. Ohtsuka, and R. Shigemoto, 2005. Differential distribution of release-related proteins in the hippocampal CA3 area as revealed by freeze-fracture replica labeling. *Journal of Comparative Neurology*, **489**, 195–216.

Haines, H. and R. J. Baerwald, 1976. Nuclear membrane changes in herpes simplex virus-infected BHK-21 cells as seen by freeze-fracture. *Journal of Virology*, **17**, 1038–42.

Haluska, C. K., K. A. Riske, V. Marchi-Artzner, *et al.*, 2006. Time scales of membrane fusion revealed by direct imaging of vesicle fusion with high temporal resolution. *Proc. Nat. Acad. Sci. USA*, **103**, 15841–6.

Hermann, R. and M. Muller, 1993. Progress in scanning electron microscopy of frozen-hydrated biological specimens. *Scanning Microscopy*, **7**, 343–9; discussion 349–50.

Hofemeister, H. and P. O'Hare, 2008. Nuclear pore composition and gating in herpes simplex virus-infected cells. *Journal of Virology*, **82**, 8392–9.

Honess, R. W. and B. Roizman, 1974. Regulation of herpes virus macromolecular synthesis. I: Cascade regulation of the synthesis of three groups of viral proteins. *Journal of Virology*, **14**, 8–19.

Hutchinson, L., C. Roop-Beauchamp, and D. C. Johnson, 1995. Herpes simplex virus glycoprotein K is known to influence fusion of infected cells, yet is not on the cell surface. *Journal of Virology*, **69**, 4556–63.

Kanaseki, T., K. Kawasaki, M. Murata, Y. Ikeuchi, and S. Ohnishi, 1997. Structural features of membrane fusion between influenza virus and liposome as revealed by quick-freezing electron microscopy. *Journal of Cell Biology*, **137**, 1041–56.

Klupp, B. G., 2001. Effect of the pseudorabies virus US3 protein on nuclear membrane localization of the UL34 protein and virus egress from the nucleus. *Journal of General Virology*, **82**, 2363–71.

Klupp, B. G., H. Granzow, and T. C. Mettenleiter. 2011. Nuclear envelope breakdown can substitute for primary envelopment-mediated nuclear egress of herpesviruses. *Journal of Virology*, **85**, 8285–8292.

Knoll, G., C. Braun, and H. Plattner, 1991. Quenched flow analysis of exocytosis in Paramecium cells: time course, changes in membrane structure, and calcium requirements revealed after rapid mixing and rapid freezing of intact cells. *Journal of Cell Biology*, **113**, 1295–304.

Kubitscheck, U., P. Wedekind, O. Zeidler, M. Grote, and R. Peters, 1996. Single nuclear pores visualized by confocal microscopy and image processing. *Biophysical Journal*, **70**, 2067–77.

Leuzinger, H., U. Ziegler, C. Fraefel, *et al.*, 2005. Herpes simplex virus 1 envelopment follows two diverse pathways. *Journal of Virology*, **79**, 13047–13059.

Lim, R. Y. and B. Fahrenkrog, 2006. The nuclear pore complex up close. *Current Opinion in Cell Biology*, **18**, 342–7.

Maul, G., 1977. The nuclear and cytoplasmic pore complex. Structure, dynamics, distribution and evolution. *International Review of Cytology, Supplement*, **6**, 75–186.

Maul, G., J. W. Price, and M. W. Lieberman, 1971. Formation and distribution of nuclear pore complexes in interphase. *Journal of Cell Biology*, **51**, 405–18.

Mettenleiter, T. C., 2002. Herpes virus assembly and egress. *Journal of Virology*, **76**, 1537–47.

Mettenleiter, T. C., 2004. Budding events in herpes virus morphogenesis. *Virus Research*, **106**, 167–80.

Mettenleiter, T. C., B. G. Klupp, and H. Granzow, 2006a. Herpes virus assembly: a tale of two membranes. *Current Opinion in Microbiology*, **9**, 423–9.

Mettenleiter, T. C., T. Minson, and P. Wild, 2006b. Egress of alpha herpes viruses. *Journal of Virology*, **80**, 1610–12.

Monaghan, P., H. Cook, P. Hawes, J. Simpson, and F. Tomley, 2003. High-pressure freezing in the study of animal pathogens. *Journal of Microscopy*, **212**, 62–70.

Mori, Y., M. Koike, E. Moriishi, *et al.*, 2008. Human herpes virus-6 induces MVB formation, and virus egress occurs by an exosomal release pathway. *Traffic*, **9**, 1728–42.

Padula, M. E., M. L. Sydnor, and D. W. Wilson, 2009. Isolation and preliminary characterization of herpes simplex virus 1 primary enveloped virions from the perinuclear space. *Journal of Virology*, **83**, 4757–65.

Pante, N. and U. Aebi, 1993. The nuclear pore complex. *Journal of Cell Biology*, **122**, 977–84.

Pante, N. and U. Aebi, 1996. Molecular dissection of the nuclear pore complex. *Critical Reviews in Biochemistry & Molecular Biology*, **31**, 153–99.

Pante, N., R. Bastos, I. McMorrow, B. Burke, and U. Aebi, 1994. Interactions and three-dimensional localization of a group of nuclear pore complex proteins. *Journal of Cell Biology*, **126**, 603–17.

Pante, N. and M. Kann. 2002. Nuclear pore complex is able to transport macromolecules with diameters of about 39 nm. *Molecular Biology of the Cell*, **13**, 425–34.

Pante, N., F. Thomas, U. Aebi, B. Burke, and R. Bastos, 2000. Recombinant Nup153 incorporates *in vivo* into Xenopus oocyte nuclear pore complexes. *Journal of Structural Biology*, **129**, 306–12.

Poon, A. P., L. Benetti, and B. Roizman, 2006. U(S)3 and U(S)3.5 protein kinases of herpes simplex virus 1 differ with respect to their functions in blocking apoptosis and in virion maturation and egress. *Journal of Virology*, **80**, 3752–64.

Purves, F. C., R. M. Longnecker, D. P. Leader, and B. Roizman, 1987. Herpes simplex virus 1 protein kinase is encoded by open reading frame US3 which is not essential for virus growth in cell culture. *Journal of Virology*, **61**, 2896–901.

Purves, F. C., D. Spector, and B. Roizman, 1991. The herpes simplex virus 1 protein kinase encoded by the US3 gene mediates posttranslational modification of the phosphoprotein encoded by the UL34 gene. *Journal of Virology*, **65**, 5757–64.

Radsak, K., M. Eickmann, T. Mockenhaupt, *et al.*, 1996. Retrieval of human cytomegalovirus glycoprotein B from the infected cell surface for virus envelopment. *Archives of Virology*, **141**, 557–72.

Ray, N. and L. W. Enquist, 2004. Transcriptional response of a common permissive cell type to infection by two diverse alpha herpes viruses. *Journal of Virology*, **78**, 3489–501.

Reynolds, A. E., B. J. Ryckman, J. D. Baines, *et al.*, 2001. U(L)31 and U(L)34 proteins of herpes simplex virus type 1 form a complex that accumulates at the nuclear rim and is required for envelopment of nucleocapsids. *Journal of Virology*, **75**, 8803–17.

Reynolds, A. E., E. G. Wills, R. J. Roller, B. J. Ryckman, and J. D. Baines, 2002. Ultrastructural localization of the herpes simplex virus type 1 UL31, UL34, and US3 proteins suggests specific roles in primary envelopment and egress of nucleocapsids. *Journal of Virology*, **76**, 8939–52.

Ritter, M., D. Henry, S. Wiesner, *et al.*, 1999. A versatile high-vacuum cryo-transfer for cryo-FESEM, cryo-SPM and other imaging techniques. *Microscopy and Microanalysis*, **5** (*Suppl. 2*), 424–5.

Roizman, B., 2001. Herpes simplex viruses and their replication. In: *Fields Virology*, Vol. 2. B. N. Fields, D. M. Knipe, and R. M. Howley, editors. Philadelphia, Lipincott-Raven Publishers, pp. 2399–460.

Schwartz, J. and B. Roizman, 1969. Concerning the egress of herpes simplex virus from infected cells: electron and light microscope observations. *Virology*, **38**, 42–9.

Schwartz, T. U., 2005. Modularity within the architecture of the nuclear pore complex. *Current Opinion in Structural Biology*, **15**, 221–6.

Schwarz-Herion, K., B. Maco, U. Sauder, and B. Fahrenkrog, 2007. Domain topology of the p62 complex within the 3-D architecture of the nuclear pore complex. *Journal of Molecular Biology*, **370**, 796–806.

Senn, C., E. Sutter, S. Leisinger, *et al.*, 2006. The Golgi complex: Crossroad in alpha-herpes virus envelopment. *Proc. 31st Herpes Workshop*, University of Washington, Seattle.

Severs, N. J., 2007. Freeze-fracture electron microscopy. *Nat. Protoc.*, **2**, 547–76.

Simpson-Holley, M., R. C. Colgrove, G. Nalepa, J. W. Harper, and D. M. Knipe, 2005. Identification and functional evaluation of cellular and viral factors involved in the alteration of nuclear architecture during herpes simplex virus 1 infection. *Journal of Virology*, **79**, 12840–51.

Skepper, J. N., A. Whiteley, H. Browne, and A. Minson, 2001. Herpes simplex virus nucleocapsids mature to progeny virions by an envelopment → deenvelopment → reenvelopment pathway. *Journal of Virology*, **75**, 5697–702.

Spring, S. B., B. Roizman, and J. Schwartz, 1968. Herpes simplex virus products in productive and abortive infection. II: Electron microscopic and immunological evidence for failure of virus envelopment as a cause of abortive infection. *Journal of Virology*, **2**, 384–92.

Stannard, L. M., S. Himmelhoch, and S. Wynchank, 1996. Intra-nuclear localization of two envelope proteins, gB and gD, of herpes simplex virus. *Archives of Virology*, **141**, 505–24.

Stoffler, D., B. Fahrenkrog, and U. Aebi, 1999. The nuclear pore complex: from molecular architecture to functional dynamics. *Current Opinion in Cell Biology*, **11**, 391–401.

Suntharalingam, M. and S. R. Wente, 2003. Peering through the pore: nuclear pore complex structure, assembly, and function. *Developmental Cell*, **4**, 775–89.

Sutter, E., A. P. de Oliveira, K., Tobler, *et al.*, 2012. Herpes simplex virus 1 induces *de novo* phospholipid synthesis. *Virology*, **429**, 124–135.

Torrisi, M. R., C. Di Lazzaro, A. Pavan, L. Pereira, and G. Campadelli-Fiume, 1992. Herpes simplex virus envelopment and maturation studied by fracture label. *Journal of Virology*, **66**, 554–61.

Wagenaar, F., J. M. Pol, B. Peeters, *et al.*, 1995. The US3-encoded protein kinase from pseudorabies virus affects egress of virions from the nucleus. *Journal of General Virology*, **76**, 1851–9.

Walther, P., Y. Chen, L. L. Pech, and J. B. Pawley, 1992. High-resolution scanning electron microscopy of frozen-hydrated cells. *Journal of Microscopy*, **168**, 169–80.

Walther, P. and M. Muller, 1999. Biological ultrastructure as revealed by high resolution cryo-SEM of block faces after cryo-sectioning. *Journal of Microscopy*, **196**, 279–87.

Walther, T. C., M. Fornerod, H. Pickersgill, *et al.*, 2001. The nucleoporin Nup153 is required for nuclear pore basket formation, nuclear pore complex anchoring and import of a subset of nuclear proteins. *EMBO Journal*, **20**, 5703–14.

Whealy, M. E., J. P. Card, R. P. Meade, A. K. Robbins, and L. W. Enquist, 1991. Effect of brefeldin A on alpha herpes virus membrane protein glycosylation and virus egress. *Journal of Virology*, **65**, 1066–81.

Wild, P., 2008. Electron microscopy of viruses and virus-cell interactions. In: *Introduction to Electron Microscopy for Biologists, Methods in Cell Biology*, T. Allen, ed., Vol. 88. Amsterdam, Elsevier, pp. 497–524.

Wild, P., M. Engels, C. Senn, *et al.*, 2005. Impairment of nuclear pores in bovine herpes virus 1-infected MDBK cells. *Journal of Virology*, **79**, 1071–83.

Wild, P., E. M. Schraner, H. Adler, and B. M. Humbel, 2001. Enhanced resolution of membranes in cultured cells by cryoimmobilization and freeze-substitution. *Microscopy Research & Technique*, **53**, 313–21.

Wild, P., E. M. Schraner, D. Cantieni, *et al.*, 2002. The significance of the Golgi complex in envelopment of bovine herpes virus 1 (BHV-1) as revealed by cryobased electron microscopy. *Micron*, **33**, 327–37.

Wild, P., C. Senn, C. L. Manera, *et al.*, 2009. Exploring the nuclear envelope of herpes simplex virus 1-infected cells by high-resolution microscopy. *Journal of Virology*, **83**, 408–19.

Wild, P., A. P. de Oliveira, S. Sonda, *et al.*, 2012. The herpes simplex virus 1 U_S3 regulates phospholipid synthesis. *Virology*, In press.

Yang, Q., M. P. Rout, and C. W. Akey, 1998. Three-dimensional architecture of the isolated yeast nuclear pore complex: functional and evolutionary implications. *Molecular Cell*, **1**, 223–34.

9 Scanning electron microscopy of chromosomes: structural and analytical investigations

Elizabeth Schroeder-Reiter and Gerhard Wanner

9.1 Introduction

Chromosomes are eukaryotic carrier units for genetic information. Although it is generally accepted that there is a common chromosome architecture for all eukaryotes, chromosomes vary in number, size, and morphology. Their function is universal, involving transcription, repair, replication, segregation, and recombination of genes during the cell cycle including mitosis and meiosis. These functions go hand-in-hand with extreme structural plasticity of chromosomes. Therefore, understanding chromosome structure in all stages of the cell cycle is particularly relevant in cytological and genetic investigations. As a dramatic example, in mitosis, chromosomes represent condensation in length from the DNA molecule (meter range) to metaphase chromosomes (micron range), shortening by a factor of several thousand (e.g. 40 000 in barley; Wanner and Formanek, 2000). These extreme changes in chromosome structure, from DNA fibrils to a complex structural network, render mitotic chromosomes appropriate for studies of different intermediates in chromosome condensation. At mitosis, chromosomes have been replicated, and are compact units of two identical sister chromatids. However, although several models exist, beyond the elementary fibril (10 nm) and solenoid (30 nm) there is still no ruling consensus on the higher order structure of chromatin, leaving dramatic compaction factors that must be achieved during condensation in mitosis and meiosis essentially unexplained.

Mitotic chromosomes in plant cells were observed with light microscopy and described in impressive structural detail as early as 1888 by Strasburger (Strasburger, 1888), while concurrently the basic function of chromosomes was being postulated (Sutton, 1902; Boveri, 1902). Much later, the elucidation of the structure of DNA (Watson and Crick, 1953) lent a molecular focus to chromatin structural studies. Its composition was determined to consist of approximately equal portions of DNA, histones, and other proteins (Earnshaw, 1988). With the application of electron microscopy, it was possible to visualize chromosome substructures such as the 10 nm elementary fibril (Thoma *et al.*, 1979), the 30 nm solenoid (Finch and Klug, 1976), and further resolution of the structure of the nucleosome (Arents *et al.*, 1991). Spanning the gap between

Scanning Electron Microscopy for the Life Sciences, edited by H. Schatten. Published by Cambridge University Press

molecular and structural evidence was difficult, however, due to preparative and instrumental limitations.

Approaches to visualizing chromosomes have changed with the development of preparation and microscopic techniques. Electron microscopy allows investigation of chromosome subunits with >100-fold higher resolution compared to light microscopy. In conventional transmission electron microscopy (TEM) of specimens (fixed with glutaraldehyde and osmium tetroxide, post-stained with lead citrate and uranyl acetate), it is difficult to distinguish between DNA and protein. This is especially true when structural elements like solenoids (30 nm) overlap due to section thickness (50–70 nm). Stereo imaging with high-voltage TEM (1 MV) was successfully applied to chromosomes, but visualization of chromosome substructures was limited by the maximal thickness of sections of 1 μm (Ris and Witt, 1981). TEM investigations of ultrathin chromosome sections after histone extraction were formative in the understanding of chromosome architecture, but could only be interpreted according to two-dimensional structural observations without information about substance–class specificity (Paulson and Laemmli, 1977; Hadlaczky *et al.*, 1981a, b; 1982).

The establishment of high-resolution scanning electron microscopy (SEM) in the field of biology in the 1980s blazed the trail for three-dimensional investigation of chromosomes (Heneen, 1980; 1981; 1982; Harrison *et al.*, 1982; 1987; Allen *et al.*, 1986; Allen and Harrison, 1988; Sumner and Ross, 1989) and their higher order chromatin structure. SEM allows visualization of chromosomes in their structural entirety, and is especially advantageous for investigation of isolated chromosomes. With labeling and analytical techniques, various kinds of signal can be implemented to identify chromosome substructures and their composition. Two signals are routinely used in SEM: secondary electron (SE) signals show topography and ultrastructural details at high resolution, and back-scattered electron (BSE) signals allow visualization of labeled regions. Labels used for SEM analysis are elements with higher atomic number, typically noble metals such as gold, platinum, and silver. For the past three decades, various preparative and analytical techniques have been developed and applied for SEM investigations of chromosomes and chromatin. Instrument developments in the field have allowed analytical refinement of our understanding of chromatin structures. The aim of this chapter is to present a current understanding of chromosome architecture (and persistent challenges) based on evidence gathered by a wide range of SEM techniques.

9.2 Technical aspects

9.2.1 Fracturing of root tips

Cryo fracture of fixed barley root tips was performed according to Tanaka (1980) modified by replacing dimethylsulfoxide with dimethylformamide, omitting osmium tetroxide thiocarbohydrazide impregnation cycles, treatment with pepsin (2 hr), and post-fixation with OsO_4.

9.2.2 TEM and FIB

Barley (*Hordeum vulgare*) and onion (*Allium cepa*) root tips were fixed with 2.5% glutardialdehyde in 75 mM cacodylate buffer with 2mM $MgCl_2$ pH 7, post-fixed with 1% OsO_4, dehydrated with a graded series of acetone, en bloc stained with 1% uranyl acetate in 20% acetone, and embedded in Spurrs low-viscosity resin. Ultrathin sections were post-stained with aqueous lead citrate.

9.2.3 SEM

Preparation of chromosome suspensions (plant and animal material): barley (*Hordeum vulgare* Steffi, 2n = 14) or woodrush (*Luzula nivea*, 2n = 6) seeds were germinated on moist filter paper. Barley root tip cells were synchronized by 18 hr incubation in hydroxy urea (1.25 mM), incubated 4 hr in water, and arrested in amiprophosmethyl (0.2 μM; 4 hr). After washing, root tips were dissected, incubated overnight in water at 0 °C, fixed in 3:1 (v/v) ethanol:acetic acid and stored at −20 °C. For chromosome isolation, fixed root tips were washed in distilled water for 30 min prior to dissection of the meristematic tips and maceration in 2.5% pectolyase and 2.5% cellulase (w/v in 75 mM KCl, 7.5 mM EDTA) 1 hr at 30 °C. The mixture was filtered through 100 μm gauze, hypotonically treated for 5 min in 75 mM KCl, washed by low-speed centrifugation in 3:1 ethanol:acetic acid fixative and stored at −20 °C.

Polytene chromosomes were dissected from salivary glands of *Drosophila melanogaster* larvae that were fixed in 2% formaldehyde (in 75 mM Tris buffer, pH 3.5). Chromosomes were dissected in 45% acetic acid (in Tris buffer), lightly squashed on a glass slide, and then frozen in liquid nitrogen. Specimens were subsequently fixed in 2% formaldehyde (in Tris buffer) for >15 min prior to further processing for SEM. Marsupial cultures were from ear biopsies of Australian wallaby (Macropodidae) *Macropus rufogriseus x Macropus agilis* hybrid individuals (Metcalfe *et al.*, 2007). Human chromosomes were isolated from lymphocytes that were cultivated and prepared according to a slightly modified version of Yunis *et al.* (1978).

9.2.4 Chromosome isolation with the drop/cryo technique

Cell/chromosome suspensions were dropped onto cold laser-marked slides (LASERMarking, Munich, Germany). Just prior to fixative evaporation, 20 μl 45% acetic acid was applied, the specimens were covered with a cover slip, and frozen upside-down on dry ice for 15 min. The cover slip was prized off, and specimens were immediately transferred into fixative (2.5% glutardialdehyde, 75 mM cacodylate buffer, 2 mM $MgCl_2$, pH 7.0) (Martin *et al.*, 1994). This technique allows investigation of complete chromosome complements relatively free (for plants) of cytoplasmic residue.

9.2.5 Carbon coated slides

For low-voltage investigations of uncoated chromatin, glass slides were carbon coated by evaporation with an electron beam gun (Balzers BAE 080T, Liechtenstein) to a layer of approx. 18 nm (monitored with quartz film thickness monitor QSG 100, BAL-TEC) resulting in light transmittance of 31% (compared to uncoated slides).

9.2.6 Chromosome isolation by mechanical dispersion

Root tips were fixed (2% formaldehyde in 10 mM Tris, 10 mM Na_2EDTA, 100 mM NaCl), mechanically dispersed (Polytron® 5 mM mixer, Kinematica, Switzerland) in isolation buffer (15 mM Tris, 2 mM Na_2EDTA, 0.5 mM Spermin, 80 mM KCl, 20 mM NaCl, 15 mM Mercaptoethanol, 1% Triton-X-100) and spun ("swing out" centrifugation) onto laser marked slides (Schubert *et al.*, 1993). This technique produces a suspension of individual chromosomes (not in their metaphase complements).

9.2.7 Metal contrasting of DNA

For DNA staining, chromosomes were stained for 30 min at room temperature with platinum blue (5 mM $[CH_3CN]_2Pt$ oligomer, in 75 mM cacodylate buffer, pH 7), and subsequently washed with buffer, distilled water and then with 100% acetone prior to critical point drying (for details see Wanner and Formanek, 1995).

9.2.8 Chromatin loosening with Tris

Chromosomes were incubated at room temperature in Tris buffer (75 mM and 2mM $MgCl_2$, pH 9 and pH 10) for 2 hr, followed by fixation with 2.5% glutaraldehyde in Tris buffer.

9.2.9 Artificial decondensation with proteinase K

Chromosomes were fixed with 2.5% glutardialdehyde in 75 mM cacodylate buffer with 2mM MgCl, pH 7, washed in buffer, blocked with 1% glycine solution in buffer, incubated with proteinase K (ICN Biochemicals, 2 mg/ml in 75 mM cacodylate buffer, pH 7) at 37 °C for 30–60 min, as judged by monitoring with phase contrast light microscopy, and washed with buffer and distilled water (for details see Wanner and Formanek, 2000).

9.2.10 Indirect immunogold labeling for scanning electron microscopy (SEM)

Chromosomes were incubated for 1hr with affinity purified rabbit anti-serine 10 phosphorylated histone H3 (diluted in blocking solution 1:250, Upstate Biotechnologies, USA) or anti-CENH3 (centromere specific histone H3, diluted 1:300; Nagaki *et al.*, 2004) followed by incubation for 1 hr with anti-rabbit Nanogold® Fab'-fragments (diluted in blocking solution 1:20; Nanoprobes, USA). Specimens were washed (PBS + 0.1% Tween) and post-fixed with 2.5% glutaraldehyde in PBS and subsequently silver enhanced (HQ Silver, Nanoprobes, USA) according to the manufacturer's instructions (for details see Schroeder-Reiter *et al.*, 2003).

9.2.11 *In situ* hybridization

45SrDNA was labeled by nick translation with biotin-16-dUTP; for ISH 20 ng probe was applied per slide. Conventional ISH procedure was shortened: enzymatic (pepsin, RNase) treatment and intermediate air-drying and dehydration steps were omitted. For immunodetection, AlexaFluor®488-FluoroNanogold™-Streptavidin (diluted 1:100, Nanoprobes, NY, USA) was applied for 1 hr, after which specimens were washed and silver enhanced (see above) (Schroeder-Reiter *et al.*, 2006).

9.2.12 Dehydration, critical point drying, and mounting for SEM

Prior to SEM, specimens were dehydrated with 100% acetone, critical point dried from liquid CO_2, cut to size, and mounted onto aluminum stubs with carbon or silver conductive adhesive.

9.2.13 Field emission scanning electron microscopy (FESEM)

Specimens examined with higher voltages were either sputter-coated to 3–5 nm with Pt (BAE S050, Balzers) or carbon-coated by evaporation (high-vacuum evaporator BAE 121, or electron beam gun BAE 080T, both Balzers) to a layer of 3–5 nm (monitored with quartz film thickness monitor QSG 100, BAL-TEC). Specimens were examined at various accelerating voltages with a Hitachi S-4100 field emission scanning electron microscope equipped with a YAG-type back-scattered electron (BSE) detector (Autrata). For BSE detection acceleration voltages from 15–30 kV were used. Secondary electron (SE) and BSE images are recorded simultaneously with Digiscan™/Digital Micrograph (Gatan, USA).

Low-voltage SEM and focused ion beam/FESEM (FIB/FESEM) investigations were performed with a dual beam Zeiss Auriga® FESEM equipped with an ion beam system, in-lens SE chamber SE, and EsB detectors. The focused ion beam consists of Ga^{+} ions accelerated by a voltage of 30 kV; milling currents (pA) were chosen according to desired milling steps. In the cut-and-view mode, "sections" were produced with the FIB and FESEM images recorded using the chamber SE and in-lens energy selective back-scattered (EsB) and/or chamber Everhard-Thornley SE detectors. Specimens were tilted to an angle of 54° and, dependent on the experimental goal, images were tilt-corrected for undistorted view of the milled surface.

9.2.14 Stereo pairs/anaglyph images

3-D anaglyph images were mounted from stereo micrograph pairs recorded after tilting the specimen stage to an angle difference of 3°.

9.2.15 Quantification and image analysis

Monte Carlo simulations of electron trajectories were performed with Casino v2.42 (Université de Sherbrooke, Canada). Marker molecules were quantified by counting the number of signal spots per segment. Marker volume was calculated according to the signal count using an average marker diameter of 15 nm. Alignment of image stacks was performed in part by ImageJ (Rasband, W. S., ImageJ, US National Institute of Health, Bethesda, Maryland, USA, http://rsb.info.nih.gov/ij/, 1997–2007). A 3-D reconstruction signal distribution, the volume of chromatin and cavities, and animations were calculated and performed using Avizo software (VSG, France).

9.3 Data collection

9.3.1 Chromosome structure *in situ* in LM, TEM, and SEM

Using the light microscope (LM) for chromosomes of semi-thin sections of resin-embedded barley root tips, different mitotic stages were readily recognizable *in situ* in phase contrast LM due to their high DNA content (Figure 9.1A). With transmission electron microscopy (TEM) of ultrathin sections, chromosomes exhibit high contrast when fixed/stained with OsO_4, uranyl acetate, and lead citrate. Their profile is irregular from prometa- to metaphase, and chromatin is granular in appearance (Figure 9.1B). The *in situ* orientation of chromosomes is limited by interpretation of single sections. It is not clear whether the total contrast originates from DNA alone, or DNA and protein, and to what extent. With SEM *in situ* investigations of cryo-fractured and critical point dried root tips it is difficult to distinguish chromosomes from the compact nucleoplasm without additional staining. In favorable cases, chromosomes are recognizable due to denser chromatin structure with respect to their surroundings (Figure 9.1C).

9.3.2 Ultrastructure of isolated mitotic plant chromosomes

Isolated chromosomes are appropriate for SEM investigations, allowing visualization of topographic features and substructural details correlated to light microscopic observations of the same specimen. Chromosome morphology can be visualized in all mitotic stages, from prophase through telophase (Figure 9.2). Chromosomes condense to recognizable units from interphase to prometaphase (Figure 9.2). At metaphase, the primary constriction at the centromere becomes visible, and the sister chromatids are distinguishable from each other at the telomeres (distal chromosome ends) (Figures 9.2 and 9.3 (color plate)). At anaphase, chromatids separate and are longer and thinner than at metaphase (Figure 9.2). From telophase to interphase, the chromosomes decondense, and are not recognizable in SEM as distinct units (Figure 9.2).

A closer look at chromosome architecture with high resolution shows that throughout all stages of the cell cycle chromosomes are composed of fibrous elements that vary in diameter and orientation. The predominant fibers observed have a diameter of approximately 30 nm (Figure 9.4; Martin *et al.*, 1996). Some of these fibers form chromomeres with diameters ranging from 200–400 nm (Figure 9.4B–C; Wanner and Formanek, 2000). The surface topography of chromosomes observed in SEM can be confirmed with scanning force microscopy of hydrated chromosomes (Schaper *et al.*, 2000).

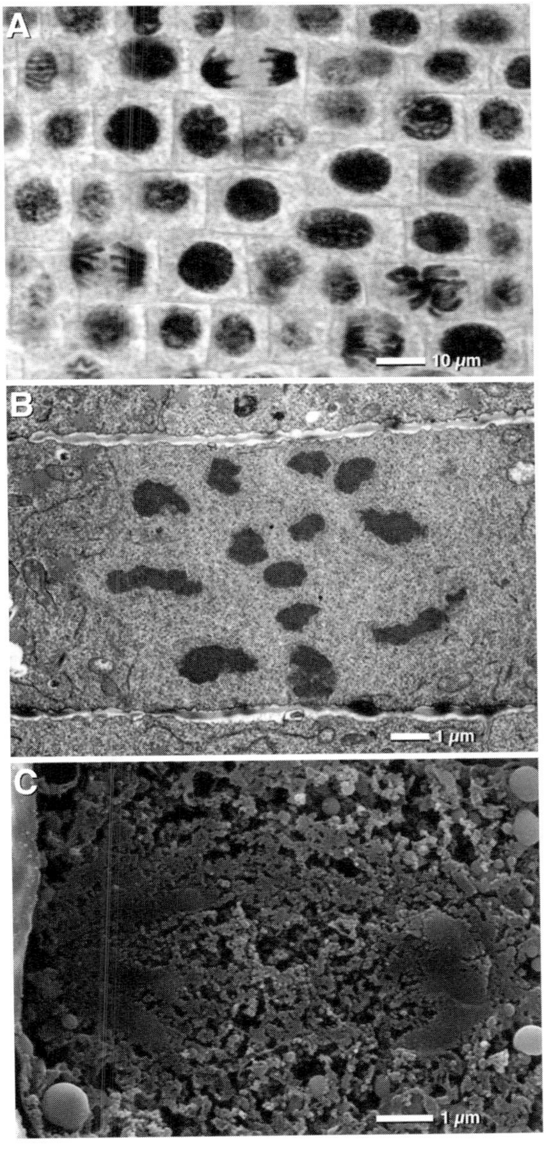

Figure 9.1 Different microscopy techniques visualizing barley mitotic chromosomes in root tip cells. (A) Semi-thin section of root tip tissue viewed with phase contrast LM exhibits chromosomes in different mitotic stages. (B) In TEM ultrathin section, chromatin appears granular, and (pro-) metaphase chromosomes are visible as irregularly outlined electron dense regions. (C) In SEM (cryo-fracture according to Tanaka), compact anaphase chromosomes are barely distinguishable from the nucleoplasm.

During condensation from prophase to prometaphase, chromomeres are interspersed with longitudinally oriented fibers (Figure 9.4B; Wanner and Formanek, 2000). At metaphase, chromosomes show two characteristic features: coherent, more or less cylindrical sister chromatids that appear "knobby" at low magnifications due to tightly

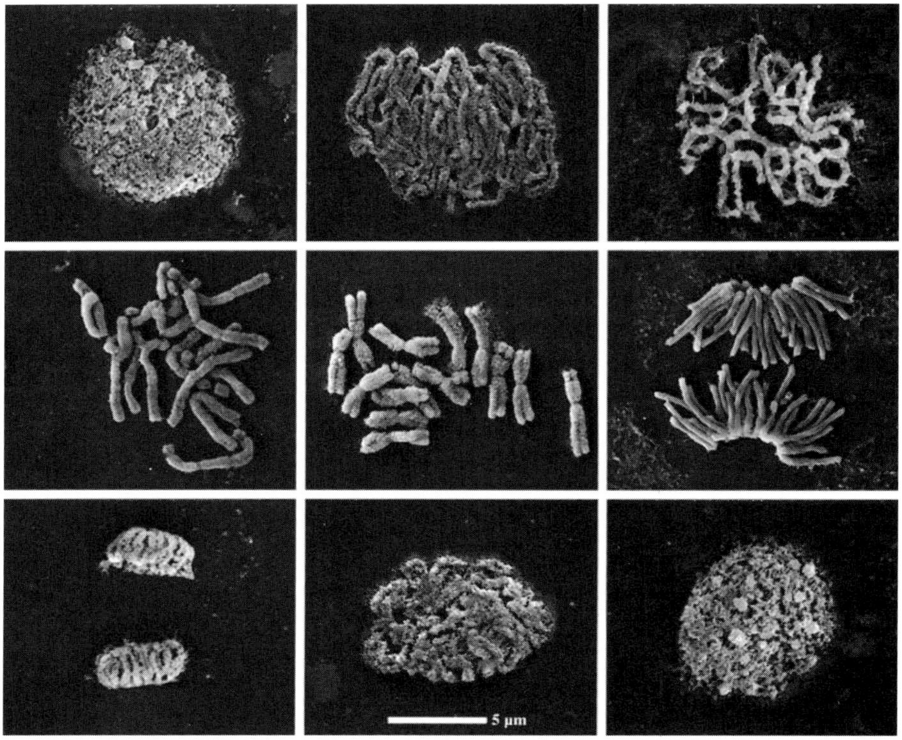

Figure 9.2 SEM micrographs of mitotic rye chromosomes (2n = 14) isolated with the drop/cryo preparation technique show chromosome morphology in all stages of condensation: (from upper left to right) interphase, early prophase showing *Rabl* configuration, prophase, prometaphase, metaphase, anaphase, telophase, transition from telophase to interphase, interphase. (Reprinted from Zoller *et al.* 2004, with permission from S. Karger AG, Basel.)

packed chromomeres, and a centromeric region, recognizable as a constriction, characterized mainly by parallel fibers and fewer, smaller chromomeres (Figure 9.4C and 9.5 (color plate)). Exceptions are holocentric chromosomes, which lack a centromeric constriction and exhibit parallel fibers interspersed with chromomeres along the entire chromosome (Schroeder-Reiter and Wanner, 2009). Secondary constrictions, generally the structural manifestation of nucleolus organizing regions (NORs), are also characterized at metaphase by parallel fibers (Wanner and Formanek, 2000).

9.3.3 Chromosome analysis in SEM

9.3.3.1 Influence of additives

Excellent structural preservation is obviously the ultimate goal in EM studies, but each step in any procedure potentially influences chromosome structure. These effects themselves, e.g. compaction, loosening, shrinkage, and stretching, are more or less reproducible and can be employed for investigations of chromosome structure.

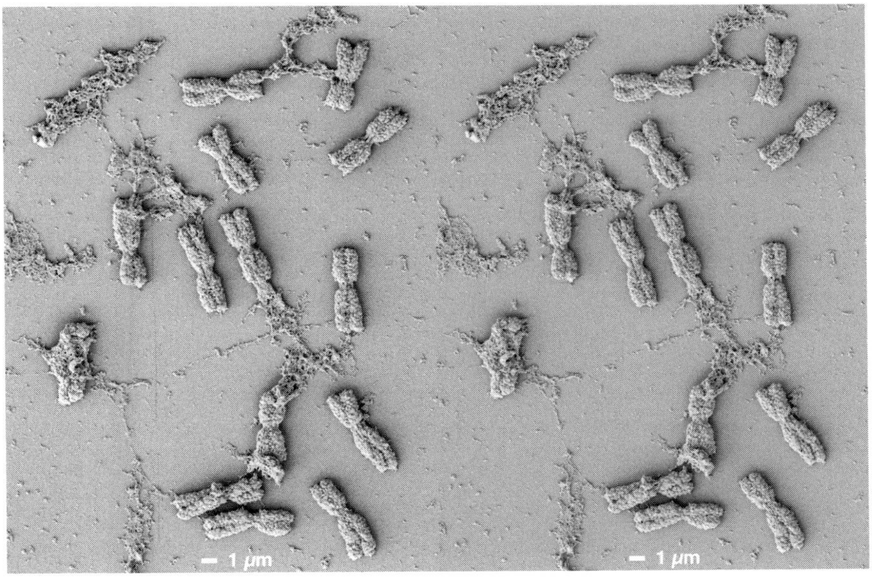

Figure 9.3 Stereo micrograph showing 3-D topography of a barley mitotic metaphase spread (2n = 14). Sister chromatids are distinguishable, and in some cases are separated at the telomeres. Centromeres exhibit a distinct primary constriction. (See color plate section for anaglyph.)

Synchronization and arrest of cells, although highly efficient in increasing mitotic chromosome yield, appear to influence chromosome morphology. Due to disruption of spindle formation, arrest causes (artificial) shortening and, consequently, compaction of the chromosomes in addition to extremely pronounced primary and secondary constrictions (Figure 9.6A). As an alternative to 3:1 (ethanol:acetic acid) fixation, classical fixation of root tips with glutardialdehyde and OsO_4 fixes the highly compact cell contents of meristematic cells to such an extent that isolation of chromosomes is not possible. With its weak cross-linking capability, brief formaldehyde fixation of root tips allows isolation of chromosomes by mechanical dispersion, filtration and centrifugation. These chromosomes do not remain in their respective metaphase complement, and are slightly stretched and often distorted (Figure 9.6B).

Studies on the influence of buffers, pH, and chemical reagents revealed that chromosomes do not react uniformly to most treatments, which limits the reproducibility for experiments. For example, incubation in phosphate buffer results in variations of chromosome structure from experiment to experiment. In the case of citrate buffer, results are quite reproducible; chromosomes typically are somewhat stretched and "fan out" at all four telomeres (Schaper *et al.*, 2000; Wanner and Schroeder-Reiter, 2008). Chromosome structure appears stable over a broad range of pH, from acidic to neutral conditions (pH3–pH7), but loosens with increasing alkalinity, effectively increasing accessibility of fibrils (Figure 9.6C, D; Wanner and Schroeder-Reiter, 2008). Incubation with dextran sulfate, as is used in standard *in situ* hybridization (ISH)

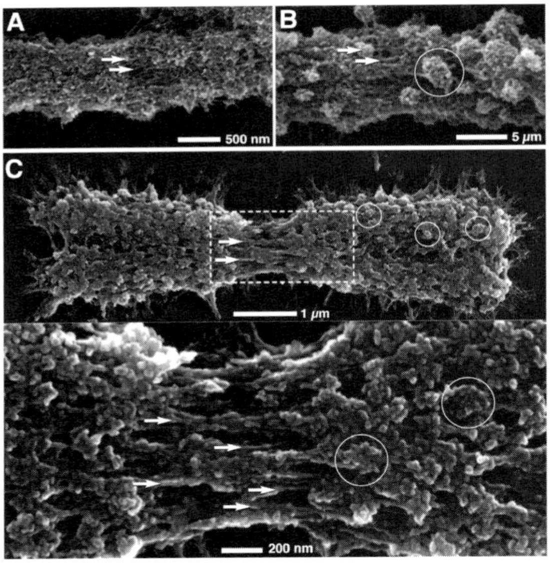

Figure 9.4 SEM micrographs of barley mitotic chromosomes in prophase (A), prometaphase (B), and metaphase (C). Two common substructural elements are characteristic for chromosomes: chromomeres (circles) and parallel fibrils (arrows). At prophase, parallel fibrils are visible along the entire chromosome (A), at prometaphase they are interspersed with chromomeres (B), and at metaphase parallel fibrils are only visible at the centromere due to a large number and density of chromomeres (C). (B reprinted from Zoller *et al.* 2004, with permission from S. Karger AG, Basel; C reprinted from Wanner and Schroeder-Reiter, 2008, with permission from Elsevier.)

assays, also results in loosening of the chromomere structure (Wanner *et al.*, 2005). Most buffer effects are accompanied by appearance of peripheral "stress fibrils," presumably due to mechanical and adhesive forces of elastic chromatin on the glass slide during preparation (Figure 9.6C).

In general, the stronger the fixation, the less the influence of subsequent treatment with buffers or detergents. Glutaraldehyde fixes proteins so well that the 3-D structure is unaffected by most reagents. If, however, glutaraldehyde-fixed chromosomes are treated with high concentrations of proteinase K (0.1–1.0 mg/ml), there is a dramatic (time-dependent) effect on the chromosomes: they stretch in length and/or fan out in breadth. This artificial loosening of the compact metaphase chromosomes is referred to as "controlled decondensation" (Wanner and Formanek, 2000). In some cases, chromosomes reach an extraordinary length of up to 200 mM (Wanner and Formanek, 2000). In such extreme stages of stretching, chromatin begins to elongate to elementary chromatin substructures: loosened chromomeres, solenoids (30 nm) and elementary fibers (10 nm) (Figures 9.7 and 9.8 (color plate)). The loosening effect results from both partial digestion of chromatin proteins, as well as physical and mechanical forces (e.g. surface tension) during preparation and dehydration.

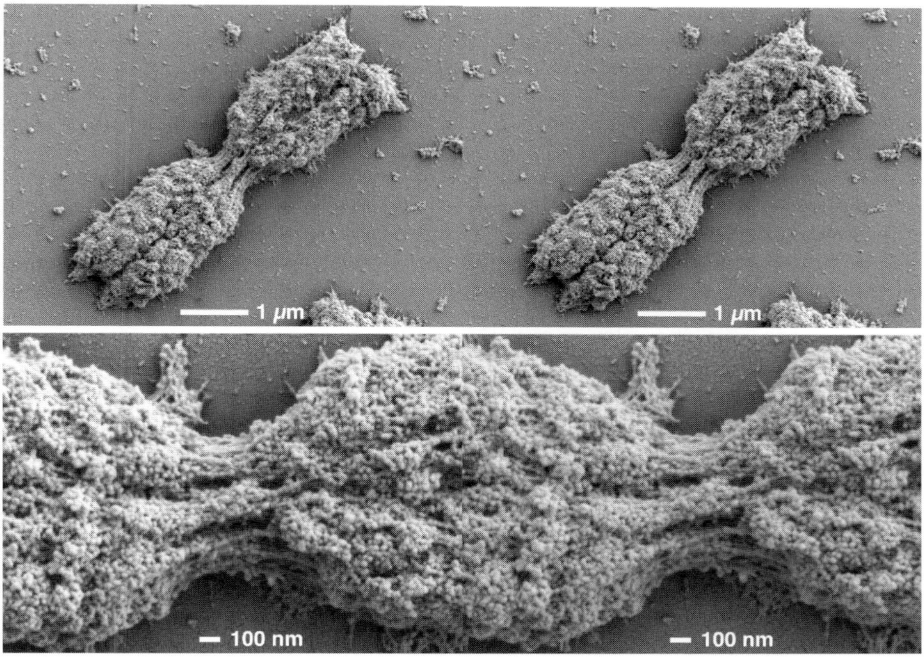

Figure 9.5 Stereo micrographs showing topographic structure of a barley mitotic metaphase chromosome with compact chromomeres along the chromosome arms and exposed parallel fibers with fewer and smaller chromomeres at the centromere. At higher magnification (lower image) 30 nm fibrils are visible along the entire chromosome, in particular at the centromere. (See color plate section for anaglyph.)

9.3.3.2 Specific staining of DNA in chromosomes with platinum blue

Platinum blue (Pt blue; $[CH_3CN]_2Pt$ oligomer; 28) and related platinum-organic compounds selectively react with nucleic acids, with especially high affinity to DNA (Aggarwal *et al.*, 1975; 1977; Lippert and Beck, 1983; Köpf-Maier and Köpf, 1986; Frommer *et al.*, 1990). When viewing Pt blue-stained chromosome preparations with back-scattered electron (BSE) microscopy with various voltages (0.5–30 kV), platinum signals from chromatin vary between fibrous networks, weakly contrasted regions, and highly contrasted "bright" regions, indicating areas of varying DNA density (Figures 9.9–9.11). In mitotic metaphase vertebrate and plant chromosomes, the BSE signal from stained DNA clearly shows regions of different signal density (Figure 9.9). Regions corresponding to chromomeres (recognizable from SE signal) have "bright" BSE signals, whereas fibers can be correlated with either strong or weak BSE signals, indicating that these structures are either DNA-rich or DNA-poor (Figure 9.9; Wanner and Formanek, 2000). This is particularly striking at the centromere: in monocentric chromosomes the BSE signal is narrower and weaker at the primary constriction than that

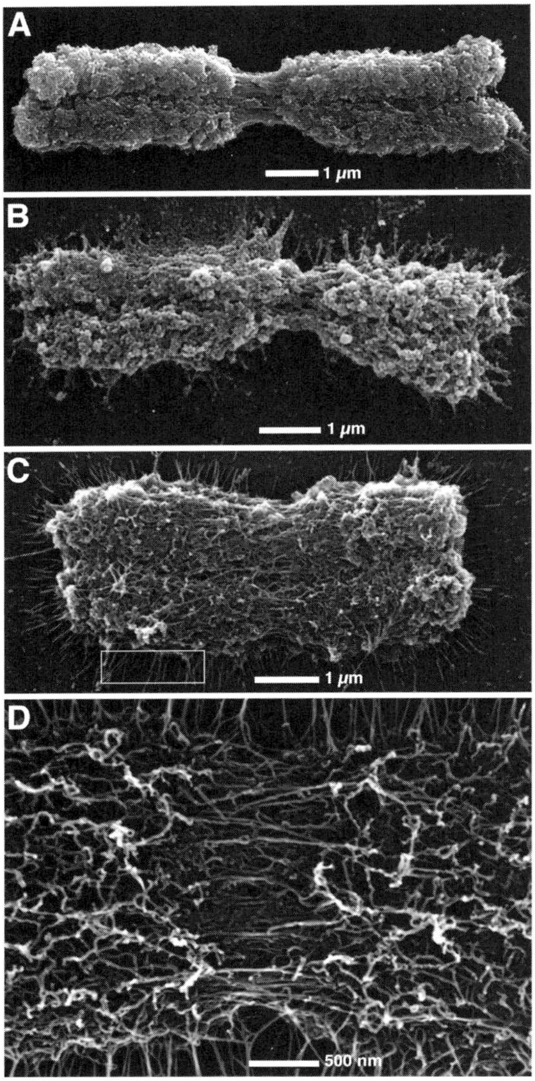

Figure 9.6 SEM micrographs of mitotic barley chromosomes show structural effects of different preparation treatments. (A) Synchronization and arrest of meristematic tissue with subsequent drop/cryo preparation results in a well-preserved but compact chromosome structure, with a particularly prominent primary constriction. (B) Formaldehyde fixation of root tip tissue and isolation of chromosomes solely by mechanical dispergation results in somewhat looser chromatin structure, but also in stretching or deformation. (C, D) Incubation in Tris buffer at high pH (pH 9–10) loosens chromomeres; at high magnifications both the centromeres and the chromosome arms clearly exhibit a fibrous network (D). "Stress fibrils" often occur after incubation with different buffers or reagents (frame). (B reprinted from Wanner *et al.*, 2005 with permission from S. Karger AG, Basel; D reprinted from Wanner and Schroeder-Reiter, 2008, with permission from Elsevier.)

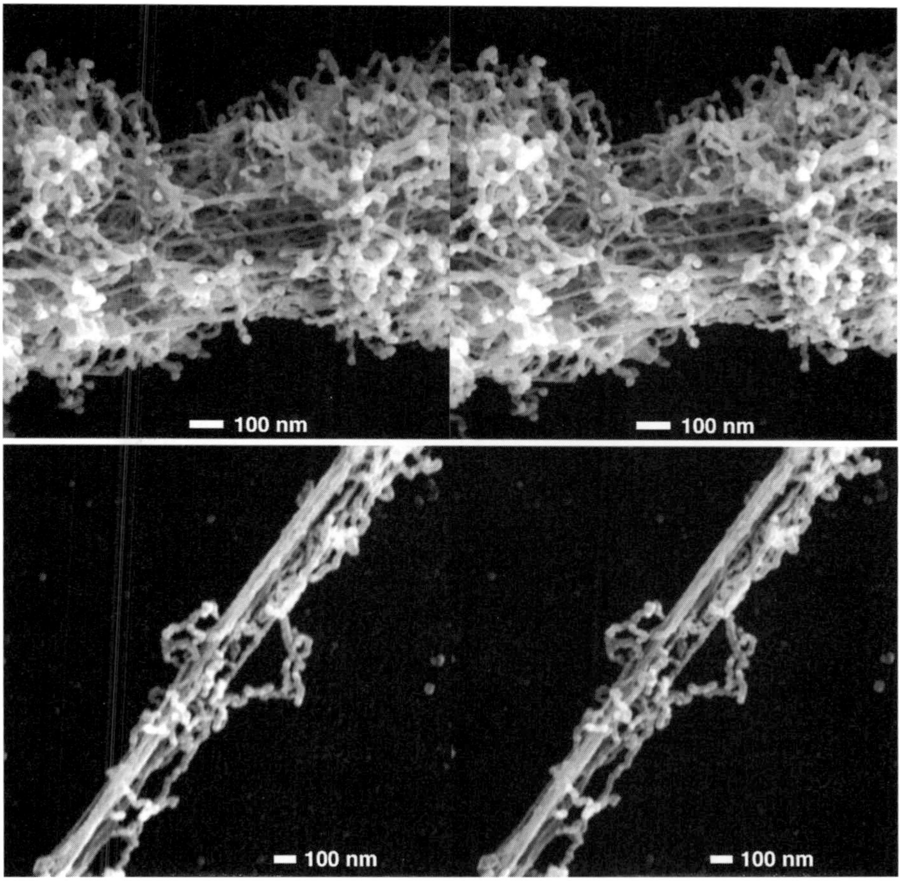

Figure 9.7 SEM stereo micrographs of barley mitotic chromosomes after controlled decondensation with proteinase K, resulting in loosening and stretching of chromatin. Loosening of the centromere and the compact chromosome arms allows insight into the centromere interior: parallel fibrils are distributed throughout the centromere and chromomeres are made up of 30 nm fibrils (upper stereo pair). Extreme stretching of the chromosomes shows that some 30 nm fibrils are oriented in a parallel fashion, while others are loosely attached to the parallel fibers (lower stereo pair). (Reprinted from Wanner and Formanek, 2000, with permission from Elsevier.)

of the chromatids (Figure 9.9 A–C; see also Metcalfe *et al.*, 2007; Schroeder-Reiter and Wanner, 2009). In the case of *Luzula nivea*, a holocentric plant species, chromosomes have no primary constriction, but exhibit regions of varying DNA density (Figure 9.9D). The efficiency of Pt blue analysis can be shown in investigations of polytene chromosomes of *Drosophila melanogaster*, for which a banding pattern, assumed to correlate to heterochromatin and euchromatin, is well known. With SEM, these bands can be investigated with higher resolution. Topography and DNA distribution show a clear orientation of parallel fibrous chromatin ("interbands"), exhibiting relatively weak platinum signals, with intermittent bands resembling densely packed chromomeres that have strikingly strong platinum signals (Figures 9.10 and 9.12 (color plate)). Even

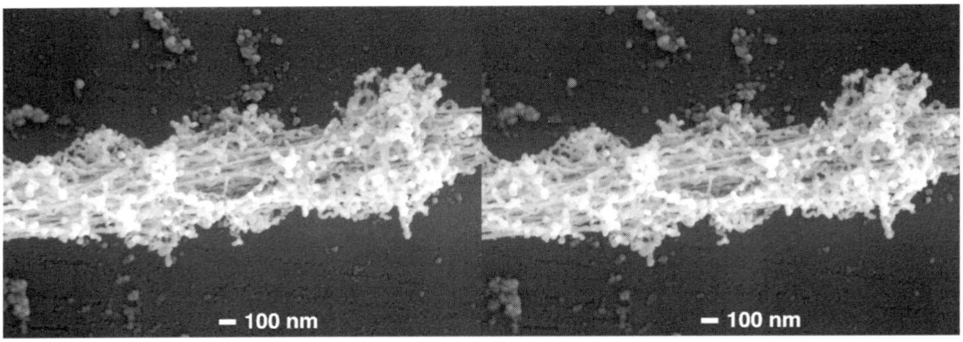

Figure 9.8 Stereo micrograph showing the stretched centromere of a barley chromosome after controlled decondensation. Parallel fibrils are distributed throughout the centromere interior, and chromomeres loosen to loops of fibrils in the range of 30 nm. (Reprinted from Wanner and Formanek; 2000, with permission from Elsevier). (See color plate section for anaglyph.)

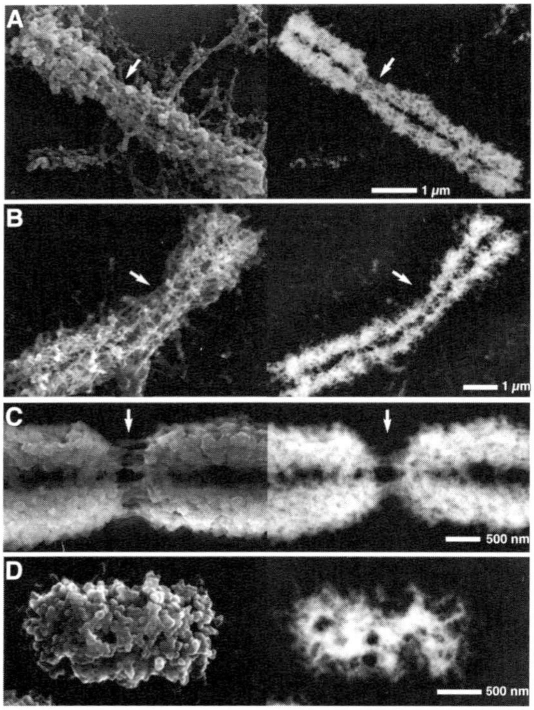

Figure 9.9 SEM micrographs of isolated mitotic metaphase chromosomes from different organisms stained with platinum blue show chromosome topography (SE images, left column) and DNA distribution (BSE images, right column). (A) Human, *Homo sapiens*; (B) wallaby, *Macropus rufogriseus*; (C) barley, *Hordeum vulgare*; (D) snowy woodrush, *Luzula nivea*. Fibrous substructural elements, parallel fibrils, and chromomeres in different stages of compaction can be seen in all chromosomes. DNA distribution is not homogeneous, showing less DNA content between sister chromatids and/or at the centromere of monocentric chromosomes (A–C, arrows). Holocentric chromosomes have no primary constriction, but show areas of higher DNA content (D).

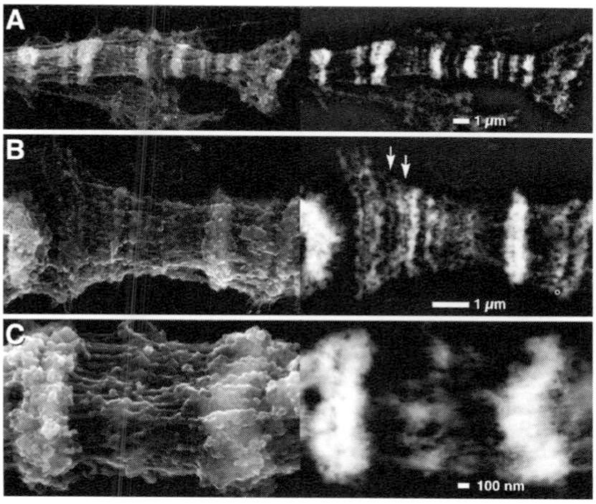

Figure 9.10 SEM micrographs of polytene chromosomes from larval salivary glands of *Drosophila melanogaster* stained with platinum blue showing chromosome topography (SE images, left column) and DNA distribution (BSE images, right column). Chromosomes are clearly organized in compact banded regions, which correspond to DNA rich regions, and interband regions with less DNA content (A–C). Enhanced material contrast with platinum blue facilitates distinction of fine bands in the order of 100–200 nm in width (arrows). At high magnifications, the fibrous nature of the polytene chromosomes is evident; the interbanded regions exhibit both DNA-rich and DNA-poor fibers interspersed with very compact DNA-rich bands (C).

extremely thin bands (<100 nm), hardly visible in SE images, are distinguishable with Pt blue staining (Figure 9.10B).

The strengths of the YAG detector lie in its high resolution and its performance at moderate and high kV (8–30 kV). This allows detection of signals from a depth of several microns, which practically covers all sizes of chromosome. By specimen tilting, BSE stereo pairs of chromosomes stained with Pt blue give important information on the bulk in-depth DNA distribution in addition to 3-D topography with SE stereo pairs (Figure 9.12). However, investigation of substructural elements is limited, as they are completely penetrated by primary electrons, and the portion of BSE from the specimen is too low compared to that from the glass slide. Any conductive coating in the nanometer range severely limits both determination of topography and BSE detection. By using conductive carbon coated glass slides, in-lens SE and energy selective back-scattered electron (EsB) detectors it is possible to examine chromosome substructures without coating at accelerating voltages in the range of 0.5 kV to 2.0 kV. SEM investigation (1 kV) of chromosomes loosened with Tris (pH 10) and stained with Pt blue demonstrates that material contrast from platinum allows EsB detection of extremely fine structural details that are otherwise barely visible even with sensitive (in-lens) SE detectors (Figure 9.11).

9.3.3.3 Immunogold labeling of specific chromosomal proteins and DNA sequences

Identification of specific elements and/or chromosomal substructures with high resolution and in a 3-D context is currently best achieved with immunogold detection.

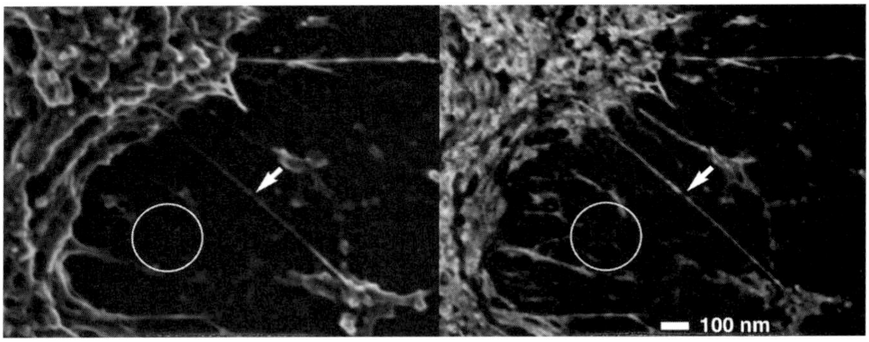

Figure 9.11 SEM micrographs of an uncoated barley chromosome loosened with Tris buffer (pH 10) and stained with platinum blue, recorded simultaneously at 1 kV with in-lens SE detector (left) and an EsB detector (right). DNA-containing substructures in the range of 30 nm can be visualized with both detectors (arrow). Fine structural elements (approx. 10 nm), hardly visible with SE due to minimal topographic contrast, are readily visible with EsB signal (circles).

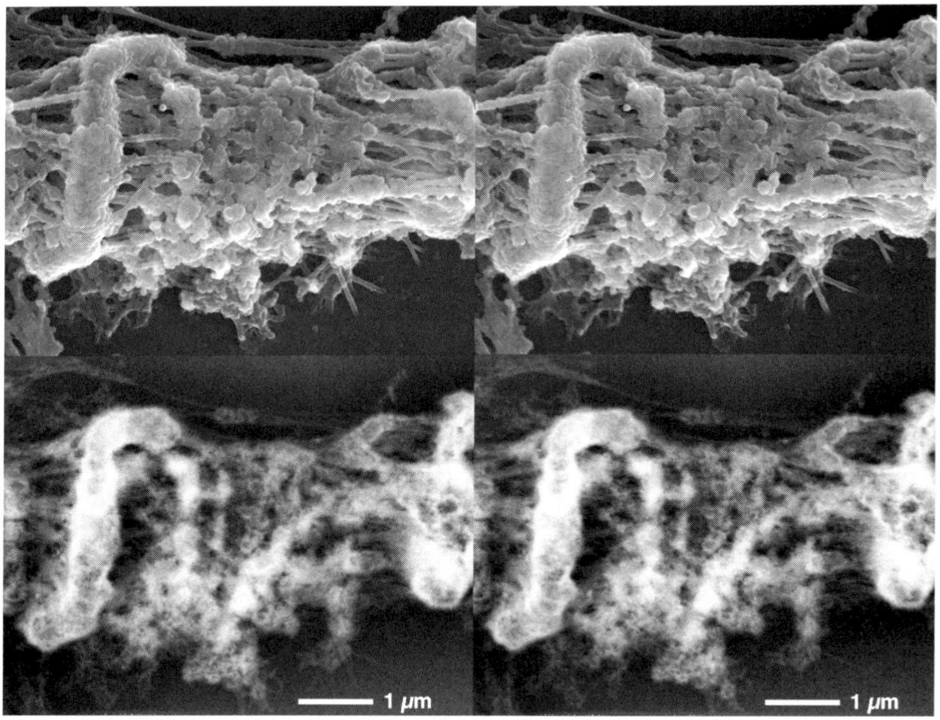

Figure 9.12 Stereo micrographs of a mutant *Drosophila melanogaster* polytene chromosome (with distortion of banding) stained with platinum blue showing structural elements (compact chromatin bands and parallel fibrils; SE, upper image) as well as corresponding 3-D DNA distribution (BSE, lower image). (See color plate section for anaglyph.)

It is critical for SEM investigations to use the smallest possible markers (e.g. Nanogold® labels, Fab' fragments, haptens when applicable). Different kinds of immunogold label have been applied for ISH and immunodetection of specific proteins: Nanogold® (NG), a 1.4 nm gold cluster, and Fluoronanogold™ (FNG), a combined fluorescein and 1.4 nm gold marker (Hainfeld and Powell, 2000). FNG allows direct correlation of fluorescent LM and SEM signals on the same specimen (Powell *et al.*, 1998; Schroeder-Reiter *et al.*, 2006). Since the size of the 1.4 nm gold clusters for both FNG and NG is at the resolution limit of the SEM, both FNG and NG must be either gold or silver enhanced, a time-dependent process of controlled metallonucleation which gradually increases the size of the bound gold cluster (Hainfeld *et al.*, 1999). Using ultra-small markers, it is currently possible to localize epitopes down to the solenoid level of chromatin.

High-resolution immunogold labeling with Nanogold® allowed quantification and three-dimensional localization of histones, histone modifications, and centromere proteins in barley (Figure 9.13; Schroeder-Reiter *et al.*, 2003, 2006; Houben *et al.*, 2007). Histone H3 exhibits strong labeling in the pericentric region, but a signal gap at the mid-centromere (Figures 9.13A and 9.14 (color plate); Schroeder-Reiter *et al.*, 2003). The centromere-specific histone H3, CENH3 (Talbert *et al.*, 2002), reveals two intense signal regions at the centromere on each sister chromatid (Figure 9.13B; Houben *et al.*, 2007). A modified ISH protocol for SEM allowed 3-D SEM analysis of an unusual terminal NOR structure and its rDNA distribution for the plant species *Ozyroë biflora* (Figures 9.13C and 9.15 (color plate); Schroeder-Reiter *et al.*, 2006). Dependent on signal size (according to time-dependent enhancement metallonucleation) and accelerating voltage, signals can be detected both on the chromosome surface and from within chromosome structures, allowing visualization of three-dimensional signal distribution with stereo pairs or anaglyph images (Figures 9.14 and 9.15).

9.3.3.4 Focused ion beam tomography with FIB/FESEM

Tomography of chromosomes is possible from both isolated chromosomes and chromosomes *in situ* of resin-embedded root tips by sequential focused ion beam (FIB) milling. Carbon coating of specimens (5–15 nm) is sufficient for conductivity and specimen stability during the milling/image acquisition process. With isolated chromosomes, milling steps of 7–20 nm, adapted to the thickness of chromatin fibrils, have allowed visualization of the inner architecture of both the centromere and the chromosome arms in the same plane of isolated chromosomes (Figure 9.16A; Schroeder-Reiter *et al.*, 2009). Parallel fibrils in the range 20–100 nm are visible throughout the entire centromere (Figure 9.16A); some milled surfaces in the centromere appear rather homogeneous, indicating intimate contact of centromere substructures (Figure 9.16A). Chromosome arms exhibited a complex network of compact chromatin areas and an extensive cavity network (Figure 9.16A). With labeled chromosomes, SE and BSE signals can be recorded simultaneously either as separate SE and BSE image files, or, if preferred, in signal-mixed images with an appropriate signal ratio for sufficient topographic information and adequate BSE signal (Figure 9.16). Marker signals from the chromosome interior are directly imaged on serial milled surfaces, and therefore detected and

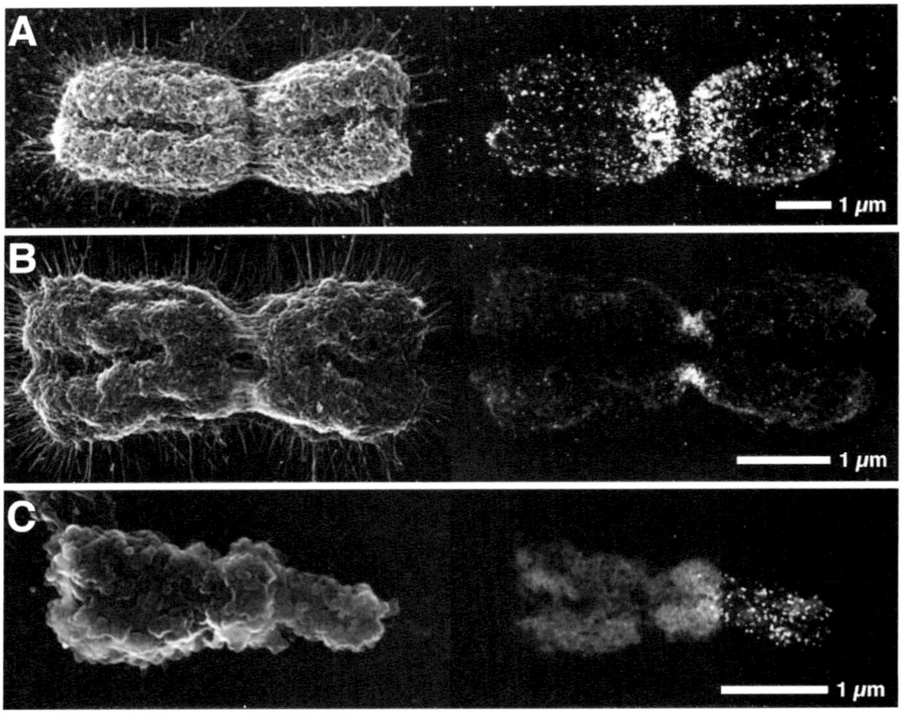

Figure 9.13 SEM micrographs of mitotic metaphase plant chromosomes immunolabeled for specific proteins and DNA sequences, showing chromosome topography (SE images, left column) and marker distribution (BSE images, right column). (A) Labeling of phosphorylated histone H3 (serine 10) on barley chromosomes shows a particularly strong signal in the pericentric region and a signal "gap" at the centromere, corresponding to the exposed parallel fibrils. (B) The histone H3 variant CENH3 shows in barley a clear specificity for the centromere, labeling a concentrated region, corresponding mainly to exposed parallel fibrils, on each sister chromatid. (C) A terminal peg-like structure observed in chromosomes of *Ozyroë biflora* (Hyacinthaceae) could be identified as a nucleolus organizing region with *in situ* hybridization of 45S rDNA. (A, C reprinted from Wanner and Schroeder-Reiter, 2008, with permission from Elsevier.)

quantifiable as signal spots (Schroeder-Reiter *et al.*, 2009). With FIB/FESEM imaging, the interior DNA distribution of metaphase chromosomes stained with Pt blue exhibits a network of cavities and DNA-rich fibers in the order of 30–50 nm in various levels of compaction (Figure 9.16B).

With FIB/FESEM imaging of embedded specimens the milled blockface is essentially smooth, and in principle both SE and BSE signals can be used for imaging. For the SE signal, brightness and contrast settings must be adjusted to the maximum in order to achieve a high contrast, which is derived from material contrast. A BSE signal gives a strong contrast, however, with a significantly reduced signal-to-noise ratio, rendering image acquisition time-consuming. However, in practice BSE detection is preferred since even minor irregularities (typical for ion beam milling) on the block face lead to

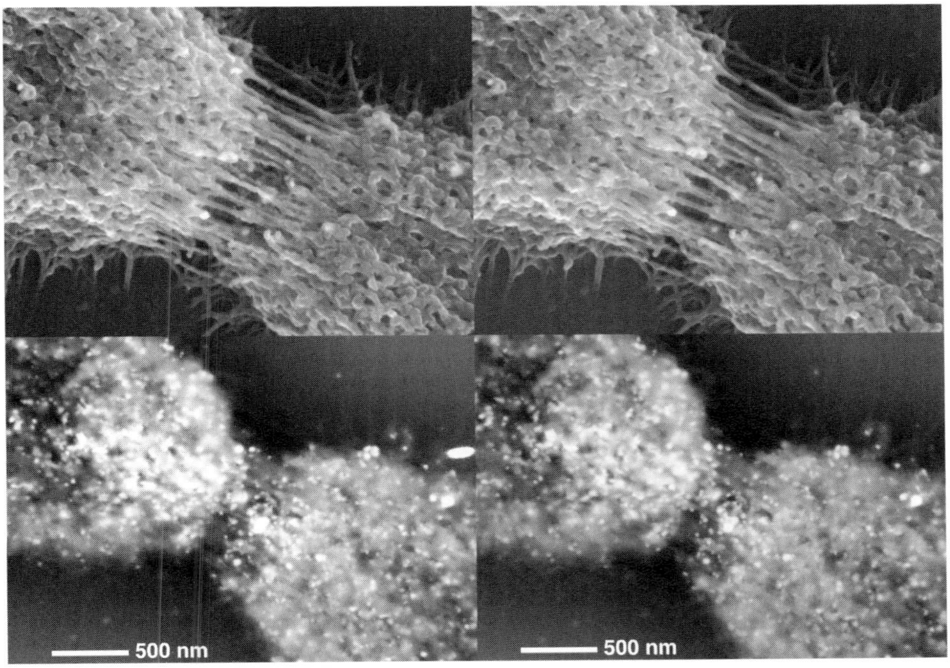

Figure 9.14 Stereo micrographs of SE (upper pair) and BSE (lower pair) signals showing 3-D distribution of markers immunolabeling phosphorylated histone H3 (serine 10). The centromere is characterized by parallel fibrils, lacking chromomeres and exhibiting very few H3P markers. The majority of the markers are localized in the pericentric chromatid arms. (See color plate section for anaglyph.)

distortion of the SE image; these are not visible in the BSE image due to larger interaction volume. Inverted BSE images of mitotic cells in fixed and embedded onion (*Allium cepa*) root tip tissue resemble TEM micrographs, with rather granular chromatin and irregularly shaped chromosomes, and exhibiting high resolution of contrasted structures (Figure 9.17). Within the chromosome regions, areas with weaker signals are visible, also indicating cavities within the dense chromatin (Figure 9.17).

9.3.3.5 High-resolution visualization of 3-D distribution of specific markers

With data from FIB/FESEM tomography series of barley metaphase chromosomes labeled with the centromere-specific histone variant CENH3 (see orthogonal view in Figure 9.13B), it is possible to analyze markers, cavities, and chromatin both individually and combined as a whole, allowing interactive analysis of marker distribution with respect to chromatin (Figure 9.18 (color plate)). 3-D reconstruction illustrates a heavily labeled region locally concentrated at the primary constriction, with very few signals extending into the pericentric region (Figure 9.18). Few markers were located directly on the surface of the chromosome, indicating that CENH3 is an interior element of the kinetochore. Chromosome surface (transparent green); chromatin (magenta); CENH3 markers (yellow).

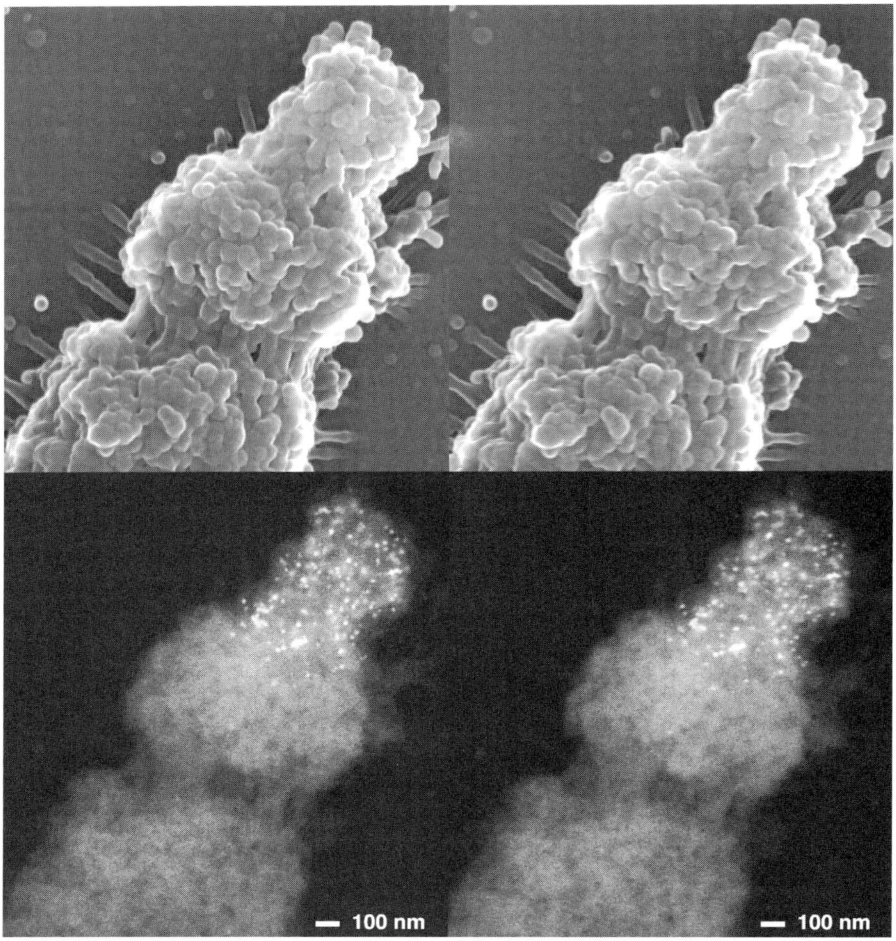

Figure 9.15 Stereo micrographs showing substructural elements (SE image pair, upper) of a plant chromosome (*Ozyroë biflora*, Hyacinthaceae), as well as 3-D marker distribution labeling 45S rDNA with *in situ* hybridization (BSE image pair, lower), identifying a peg-like terminal structure as a NOR. (See color plate section for anaglyph.)

9.4 Data analysis and discussion

Visions for chromosome architecture, influenced by a "helical" perspective starting with the double helix DNA molecule, the elementary fibril with helical winding of DNA around nucleosomes, up to the solenoid (the helical winding of the elementary fibril) have resulted in a variety of helical and folded models for chromosome higher order structure, some still currently in circulation (Dupraw, 1965; Rattner and Lin, 1985; Manuelidis and Chen, 1990). Although the technical resources were applied for confirmation of tertiary structures (Marsden and Laemmli, 1979), and formative TEM

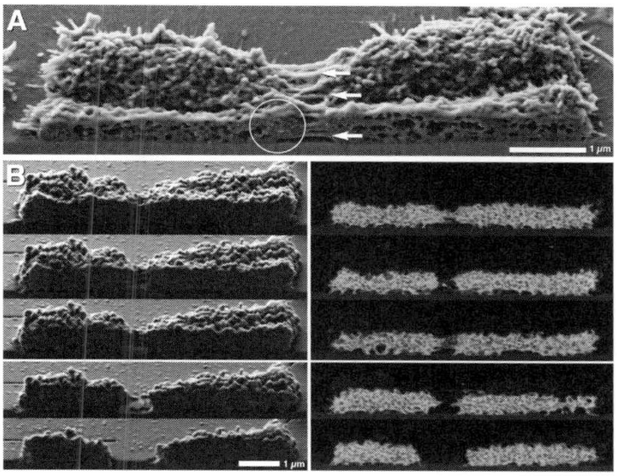

Figure 9.16 Selected images from a FIB/FESEM series, by which barley isolated chromosomes are milled in steps, allowing direct visualization of chromosome interior. (A) Chromosomes immunolabeled for phosphorylated histone H3 (ser 10) and milled longitudinally exhibit dense chromatin regions and an extensive cavity network along the chromosome arms, as well as parallel fibrils in the centromeric region (arrows). Imaging with signal mixing (50/50 SE/BSE) results in strong topographic contrast for optimal visualization of cavities; markers, however, are barely visible at low magnifications (circle). In an FIB series of simultaneously recorded SE (B, left column) and EsB (B, right column) images of a chromosome stained with platinum blue, the milled surface shows a network of "bright" DNA-rich areas as well as DNA-free areas that correspond to cavities (series shows every 10th image in a series with 7 nm milling steps).

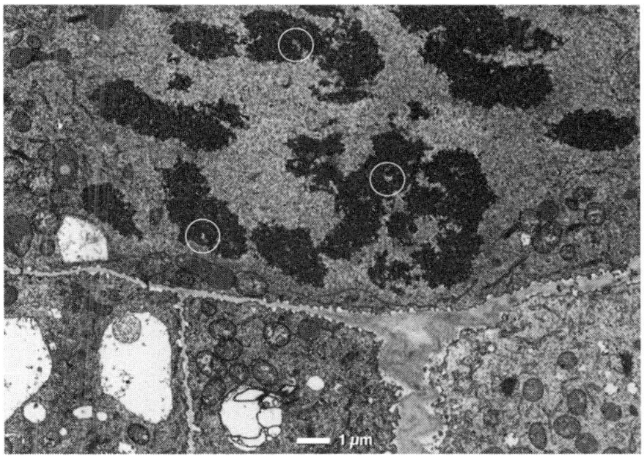

Figure 9.17 FIB/FESEM EsB image (inverted) of the milled blockface of an onion (*Allium cepa*) root tip cell (fixed, contrasted and resin-embedded according to standard TEM preparation). This technique allows *in situ* investigation of chromosomes with high resolution of contrasted structural details in large-scale image series (e.g. of entire cell). Chromatin appears granular and exhibits both dense regions as well as small, less dense areas within the chromosomes (circles).

evidence for a "scaffold" was introduced (Paulson and Laemmli, 1977; Hadlaczky *et al.*, 1981b), the superhelix model could not be proven, and comprehensive EM investigations extending to large-scale chromosome architecture were rare. The main problem was achieving reproducible, routine chromosome preparations. The era of high-resolution

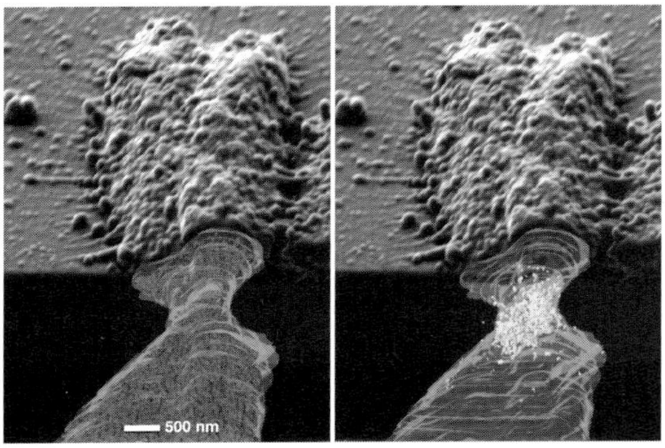

Figure 9.18 3-D reconstruction of an isolated barley chromosome immunolabeled for the centromere-specific histone H3 variant CENH3. The interior chromatin exhibits a dense network. The majority of the CENH3 signals localize to the centromere interior, indicating that it is a component of the inner kinetochore. (Reconstructed from a FIB/FESEM series of 162 images with 15 nm milling steps.) (See plate section for color version.)

SEM (1980s–1990s) initiated quality pioneer investigations (Heneen, 1980; 1981; 1982; Allen *et al.*, 1986; Allen and Harrison, 1988; Sumner, 1991). The majority of these investigations focused on mammalian chromosomes, which are well suited for LM, but are difficult for SEM due to artificial layers (Sumner, 1996; Wanner *et al.*, 2005; Wanner and Schroeder-Reiter, 2008) and for studies of different stages of condensation. The development of the "drop/cryo technique," an efficient chromosome isolation method from plant root tips (Martin *et al.*, 1994), opened the door for large-scale routine structural investigations.

9.4.1 Dynamic Matrix Model for chromosome condensation

The drop/cryo technique in combination with development of analytical applications established the foundation for a simple, structurally based chromosome model that is able to accommodate contemporary data and pre-existing evidence as well as explain some apparent contradictions. The "Dynamic Matrix Model" explains chromosome condensation and higher order chromatin states (Figure 9.19; Wanner and Formanek, 2000). From prophase to prometaphase, solenoids (30 nm) form chromomeres which are interspersed with parallel matrix fibers (Figure 9.19). The number of chromomeres increases with further condensation. Constrictions first occur at prometaphase. At maximum condensation, chromomeres are tightly compacted and parallel fibers are exposed at the centromere. The motor for chromosome condensation is postulated to be an antiparallel movement of the matrix fibers, an energy-dependent process which is limited sterically in binding sites, and which is counteracted by elastic tension (potential energy) in the condensed chromomeres. Under normal conditions these two forces reach an equilibrium, which can be shifted depending on preparative treatment (such as disruption of spindle apparatus formation during arrest). In telophase, constrictions disappear and decondensation starts at the telomeres by successive separation and loosening of chromomeres. In physical terms, decondensation of

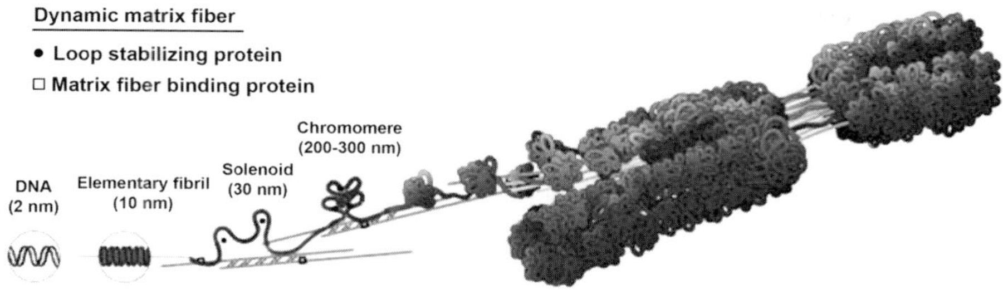

Figure 9.19 Dynamic Matrix Model for chromosome condensation according to Wanner and Formanek (2000). DNA assembles with histone octamers to form nucleosomes and elementary fibrils which wind up to solenoids. Solenoids attach to matrix fibers by matrix fiber binding proteins. Dynamic matrix fibers move in an antiparallel fashion, possibly assisted by linker proteins. As condensation progresses, chromomeres are formed by loop stabilizing proteins, and the chromosomes become shorter and thicker as more chromomeres are formed. At late stages of condensation the centromere becomes visible as a constriction with few chromomeres and exposed parallel fibrils. Condensation causes a tension vertical to the axial direction which forces the chromatids apart. (Reprinted from Wanner *et al.* 2005, with permission from S. Karger AG, Basel.)

chromosomes may be described as a release of potential energy in the chromomeres and matrix fibers. To date, the Dynamic Matrix Model remains a viable and comprehensive explanation for chromatin condensation, taking into account structural evidence along with implicit chromatin accessibility and functional plasticity (accommodating transcription, replication, segregation) of chromosomes during the cell cycle.

Many tenets of the model have been upheld for the past decade of routine structural and analytical investigations: i) chromosomes condense in a linear, non-helical, fashion, ii) the dominant chromosome substructures are solenoid fibrils, which form intermediate structures up to chromomeres, iii) chromomeres exhibit high DNA content, iv) centromeres are structured differently compared to chromosome arms, v) in metaphase parallel fibrils are typical for centromere and satellite regions. However, analysis, by definition (*ana lyse* = *Gr* to take apart), can have a severe impact on structural preservation, and therefore analytical techniques often represent a compromise based on the goals of a particular investigation. It is therefore a justifiable question, whether substructures observed in isolated chromosomes may be altered by preparation and/or fixation.

9.4.2 Preparation/isolation

In principle, fibrillar and chromomeric substructures are observed with all fixations/preparation techniques applied, and are supported by numerous publications in the field of high-resolution SEM (Allen *et al.*, 1986; Sumner, 1991; Wanner *et al.*, 2005; Schroeder-Reiter *et al.*, 2006; Wanner and Schroeder-Reiter, 2008). These substructures are found with both formaldehyde and 3:1 (ethanol:acetic acid) fixations in all organisms studied (plants, insects, birds, mammals). Proving the dynamic mode of condensation, identifying protein matrix fibrils, as well as the structures involved in spindle attachment and chromosome separation remain elusive due to limitations in preparation techniques. The fundamental problem is distinguishing chromosomes from their surroundings, which is why isolation techniques have been preferred. Mitotic chromosomes are surrounded by compact nucleoplasm (approximately the viscosity of a 20% protein solution), they are themselves compact and consist possibly *in vivo* of a two-phase system of a chromatin network and a sol-like phase that represents either the nucleoplasm imbibing the interchromatin space or a specific interior component. This is supported by FIB data showing cavities in isolated chromosomes (Figure 9.16) and, in addition, by low-contrast interior areas in chromosomes *in situ* in root tip cells seen both in TEM and in FIB milled blockfaces (Figure 9.17). Obviously, successful isolation of chromosomes excludes preservation of all components, but is compensated by visualization of individual substructures like chromatin fibrils. Preservation and visualization of the 3-D structure of chromosomes *in situ* in mitotic cells, in addition to kinetochore structures, microtubuli, and the spindle apparatus, still remains a challenge.

9.4.3 Labeling of DNA and proteins

Contrasting DNA with Pt blue in chromosomes is beneficial for investigation of 3-D DNA distribution at higher voltages. With very small chromosomes and decondensed chromatin to solenoid levels, detection of Pt becomes critical: accelerating voltage must be lowered, and consequently the thickness of C-coating must be reduced, which drastically decreases the BSE-signal. The determination of signal specificity from Pt-labeled structures becomes increasingly difficult. An adequate alternative, i.e. high-resolution labeling of proteins as a substance class, is not currently available; at present there is still no highly sensitive method for protein staining of small structures.

Specific labeling with immuno-Nanogold® has been successfully applied for investigation of 3-D distribution at moderate magnifications. Enhanced NanoGold labels can be localized to chromatin structures or areas in the range 15–30 nm. Use of 3–5 nm NanoGold would be desirable for direct detection of signals and higher local resolution.

9.4.4 Resolution

For many years, the resolution limit of biological specimens was primarily determined by the resolution power of the SEM. With improvement of all instrumental components, the limiting resolution factor has shifted to the quality of specimen preparation. With the best instruments, resolution of ±1 nm at *c.* 1 kV is achievable. For chromosome specimens, visualizing substructures to 10 nm currently can be achieved for routine investigation; in order to visualize structures down to 5 nm the difficulties increase exponentially. Beam damage and contamination are the main, and immediately obvious, obstacles because they alter topographic details.

Aside from alterations resulting from microscopy, it is unclear whether the applied preparation is adequate for preservation of the fine structures expected. For example, one would expect that the 30 nm fibril would show substructures with a 10 nm periodicity resulting from helical winding of elementary fibrils, but the fibrils appear smooth. One could argue that the substructures are concealed even by the thinnest possible sputter coating (1 nm), a well-known problem, but even fibrils from uncoated specimens (mounted on conductive slides) appear smooth. The following possibilities could be considered: i) the groove between elementary fibrils may be too small to give enough topographic contrast for detection; ii) compaction of chromatin could occur during critical point drying; iii) during chromosome preparation putative grooves are concealed, possibly by unspecific absorption of proteins; iv) contamination may occur during transfer between critical point drying and the microscope chamber.

9.4.5 Detection of labels from the interior: possibilities with FIB/FESEM

Application of FIB/FESEM techniques for tomography of chromosomes is a relatively new approach, and is promising for new insights into 3-D chromosome architecture. FIB tomography of isolated chromosomes has the advantage of directly visualizing the uncoated milled surface, enabling investigation of interior structures and 3-D distribution of different kinds of label. However, the exposed organic material is very sensitive to beam damage from both the ion beam and, to a lesser extent, the electron beam. In addition, there is some redeposition, which has little effect on BSE signals, but influences topographic information (e.g. milled material can fill small cavities). Since redeposition increases with ion beam current, it is important to apply low currents (10–50 pA), which increases milling rate and, consequently, total milling time.

Efforts are currently aimed toward high-resolution 3-D reconstruction of chromosomes *in situ* from large-scale FIB tomography data from resin-embedded mitotic cells. Embedded specimens are very stable compared to isolated specimens, show much less redeposition, and have the advantage of increased resolution of stained structures if the contrast is high enough. Very small milling steps (between 5 and 8 nm) can be routinely achieved in series up to several hundred "sections." However, embedding cells in epoxy resins reintroduces the problem of differentiating between chromatin and surrounding nucleoplasm in mitotic cells, emphasizing the importance of finding efficient ways of specific contrasting of DNA and protein as substance classes.

Efficient *in situ* hybridization of DNA sequences or immunolabeling of specific proteins in embedded specimens is at present very restricted due to limited accessibility of markers in well-fixed cells, but will become more feasible as markers become smaller. Small markers (Fab' fragments, nanobodies; Rothbauer and Leonhardt, 2008) with covalently bound 3–10 nm (Fluoro)Nanogold particles, that are large enough to be detected at relevant magnifications without additional metalloenhancement, would be desirable. This would allow combinations of high-resolution techniques, both LM and SEM, to refine present models for tertiary chromatin structures and chromosome architecture.

9.5 Acknowledgments

The authors express deep gratitude to Sabine Steiner for excellent chromosome preparations starting from pioneering work up to sophisticated state-of-the-art techniques. Our thanks are extended to Silvia Dobler, Cornelia Niemann, Jennifer Grünert, and Martina Reymers for excellent technical assistance. Graduate work by Ursula Wengenroth (human chromosomes) and Stephanie Fritsch (Tanaka fracture), and collaboration with Professor Dr. Alberto Ferrús of the Cajal Institute in Madrid, Spain (*Drosophila melanogaster* larvae), have been greatly appreciated. This work was supported by the Deutsche Forschungsgemeinschaft and the LMU Mentoring program.

9.6 References

Aggarwal, S., Wagner, R., McAllister, P., and Rosenberg, P. (1975). Cell-surface-associated nucleic acid in tomorigenic cells made visible with platinum-pyrimidine complexes by electron microscopy. *Proceedings of the National Academy of Sciences*, **72**, 928–932.

Aggarwal, S., Ofosu, G., and Waku, Y. (1977). Cytotoxic effects of platinum-pyrimidine complexes on the mammalian cell *in vitro*: a fine structural study. *Journal Clin. Hematol. Oncol.*, **75**, 547–561.

Allen, T. D., Jack, E. M., Harrison, C. J., and Claugher, D. (1986). Scanning electron microscopy of human metaphase chromosomes. *Scanning Electron Microscopy*, **1**, 301–308.

Allen, E. M. and Harrison, C. J. (1988). *The three dimensional structure of human metaphase chromosomes determined by scanning electron microscopy.* Boca Raton, FL, CRC Press, Inc., pp. 51–72.

Arents, G., Burlingame, R. W., Wang, B. C., Love, W. E., and Moudrianakis, E. N. (1991). The nucleosomal core histone octamer at 3.1Å resolution: A tripartite protein assembly and a left-handed superhelix. *Biochemistry*, **88**, 10148–10152.

Boveri, T. (1902). Über mehrpolige Mitosen als Mittel zur Analyse des Zellkerns. *Verhandlungen der physikalisch-medizinischen Gesellschaft zu Würzburg*, **35**, 67–90.

Dupraw, E. J. (1965). Macromolecular organization of nuclei and chromosomes: A folded fibre model based on whole-mount electron microscopy. *Nature*, **206**, 338–343.

Earnshaw, W. C. (1988). Mitotic chromosome structure. *Bioessays*, 147–150.

Finch, J. T. and Klug, A. (1976). Solenoidal model for superstructure in chromatin. *Proceedings of the National Academy of Sciences USA*, **73**, 1897–1901.

Frommer, G., Schöllhorn, H., Thewalt, U., and Lippert, B. (1990). Platinum(II) binding to N7 and N1 of guanine and a model for a purine-N^1, pyrimidine-N^3 cross-link of cis-platin in the interior of a DNA duplex. *Inorg. Chem.*, **29**, 1417–1422.

Hadlaczky, G., Sumner, A. T., and Ross, A. (1981a). Protein-depleted chromosomes I: Structure of isolated protein-depleted chromosomes. *Chromosoma*, **81**, 537–555.

Hadlaczky, G., Sumner, A. T., and Ross, A. (1981b). Protein-depleted chromosomes II: Experiments concerning the reality of chromosome scaffolds. *Chromosoma*, **81**, 537–555.

Hadlaczky, G., Praznovszky, T., and Bisztray, G. (1982). Structure of isolated protein-depleted chromosomes of plants. *Chromosoma*, **86**, 643–659.

Hainfeld, J. F., Powell, R. D., Stein, J. K., and Hacker, G. W. (1999). Gold-based autometallography. Proceedings of the 57th annual meeting, Microscopy Society of America. New York, Springer, pp. 486–487.

Hainfeld, J. F. and Powell, R. D. (2000). New frontiers in gold labeling. *J. Histochem. Cytochem.*, **48**, 471–480.

Harrison, C. J., Allen, T. D., Britch, M., and Harris, R. (1982). High-resolution scanning electron microscopy of human metaphase chromosomes. *J. Cell Sci.*, **56**, 409–422.

Harrison, C. J., Jack, E. M., and Allen, T. D. (1987). Light and scanning electron microscopy of the same metaphase chromosomes. In: *Correlative Microscopy in Biology Instrumentation and Methods*. Orlando, FL, Academic Press, Inc.

Heneen, W. K. (1980). Scanning electron microscopy of the intact mitotic apparatus. *Biol. Cellulaire*, **37**, 13–22.

Heneen, W. K. (1981). Mitosis as discerned in the scanning electron microscope. *European Journal of Cell Biology*, **25**, 242–247.

Heneen, W. K. (1982). The centromeric region in the scanning electron microscope. *Hereditas*, **97**, 311–314.

Houben, A., Schroeder-Reiter, E., Nagaki, E., *et al.* (2007). CENH3 interacts with the centromeric retrotransposon *cereba* and GC-rich satellites and locates to centromeric substructures in barley. *Chromosoma*, **116** (*3*), 275–283.

Köpf-Maier, P. and Köpf, H. (1986). Cystostatische Platinkomplexe: Eine unerwartete Entdeckung mit weitreichenden Konsequenzen. *Naturwissenschaften*, **73**, 239–247.

Lippert, B. and Beck, W. (1983). Platinkomplexe in der Krebstherapie. *Chemie u Z.*, **17**, 190–197.

Manuelidis. L. and Chen, T. L. (1990). A unified model of eukaryotic chromosomes. *Cytometry*, **11**, 8–25.

Marsden, M. P. F. and Laemmli, U. K. (1979). Metaphase chromosome structure: evidence for a radial loop model. *Cell*, **17**, 849–858.

Martin, R., Busch, W., Herrmann, R. G., and Wanner, G. (1994). Efficient preparation of plant chromosomes for high-resolution scanning electron microscopy. *Chrom. Research*, **2**, 411–415.

Martin, R., Busch, W., Herrmann, R. G., and Wanner, G. (1996). Changes in chromosomal ultrastructure during the cell cycle. *Chrom. Research*, **4**, 288–294.

Metcalfe, C. J., Bulazel, K. V., Ferreri, G. C., *et al.* (2007). Genomic instability within centromeres of interspecific marsupial hybrids. *Genetics*, **177**, 2507–2517.

Nagaki, K., Cheng, Z., Ouyang, S., *et al.* (2004). Sequencing of a rice centromere uncovers active genes. *Nat. Geet.*, **36**, 138–145.

Paulson, J. R. and Laemmli, U. K. (1977). The structure of histone-depleted metaphase chromosomes. *Cell*, **12**, 817–828.

Powell, R. D., Halsey, C. M., and Hainfeld, J. F. (1988). Combined fluorescent and gold immunoprobes; Reagents and methods for correlative light and electron microscopy. *Microsc. Res. Tech.*, **42** (*1*), 2–12.

Rattner, J. B. and Lin, C. C. (1985). Radial loops and helical coils coexist in metaphase chromosomes. *Cell*, **42**, 291–296.

Ris, H. and Witt, P. L. (1981). Structure of the mammalian kinetochore. *Chromosoma*, **82**, 153–170.

Rothbauer, U. and Leonhardt, H. (2008). Connecting biochemistry and cell biology with nanobodies. *Zellbiologie aktuell*, **34**, 9–12.

Schaper, A., Rößle, M., Formanek, H., Jovin, T. M., and Wanner, G. (2000). Complementary visualization of mitotic barley chromatin by field-emission scanning electron microscopy and scanning force microscopy. *Journal of Structural Biology*, **129**, 17–29.

Schroeder-Reiter, E., Houben, A., and Wanner, G. (2003). Immunogold labeling of chromosomes for scanning electron microscopy: A closer look at phosphorylated histone H3 in mitotic metaphase chromosomes of *Hordeum vulgare*. *Chrom. Res.*, **11**, 585–596.

Schroeder-Reiter, E., Houben, A., Grau, J., and Wanner, G. (2006). Characterization of a peg-like terminal NOR structure with light microscopy and high-resolution scanning electron microscopy. *Chromosoma*, **115**, 50–59.

Schroeder-Reiter, E. and Wanner, G. (2009). Chromosome centromeres: structural and analytical investigations with high resolution scanning electron microscopy in combination with focused ion beam milling. *Cytogenet Genome Research*, **124**, 239–250.

Schroeder-Reiter, E., Pérez-Willard, F., Zeile, U., and Wanner, G. (2009). Focused ion beam (FIB) combined with high resolution scanning electron microscopy: A promising tool for 3D analysis of chromosome architecture. *Journal Structural Biology*, **165** (*2*), 97–106.

Schubert, I., Dolezel, J., Houben, A., Scherthan, H., and Wanner, G. (1993). Refined examination of plant metaphase chromosome structure at different levels made feasible by new isolation methods. *Chromosoma*, **102**, 96–101.

Strasburger, E. (1988). *Über Kern- und Zelltheilung im Pflanzenbereich, nebst einem Anhang über Befruchtung*. Jena, Gustav Fischer.

Sumner, A. T. and Ross, A. (1989). Factors affecting preparation of chromosomes for scanning electron microscopy using osmium impregnation. *Scanning Microscopy*, **3**, 87–99.

Sumner, A. T. (1991). Scanning electron microscopy of mammalian chromosomes from prophase to telophase. *Chromosoma*, **100**, 410–418.

Sumner, A. T. (1996). Problems in preparation of chromosomes for scanning electron microscopy to reveal morphology and to permit immunocytochemistry of sensitive antigens. *Scanning Microscopy Suppl*, **10**, 165–176.

Sutton, W. S. (1902). The chromosomes in heredity. *Biol. Bulletin*, **4**, 231–251.

Talbert, P. B., Masuelli, R., Tyagi, A. P., Comai, L., and Henikoff, S. (2002). Centromeric localization and adaptive evolution of an Arabidopsis histone H3 variant. *Plant Cell*, **14**, 1053–1066.

Tanaka, K. (1980). Scanning electron microscopy of intracellular structures. *International Review of Cytology*, **68**, 97–125.

Thoma, F., Koller, T., and Klug, A. (1979). Involvement of histone H1 in the organization of the nucleosome and the salt-dependent superstructures of chromatin. *Journal of Cell Bio*logy, **83**, 402–427.

Wanner, G. and Formanek, H. (1995). Imaging of DNA in human and plant chromosomes by high-resolution scanning electron microscopy. *Chrom. Research*, **3**, 368–374.

Wanner, G. and Formanek, H. (2000). A new chromosome model. *Journal Struct. Biology*, **132**, 147–161.

Wanner, G., Schroeder-Reiter, E., and Formanek, H. (2005). 3D analysis of chromosome architecture: advantages and limitations with SEM. *Cytogenet Genome Research*, **109**, 70–78.

Wanner, G. and Schroeder-Reiter, E. (2008). Scanning electron microscopy of chromosomes. *Methods in Cell Biology*, **88**, 451–474.

Watson, J. D. and Crick, F. H. C. (1953). A structure for deoxyribose nucleic acid. *Nature*, **171**, 737–738.

Yunis, J. J., Sawyer, J. R., and Ball, D. W. (1978). The characterization of high-resolution G-banded chromosomes of man. *Chromosoma*, **67**, 293–307.

Zoller, J. F., Herrmann, R. G., and Wanner, G. (2004). Chromosome condensation in mitosis and meiosis of rye (*Secale cereale* L.). *Cytogenet Genome Res*earch, **105**, 134–144.

10 A method to visualize the microarchitecture of glycoprotein matrices with scanning electron microscopy

Giuseppe Familiari, Rosemarie Heyn, Luciano Petruzziello, and Michela Relucenti

10.1 Introduction

10.1.1 The extracellular matrix

The extracellular matrix (ECM) is an intricate network of macromolecules that binds cells together and affects several cell processes: development, proliferation, differentiation, migration, survival, behavior, shape, and polarity (Alberts *et al.*, 2002; Hynes, 2009). This matrix is composed of a variety of proteins and polysaccharides that are locally secreted and assembled into an organized meshwork (Alberts *et al.*, 2002).

The ECM is a dynamic structure that interacts with cells and generates signals through feedback loops (Järveläinen *et al.*, 2009). Traditionally this matrix has been thought to provide a support function, providing strength to cells within tissues. Recent literature about the ECM focuses on its role in inducing pleiotropic effects during development and growth (Tsang *et al.*, 2010).

Specific genetic alterations of the ECM components cause phenotypic and developmental alterations. The ECM interacts with signaling molecules and morphogens, modulating their activities, thereby modifying the regulation of key processes in morphogenesis and organogenesis (Tsang *et al.*, 2010). Specific structural features of ECMs are recognized from cell surface receptors, thus ECMs represent overall physical properties of the acellular environment (Bruckner, 2010).

The ECM structures are subject to hierarchic organizations; its isolated macromolecules accomplish only few specialized tasks. ECM molecular components attain their prominent functions only after polymerization into insoluble suprastructural elements (fibrils, microfibrils, or networks) that, in turn, are assembled into regional tissue structures, such as fibers or basement membranes. ECM suprastructures resemble composite biological amalgamates, so that structural and functional characteristics of each ECM composite are different (Bruckner, 2010).

The ECM components are produced by resident cells and secreted by exocytosis into the extracellular environment where they aggregate with the existing matrix

Scanning Electron Microscopy for the Life Sciences, edited by H. Schatten. Published by Cambridge

(Sorokin, 2010). Two main classes of extracellular macromolecules are recognized: polysaccharide chains called glycosaminoglycans (GAGs), which are usually found covalently linked to protein forming proteoglycans, and fibrous proteins, such as collagen, elastin, fibronectin, and laminin, with both structural and adhesive functions (Varki *et al.*, 1999). The ECM is highly insoluble and voluminous and consists of proteins with independent structural glycosylated domains, whose sequences and arrangements are highly conserved. The conserved domain structure of ECM molecules may reflect a conserved structure–function relationship. The domains frequently contain glycosaminoglycan sulfated chains, which give a total negative charge (Sorokin, 2010).

Glycosaminoglycans are the characteristic long unbranched polysaccharide chains of proteoglycans (hyaluronic acid is a notable exception) (Varki *et al.*, 1999). GAGs consist of a repeating disaccharide unit (hexose or hexuronic acid, linked to hexosamine) and they are negatively charged under physiological conditions, due to the occurrence of sulfate and uronic acid groups. The negative charge attracts positively charged sodium ions (Na^+), which carry on water molecules, thus creating a water-rich gel in the extracellular space. Proteoglycans help to trap and store growth factors within the ECM. Proteoglycans are also found on the surface of cells, where they function as co-receptors to help cells respond to secreted signal proteins.

Hyaluronic acid (also called hyaluronan or hyaluronate) is an anionic, nonsulfated glycosaminoglycan, which is not found as a proteoglycan. Hyaluronic acid consists of alternative residues of D-glucuronic acid and *N*-acetylglucosamine, and in the extracellular space it confers the ability to resist compression by absorbing significant amounts of water. It is also a chief component of the interstitial gel (Kogan *et al.*, 2007).

10.1.2 Two special kinds of extracellular matrix

10.1.2.1 Zona pellucida

Mammalian zona pellucida (ZP) is the peculiar extracellular matrix that surrounds the oocyte during folliculogenesis up to blastocyst hatching. The ZP is often defined as an "extracellular coat," although its functions are not only those of a strictly passive envelope. In fact, ZP plays an important role in fertilization because species-specific recognition between gametes is the first nodal point in this process (Monné *et al.*, 2008).

ZP is an agonist of the acrosomal reaction, playing an important role in preventing polyspermy (Wassarman and Litscher, 2008). The human ZP matrix is formed by four glycoproteins called ZP1, ZP2, ZP3, and ZP4 (Hughes and Barratt, 1999; Lefievre *et al.*, 2004; Ganguly *et al.*, 2008), of which ZP3, ZP4, and also ZP1 bind to capacitated sperm, inducing acrosomal reaction (Ganguly *et al.*, 2010). Recently, Bhandari *et al.* (2010) have demonstrated that the human zona pellucida-mediated acrosome reaction induces a downstream signaling. The mouse ZP is made up of three species-specific glycoproteins: mZP1, mZP2, mZP3 arranged in a delicate filamentous matrix (Wassarman, 2008; Litscher *et al.*, 2009). The orthologue of the human Zp4 gene is present in the mouse genome as a pseudo gene (Ganguly *et al.*, 2010). Polymers of mZP2 and mZP3 form long interconnected filaments, which compose the ZP (Greve and Wassarman, 1985). An mZP2–mZP3 dimer is located every 14 nm or so along the filaments, and its presence causes the structural periodicity visible in dissolved ZP. mZP1 cross-links the filaments, thus creating a 3-D scaffold in the matrix (Greve and Wassarman, 1985). Therefore, each of the ZP glycoproteins plays a structural role during ZP assembly.

Mechanical properties of ZP change between the immature, mature, and fertilized conditions. The "zona hardening" process allows a recovery of elasticity (as it is in the immature ZP), and this phenomenon causes an increased resistance to proteolytic digestion by an actual stiffening of its structure (Papi *et al.*, 2010).

10.1.2.2 Mucus

Mucus is a dynamic, semipermeable barrier, which enables exchanging of nutrients, water, gases, odorants, hormones, and gametes, but at the same time it does not allow the passage of most bacteria and many pathogens (Cone, 2009). The major gel-forming glycoprotein components called mucins supply the viscoelastic, polymer-like properties of the mucus (Forstner *et al.*, 1995). Mucins consist of a peptide backbone containing alternating glycosylated and nonglycosylated domains, with *O*-linked glycosylated regions comprising 70–80% of the polymer. The four primary mucin oligosaccharides are *N*-acetylglucosamine, *N*-acetylgalactosamine, fucose, and galactose (Forstner *et al.*, 1995). Mucin oligosaccharide chains are often terminated with sialic acid or sulfate groups, which account for the polyanionic nature of mucins at a neutral pH (Forstner *et al.*, 1995).

At the chemical level, mucus can be described as an integrated structure of biopolymers with a complex (non-Newtonian) physical behavior. It owns properties between those of a viscous liquid and an elastic solid. In order to describe the consistency of mucus, rheological measurements are often used, including viscosity (resistance to flow) and elasticity (stiffness) (Lai *et al.*, 2009). Changes in mucus rheological properties affect its ability to function as a lubricant, a selective barrier, and the body's first line of defense against infection (Slomiany and Slomiany, 1991; Girod *et al.*, 1992; Randell and Boucher, 2006). In the human small intestine, the thickness of the mucus blanket varies greatly depending on digestive activity, being thicker in the stomach (180 μm; range 50–450 μm) than in the colon (110–160 μm) (Copeman *et al.*, 1994; Kerss *et al.*, 1992; Sandzen *et al.*, 1988).

10.2 Cationic dye ruthenium red

Ruthenium red (ammoniated ruthenium oxychloride) is an inorganic, synthetically prepared and intensely colored, crystalline compound. Ruthenium, atomic number 44, is moderately heavy. This reagent was successfully used by Luft (Luft, 1971a, b) as an electron stain in animal histology to demonstrate the presence of glycosaminoglycans and it is still widely used when surface mucin-like glycoproteins have to be detected (Ferreira *et al.*, 2008; Chatterjee *et al.*, 2010). Glycosaminoglycans stained with ruthenium red during fixation are resistant to elastase digestion, and persist as a distinct meshwork in the regions of digested aggregates (Kádár *et al.*, 1972). Ruthenium red has strong oxidant properties, acting not only as a stain, but also as a fixative. Applying this dye during fixation, it preserves and sharply defines membranes and myofilaments without post-staining. Ruthenium red binds to its substrates electrostatically (salt linkage); the active group in ruthenium red is the ruthenium ion and its associated four ammonia molecules. The staining is considered to be accomplished when the host molecule has two negative charges 0.42 nm apart to accommodate the staining group (Sterling, 1970). Use of ruthenium red together with OsO_4 produces RO_4, which reacts with some of the more polar lipids, proteins, glycogen, and common oligosaccharides. The use of OsO_4 together with ruthenium red permits *en bloc* staining and allows achieving a higher contrast. The stain can be added to water or buffered solutions of the fixative. It has been noted that phosphate buffers are not recommended for ruthenium red-OsO_4 procedures; chloride-free cacodylate buffer is considered the most suitable (Hayat, 1989).

10.3 Saponin

Saponins are complex compounds that are formed by a steroid attached to a carbohydrate moiety (triterpenoid glycosides). They are natural surfactants and detergents and have shown membrane-permeabilizing and other biological effects (Williams and Gong, 2007; Desai *et al.*, 2009). Saponin is a weak detergent that forms pores in the plasma membrane by preferentially extracting cholesterol, leaving intact much of the plasma membrane phospholipid bilayer while allowing soluble proteins to diffuse away from the cell (Penman, 1995). Pretreatment with saponin, or other detergents, may be used in immunohistochemical procedures, even if some artifacts may occur when lipid-linked molecules are studied (Heffer-Lauc *et al.*, 2007).

Use of saponin permits study of cellular cytoskeletons without destroying the detailed morphology of cell structures (Penman, 1995). The mechanism of detergent extraction is often misunderstood (Penman, 1995). In fact, nonionic detergents, in contrast to ionic types, do not affect proteins so they do not extract in the usual sense of this term, allowing soluble proteins to diffuse away passively after destroying the plasma membrane barrier. Use of saponin combined with ruthenium red probably helps in stabilization of glycoproteins and protoglycans and favors the diffusion of soluble proteins, revealing the 3-D filamentous nature of the extracellular matrix.

10.4 Saponin and ruthenium red preserve the extracellular matrix

It is very difficult to preserve the 3-D structure of the extracellular matrix, including glycoprotein matrices such as the ZP or the mucus, during procedures for electron microscopy, because they are extremely hydrated. Solutions of ethanol or glutaraldehyde used for electron microscopy may cause shrinkage of the glycoprotein structure and may produce artifacts, thus explaining why only fragments of the mucus layer are observed on fixed sections (Allen and Leonard, 1985). Moreover, freezing and drying procedures may stress the filaments, causing their alignment in the direction of shear or flow, influencing in that way the organization of molecules (Sturgess, 1988).

For all the above reasons, development of a high-resolution scanning electron microscopy (SEM) technique in combination with saponin, ruthenium red (RR), and osmium-thiocarbohydrazide impregnation is required really to assess the 3-D structure that glycoprotein filaments assume in the matrices. Therefore, an analysis of the 3-D microarchitecture of the ZP and the intestinal (jejunum) mucus is herein proposed.

10.5 Conventional SEM technique

In the standard method (Familiari *et al.*, 1992a; Familiari *et al.*, 2006), specimens were fixed in a 3.0% glutaraldehyde solution in 0.1 M cacodylate buffer at pH 7.4 for two to five days. The specimens were carefully washed twice, each for 10 min, without stirring, in 0.1 M cacodylate buffer at pH 7.4. Post-fixation was performed with a solution of 1.0% osmium tetroxide in 0.1 M cacodylate buffer at pH 7.4. Samples were rinsed gently in distilled water twice, 10 min each, to remove excess osmium tetroxide, dehydrated in a series of ascending concentrations of acetone or ethanol (30%–50%–70%–95%–100%–100%–100%, 20 min each) and critical point-dried in a CPD020 Balzers device (Balzers, Liechtenstein). Dehydrated specimens were mounted on aluminum stubs with silver paint and sputtered with platinum at 10–15 mA for 1 min (obtaining a 3 nm thickness film) in a sputter coater device (EMITECH K550, EMITECH Ltd, Ashford, Kent, England). Observations were made with a Hitachi S-4000 field-emission scanning electron microscope operating at 5–10 kV using secondary electrons and with a working distance of 5–9 mm.

10.6 High-resolution SEM technique for glycoproteins

The authors have developed and applied the following eight-step schedule (Familiari *et al.*, 1992a; Familiari *et al.*, 2006):

1) The extraction-stabilization procedure was made with 0.02% saponin and 1.0% RR in cacodylate buffer 0.1 M at pH 7.4 for 15–45 min;
2) Fixation was obtained by immersion in 3.0% glutaraldehyde plus 0.02% saponin and 1.0% RR in cacodylate buffer overnight;
3) Four washes, 25 min each, using a solution of 0.02% saponin and 1.0% RR in cacodylate buffer 0.1 M at pH 7.4 for 1 hr;
4) Post-fixation was performed with a solution of 1.0% osmium tetroxide containing 0.02% saponin and 0.75% RR in cacodylate buffer for 2 hr;
5) Impregnation with two solutions, 1% thiocarbohydrazide in distilled water (T), 1.0% osmium tetroxide (O) in distilled water. The solutions were used alternately in four steps (T/O/T/O for 20, 60, 20, and 60 min), modified from Kelley *et al.* (1973);
6) Samples were washed four times 10 min each in distilled water and then dehydrated in a series of ascending concentrations of acetone or ethanol (30%–50%–70%–95%–100%–100%–100%, 20 min each);

7) The drying procedure was carried out with a critical point dryer device (CPD020, Balzers, Liechtenstein) in liquid carbon dioxide; samples were then mounted with silver paint on aluminum stubs.
8) Specimens were finally observed in a Hitachi S-4000 field emission scanning electron microscope operating at 5–10 kV using secondary electrons, with a working distance of 5–9 mm. Note that no sputter coating was performed.

10.7 Zona pellucida visualized by HR-SEM

After standard SEM preparation the outer surface of the ZP was usually characterized by a spongy appearance, due to the presence of numerous branches surrounding fenestrations (Figure 10.1). In control specimens, at high magnification, the surface of the branches as well as fenestrations of the outer ZP displayed a granular appearance (Figure 10.2).

Specimens treated with the RR-saponin-T/O/T/O method showed the outer zona surface consisting of fine filaments arranged in a regular alternation of tight and large meshed networks (Figure 10.3). These filaments were 22–28 nm in thickness and showed a globule-bearing structure (Figure 10.4). The tight meshed arrangement of filaments was observed in correspondence of branches whereas a large meshed network of filaments was seen within the fenestrations observed by conventional SEM.

10.8 Mucus covering the jejunum surface visualized by HR-SEM

After standard SEM preparation, jejunum samples evidenced large areas of the villous surface lacking the mucus layer (Figure 10.5). In the same specimens, at high magnification, the residual mucus appeared as small patches consisting of a granular material (Figure 10.6).

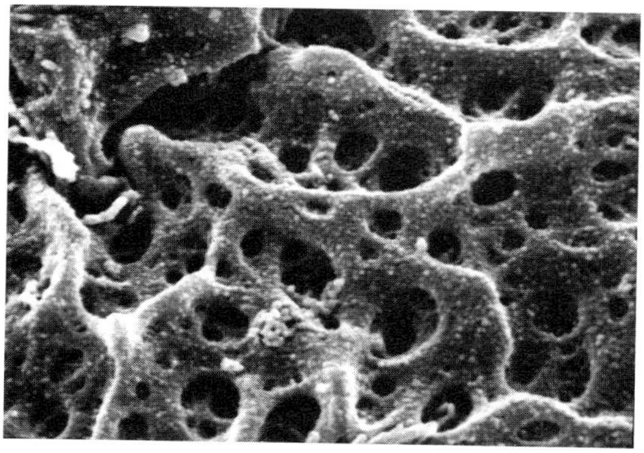

Figure 10.1 Human mature oocyte. Conventional SEM with platinum coating. Note the spongy appearance of the ZP (2000×).

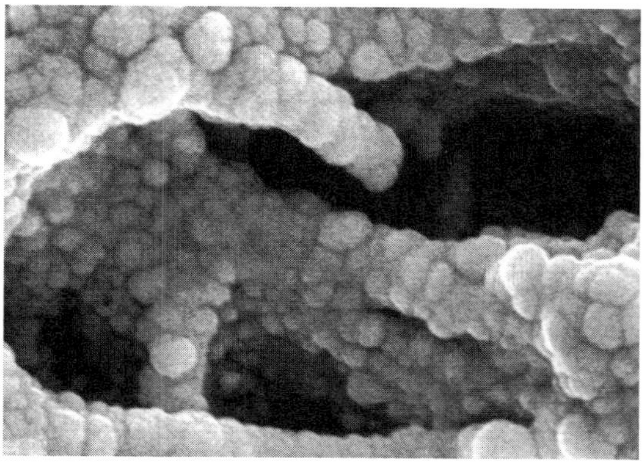

Figure 10.2 Conventional SEM with platinum coating. Human mature oocyte. Higher magnification of the ZP. Note the granular appearance of its outer aspect (40000×).

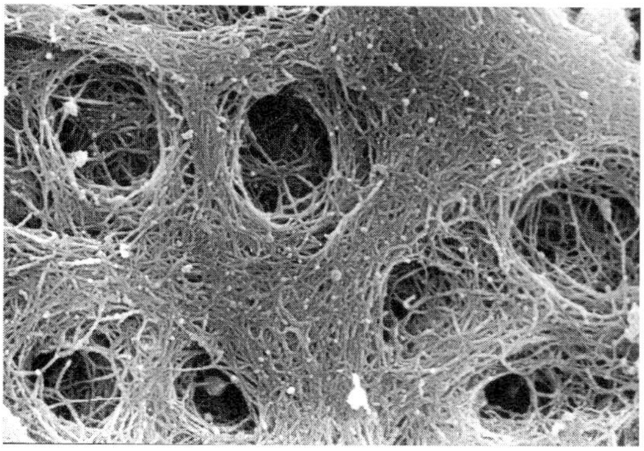

Figure 10.3 Human mature oocyte. SEM: Saponin, RR, T/O/T/O treatment. The outer ZP surface consists of fine filaments arranged in a regular alternation of tight and large meshed networks (10000×).

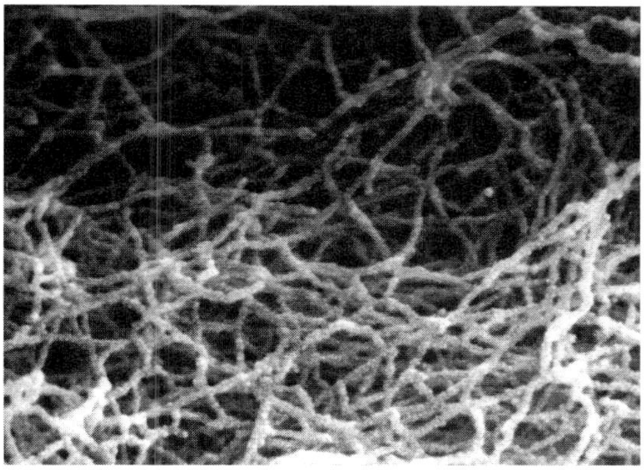

Figure 10.4 Human mature oocyte. SEM: Saponin, RR, T/O/T/O treatment. At higher magnification, note the presence of filaments arranged in complex networks at the outer aspect of the ZP (50000×).

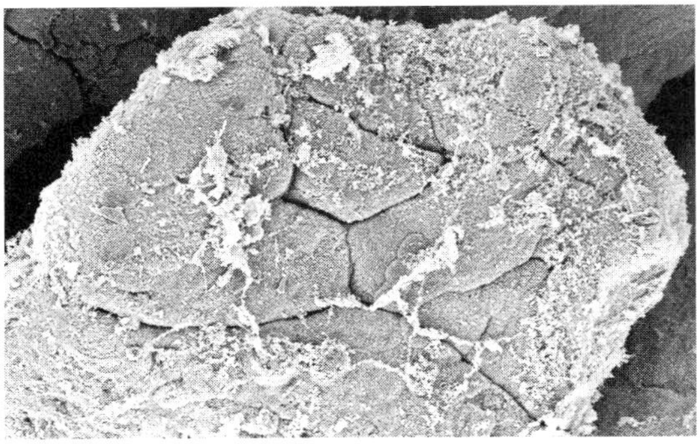

Figure 10.5 Mouse jejunum. Conventional SEM with platinum coating. The mucus forms patches of amorphous material on the villous surface (200×).

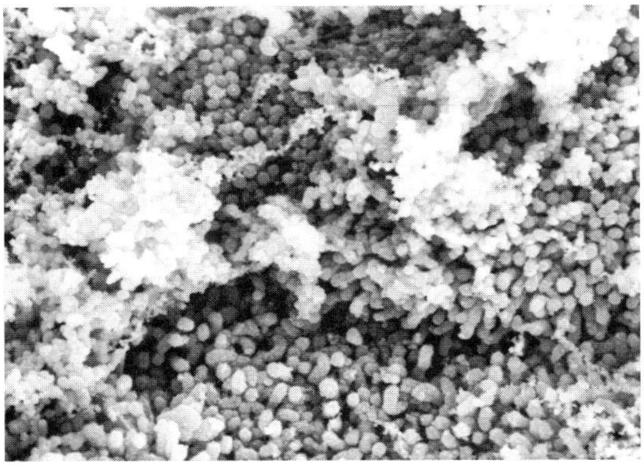

Figure 10.6 Mouse jejunum. Conventional SEM with platinum coating. At higher magnification, note the granular appearance of the residual mucus (9000×).

Specimens treated with the RR-saponin-T/O/T/O method evidenced the mucus forming a continuous layer covering almost the entire surface of the jejunal villi (Figure 10.7). When seen at high magnification, the mucus was characterized by filaments, measuring 30–50 nm in thickness, arranged in regular meshed networks (Figure 10.8).

10.9 Discussion and data analysis

Several ultrastructural methods are available to study the glycoprotein matrix with transmission (Greve and Wassarman, 1985) and scanning electron microscopy (Sturgess, 1988). All the techniques characterize the glycoprotein matrix as a filamentous structure. In particular, methods comprising rapid freeze-drying or freeze substitution (Sturgess, 1988) show fibrils ranging from 50 to 500 nm in diameter, forming interconnected layers; therefore, the relevant diameter of filaments observed using these methods was likely related to packing of thin filaments, not properly stabilized.

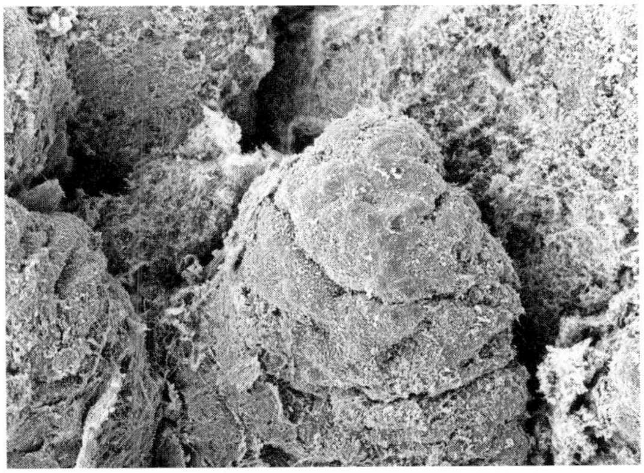

Figure 10.7 Mouse jejunum. SEM: Saponin, RR, T/O/T/O treatment. The mucus layer forms a continuous sheet covering the entire surface of the villi (200×).

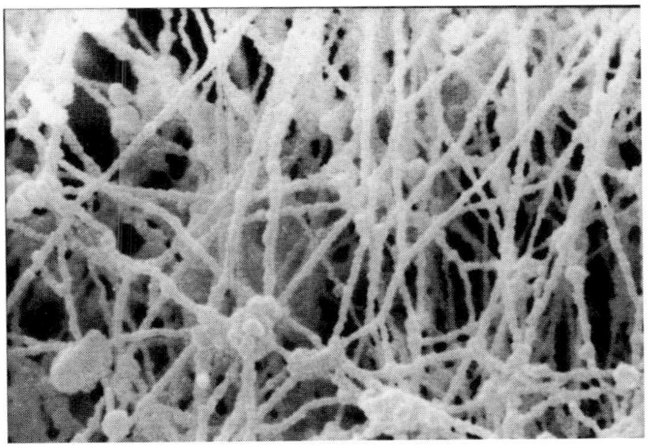

Figure 10.8 Mouse jejunum. SEM: Saponin, RR, T/O/T/O treatment. At higher magnification, the mucus displays filaments that are arranged in a regular meshwork, revealing a globule-bearing structure (15000×).

10.9.1 Mucus

The structural arrangement of mucin fibers is difficult to capture because of its dynamic features. Besides, the molecular structure of a mucus gel is difficult to observe, due to the high hydration state. Many attempts have been made with the aim of preparing mucus for electron microscopy, in order to produce minimal artifacts to the actual structural array of mucin fibers. Most fixation methods for fresh mucus for electron microscopy produce a random mesh of thick individual fibers about 30–100 nm in diameter (Chretien *et al.*, 1975; Poon and McCoshen, 1985; Yudin *et al.*, 1989), about 10-times thicker than the diameter of 3–10 nm determined biochemically. Smooth and curled fibers with a diameter of about 5–7 nm are visible when individual purified mucin fibers are placed onto a grid and then lightly shadowed with heavy metal (Sheehan *et al.*, 1986; Slayter *et al.*, 1991).

These features appear consistent with the ~15 nm persistent length deduced from the behavior of mucin fibers as random coils in solution (Shogren *et al.*, 1989). Images of mucus gels show thickened and straightened fibers, probably because of the tendency of antibodies, lysozyme, lactoferrin, albumin, present in the gel, to adsorb to the mucin fibers, thickening themselves during preparation of the gel for electron microscopy. In addition, the heavy metals used to enhance electron contrast also increase the apparent fiber diameter. The apparent thickness of mucin fibers can be at most double due to these two effects. Moreover, usual fixation methods and even fast freezing followed by freeze substitution may produce mucin fiber aggregation, forming a "cord" 3–10 times thicker than an individual fiber. Images obtained in fast-frozen preparations fixed by freeze substitution show a ribbon-like appearance (Yudin *et al.*, 1989).

Fixed mucin fibers probably thicken by condensing longitudinally with zigzag folds, and hydrophobic beads tend to cluster together with the interspersed glycosylated regions folding into hairpin loops.

The RR-saponin-T/O/T/O method used in this study offers several advantages in terms of stabilization and preservation of the structure, and, consequently, of the morphology of thin filaments.

10.9.2 Zona pellucida

Study of the ZP by means of classical SEM or TEM yields a microstructural ZP appearance, which is not consistent with the molecular model in which ZP2 and ZP3 form long filaments anastomosed by means of ZP1 (mice) and ZP1/ZP4 (human) proteins (Florman and Ducibella, 2006). In fact, using these methods, it is possible to visualize the surface of the ZP with small or large fenestrae or smooth appearance. Interpretation of the precise ZP structure seen with traditional SEM is difficult, because the ZP is much hydrated and thus shrinks considerably when dried, even if a critical-point drying apparatus is used.

A true detailed description of the filaments contained within the multilayered lattice of the ZP was obtained by high-resolution SEM using the RR-saponin-T/O/T/O method that offers several advantages in terms of ZP stabilization and preservation of the morphology of its delicate filaments (Familiari *et al.*, 1992a; Familiari *et al.*, 2006). Moreover, RR acts as a staining and a stabilizing agent of structural glycoproteins and polyanionic carbohydrates by preventing their dissolution and/or alteration induced by aqueous fixatives. Even if use of cationic dyes such as RR results in a molecular collapse of polyanionic chains causing these to appear as condensed granules (Hunziker and Schenk, 1984), addition of saponin avoids the formation of globular artifacts (Familiari *et al.*, 1992b). Osmium-thiocarbohydrazide treatment increases the molecular weight and yields a very fine structure under the electron beam. Besides, it hardens and preserves the glycoprotein matrix filaments from the mechanical stress induced by dehydration and critical point drying, and thus reduces filaments packing and shrinkage.

This method allowed observation of thin anastomosed, globule-bearing filaments, 0.1 to 0.4 mm in length and 22 to 28 nm thick as seen by SEM. The filaments observed by means of this particular technique probably corresponded to ZP2/ZP3 filaments, and the intersections corresponded to ZP1/ZP4 bounds. The filament arrangement was remarkably different between the inner and outer surfaces of the ZP and among the various maturation stages of the oocytes studied (Familiari *et al.*, 2008a). The outer surface of the mature oocyte mainly consisted of filaments arranged in a multilayered network that appears compact in the parts delimiting the fenestrations of the ZP, whereas it appears loose in correspondence with the fenestrations. The latter seemed empty when observed with conventional SEM. These were, in turn, made up of a loose filamentous arrangement when observed with the Sap-RR-Os-Tc method. In immature and atretic oocytes almost a tight meshed network of filaments was displayed. The inner surface of the ZP belonging to unfertilized oocytes at any stage was arranged in repetitive structures characterized by numerous short and straight filaments that anastomosed with each other, sometimes forming small, rounded structures at the intersections (Familiari *et al.*, 2008a, b).

10.10 Concluding remarks

The advantages of the RR-saponin-T/O/T/O method can be summarized as follows. RR stabilizes glycoproteins and polyanionic carbohydrates, avoiding their dissolution and/or alteration induced by aqueous fixatives. Saponin removes soluble proteins and avoids formation of globular artifacts due to the use of RR. Treatment with osmium-thiocarbohydrazide increases molecular weight, yielding a very fine structure under the electron beam, but also hardens and protects the glycoprotein matrix filaments from the mechanical stress induced by dehydration and critical point drying, thus decreasing filament shrinkage and packing.

The authors' technique introduces particular improvements that have made it possible to show that: 1) the filaments' network is arranged in a very regular way; 2) the thickness of mucus filaments is smaller and more precise than that obtained with other methods (Sturgess, 1988); and 3) the thickness of ZP filaments is homogeneous, according to transmission electron microscopy data obtained by dissolution of the ZP matrix and observation by means of negative staining (Greve and Wassarman, 1985).

In conclusion, performing the high-resolution SEM technique developed in the authors' laboratory for the study of glycoprotein matrices (ZP or mucus), a more detailed view of the structural organization of these special structures is obtained.

ECM molecular components attain their prominent role in many important biological functions only after polymerization into insoluble suprastructural elements (fibrils, microfibrils, or networks). These, in turn, are assembled into regional tissue structures (Bruckner, 2010), and, this being the case, this technique can be usefully applied to investigate further the morphodynamic changes of these and other similar structures in various physiological and pathological conditions.

10.11 Acknowledgments

Thanks are due to Ezio Battaglione, Gianfranco Franchitto, and Antonio Familiari for their technical contribution. Funds were provided by Italian MIUR (PRIN-COFIN and Sapienza University of Rome grants).

10.12 References

Alberts, B., Johnson, A., Lewis, J., *et al.* (2002). *Molecular Biology of the Cell*, 4th ed. New York, Garland Science Publ.

Allen, A. and Leonard, A. (1985). Mucus structure. *Gastroenterol. Clinical Biol.*, **9**, 9–12.

Bhandari, B., Bansal, P., Talwar, P., and Gupta, S. K. (2010). Delineation of downstream signaling components during acrosome reaction mediated by heat solubilized human zona pellucida. *Reprod. Biol. Endocrinology*, **8**, 7.

Bruckner, P. (2010). Suprastructures of extracellular matrices: paradigms of functions controlled by aggregates rather than molecules. *Cell and Tissue Research*, **339**, 7–18.

Chatterjee, A., Banerjee, S., Steffen, M., *et al.* (2010). Evidence for mucin-like glycoproteins that tether sporozoites of Cryptosporidium parvum to the inner surface of the oocyst wall. *Eukaryot Cell*, **9**, 84–96.

Chretien, F. C., Cohen, J., Borg, V., and Psychoyos, A. (1975). Human cervical mucus during the menstrual cycle and pregnancy in normal and pathological conditions. *Journal Reprod. Med.*, **14**, 192–196.

Cone, R. A. (2009). Barrier properties of mucus. *Adv. Drug Deliv. Rev.*, **61**, 75–85.

Copeman, M., Matuz, J., Leonard, *et al.* (1994). The gastroduodenal mucus barrier and its role in protection against luminal pepsins: the effect of 16,16 dimethyl prostaglandin E2, carbopolyacrylate, sucralfate and bismuth subsalicylate. *Journal Gastroenterol. Hepatol.*, **9** (*Suppl 1*), S55–S59.

Desai, S. D., Desai, D. G., and Kaur, H. (2009). Saponins and their biological activities. *Pharma Times*, **41**, 13–16.

Familiari, G., Nottola, S. A., Macchiarelli, G., Familiari, A., and Motta, P. M. (1992a). A technique for exposure of the glycoproteic matrix (zona pellucida and mucus) for scanning electron microscopy. *Microscopy Research and Technique*, **23**, 225–229.

Familiari, G., Nottola, S. A., Macchiarelli, G., *et al.* (1992b). Human zona pellucida during *in vitro* fertilization, an ultrastructural study using saponin, ruthenium red, and osmium-thiocarbohydrazide. *Molecular Reproduction and Development*, **32**, 51–61.

Familiari, G., Relucenti, M., Heyn, R., Micara, G., and Correr, S. (2006). Three dimensional structure of the zona pellucida at ovulation. *Microscopy Research and Technique*, **69**, 415–426.

Familiari, G., Heyn, R., Relucenti., M., and Sathananthan H. (2008a). Structural changes of the zona pellucida during fertilization and embryo development. *Frontiers in Bioscience*, **13**, 6730–6751.

Familiari, G., Heyn, G., Relucenti, M., Nottola, S. A., and Sathananthan, A. H. (2008b). Ultrastructural dynamics of human reproduction, from ovulation to fertilization and early embryo development. *International Reviews of Cytology*, **249**, 53–142.

Ferreira Ede, O., Yates, E. A., Goldner, M., *et al.* (2008). The redox potential interferes with the expression of laminin binding molecules in Bacteroides fragilis. *Mem. Inst. Oswaldo Cruz*, **103**, 683–689.

Florman, H. and Ducibella, M. T. (2006). Fertilization in mammals. In: *Physiology of Reproduction*, 3rd ed., Neill, J. D. ed. San Diego, CA, Academic Press, Inc., pp. 55–112.

Forstner, J. F., Oliver, M. G., and Sylvester, F. A. (1995). Production, structure and biologic relevance of gastrointestinal mucins. In: *Infections of the Gastrointestinal Tract*, Blaser, M. J., Smith, P. D., Ravdin, J. I., Greenberg, H. B., and Guerrant, R. L. ed. New York, Raven Press, pp. 71–88.

Ganguly, A., Sharma, R. K., and Gupta, S. K. (2008). Bonnet monkey (Macaca radiata) ovaries, like human oocytes, express four zona pellucida glycoproteins. *Molecular Reproduction and Development*, **75**, 156–166.

Ganguly, A., Bukovsky, A., Sharma, *et al.* (2010). In humans, zona pellucida glycoprotein-1 binds to spermatozoa and induces acrosomal exocytosis. *Human Reproduction*, **25**, 1643–1656.

Girod, S., Zahm, J. M., Plotkowski, C., Beck, G., and Puchelle, E. (1992). Role of the physiochemical properties of mucus in the protection of the respiratory epithelium. *Eur. Respir. Journal*, **5**, 477–487.

Greve, J. M. and Wassarman, P. M. (1985). Mouse egg extracellular coat is a matrix of interconnected filaments possessing a structural repeat. *Journal of Molecular Biology*, **181**, 253–264.

Hayat, M. A. (1989). *Principles and Techniques of Electron Microscopy, Biological Applications*, 3rd ed. Boca Raton, Florida, CRC Press, Inc.

Heffer-Lauc, M., Viljetic, B., Vajn, K., Schnaar, R. L., and Lauc, G. (2007). Effects of detergents on the redistribution of gangliosides and GPI-anchored proteins in brain tissue sections. *Journal Histochem. Cytochem.*, **55**, 805–812.

Hughes, D. C. and Barratt, C. L. (1999). Identification of the true human orthologue of the mouse Zp1 gene: evidence for greater complexity in the mammalian zona pellucida? *Biochim. Biophys. Acta*, **1447**, 303–306.

Hunziker, E. B. and Schenk R. K. (1984). Cartilage ultrastructure after high pressure freezing, freeze-substitution, and low temperature embedding. Intercellular matrix ultrastructure – preservation of proteoglycans in their native state. *Journal of Cell Biology*, **98**, 277–282.

Hynes, R. O. (2009) The extracellular matrix: not just pretty fibrils. *Science*, **326**, 216–219.

Järveläinen, H., Sainio, A., Koulu, M., Wight, T. N., and Penttinen, R. (2009). Extracellular matrix molecules: potential targets in pharmacotherapy. *Pharmacol. Rev.*, **61**, 198–223.

Kádár, A., Gardner, D. L., and Bush, V. (1972). Glycosaminoglycans in developing chick-embryo aorta revealed by ruthenium red: an electron-microscope study. *Journal Pathol.*, **108**, 275–280.

Kelley, R. O., Deker, R. A. F., and Bluemink, J. G. (1973). Ligand-mediated osmium binding: Its application in coating biological specimens for scanning electron microscopy. *Journal of Ultrastructural Research*, **45**, 254–258.

Kerss, S., Allen, A., and Garner, A. (1982). A simple method for measuring thickness of the mucus gel layer adherent to rat, frog, and human gastric mucosa: influence of feeding, prostaglandin, N-acetylcysteine, and other agents. *Clin. Sci. (Lond.)*, **63**, 187–195.

Kogan, G., Soltés, L., Stern, R., and Gemeiner, P. (2007). Hyaluronic acid: a natural biopolymer with a broad range of biomedical and industrial applications. *Biotechnol. Lett.*, **29**, 17–25.

Lai, S. K., Wang, Y. Y., Wirtz, D., and Hanes, J. (2009). Micro- and macro-rheology of mucus. *Adv. Drug Deliv. Rev.*, **61**, 86–100.

Lefievre, L., Conner, S. J., Salpekar, A., *et al.* (2004). Four zona pellucida glycoproteins are expressed in the human. *Human Reproduction*, **19**, 1580–1586.

Litscher, E. S., Williams, Z., and Wassarman, P. M. (2009). Zona pellucida glycoprotein ZP3 and fertilization in mammals. *Molecular Reproduction and Development*, **76**, 933–941.

Luft, J. H. (1971a). Ruthenium red and violet. I: Chemistry, purification, methods of use for electron microscopy and mechanism of action. *Anat. Rec.*, **171**, 347–368.

Luft, J. H. (1971b). Ruthenium red and violet. II: Fine structural localization in animal tissues. *Anat. Rec.*, **171**, 369–415.

Monné, M., Han, L., Schwend, T., Burendahl, S., and Jovine, L. (2008). Crystal structure of the ZP-N domain of ZP3 reveals the core fold of animal egg coats. *Nature*, **456** (*7222*), 653–657.

Papi, M., Brunelli, R., Sylla, L., *et al.* (2010). Mechanical properties of zona pellucida hardening. *European Biophys. Journal*, **39**, 987–992.

Penman, S. (1995). Rethinking cell structure. *Proc. Natl Acad. Sci. USA*, **92**, 5251–5257.

Poon, W. W. and McCoshen J. A. (1985). Variances in mucus architecture as a cause of cervical factor infertility. *Fertility and Sterility*, **44**, 361–365.

Randell, S. H. and Boucher, R. C. (2006). Effective mucus clearance is essential for respiratory health. *Am. J. Respir. Cell Mol. Biol.*, **35**, 20–28.

Sandzen, B., Blom, H., and Dahlgren, S. (1988). Gastric mucus gel layer thickness measured by direct light microscopy. An experimental study in the rat. *Scand. J. Gastroenterol.*, **23**, 1160–1164.

Sheehan, J. K., Oates, K., and Carlstedt, I. (1986). Electron microscopy of cervical, gastric, and bronchial mucus glycoproteins. *Biochem. Journal*, **239**, 147–153.

Shogren, R., Gerken, T. A., and Jentoft, N. (1989). Role of glycosylation on the conformation and chain dimensions of O-linked glycoproteins: light-scattering studies of ovine submaxillary mucin. *Biochemistry*, **28**, 5525–5536.

Slayter, H. S., Wold, J. K., and Midtvedt, T. (1991). Intestinal mucin of germ-free rats. Biochemical and electron-microscopic characterization. *Carbohydr. Research*, **222**, 1–9.

Slomiany, B. L. and Slomiany, A. (1991). Role of mucus in gastric mucosal protection. *Journal Physiol. Pharmacology*, **42**, 147–161.

Sorokin, L. (2010). The impact of the extracellular matrix on inflammation. *Nat. Rev. Immunol.*, **10**, 712–723.

Sterling, C. (1970). Crystal-structure of ruthenium red and stereochemistry of its pectic stain. *Am. J. Bot.*, **57**, 172–175.

Sturgess, J. M. (1988). Electron microscopy investigation of mucus. In: *Methods in Bronchial Mucology*, Braga, P. C. and Allegra L., ed. New York, Raven Press, pp. 245–253.

Tsang, K. Y., Cheung, M. C., Chan, D., and Cheah, K. S. (2010). The developmental roles of the extracellular matrix: beyond structure to regulation. *Cell and Tissue Research*, **339**, 93–110.

Varki, A., Cummings, R., Esko, J., *et al.* (1999). *Essentials of Glycobiology*. Cold Spring Harbor, NY, Cold Spring Harbor Laboratory Press.

Wassarman, P. M. (2008). Zona pellucida glycoproteins. *Journal Biol. Chem.*, **283**, 24285–24289.

Wassarman, P. M. and Litscher, E. S. (2008). Mammalian fertilization: the egg's multifunctional zona pellucida. *Int. J. Dev. Biol.*, **52**, 665–676.

Williams, J. R. and Gong, H. (2007). Biological activities and syntheses of steroidal saponins: the shark-repelling pavoninins. *Lipids*, **42**, 77–86.

Yudin, A. I., Hanson, F. W., and Katz, D. F. (1989). Human cervical mucus and its interaction with sperm: a fine-structural view. *Biology of Reproduction*, **40**, 661–671.

11 Scanning electron microscopy of cerebellar intrinsic circuits

Orlando J. Castejón

11.1 Introduction

Scanning electron microscopy (SEM) and diverse fracture methods have shown the possibility of studying three-dimensional nerve cell interrelationship *in situ* taking advantage of SEM depth of focus, high magnification and resolution. By means of conventional SEM, and the SEM ethanol-cryofracturing technique designed by Humphreys *et al.* (1975), we have described the course of mossy fibers in the granular layer, and the climbing fiber pathways through the granule cell, Purkinje cell, and molecular layers of teleost fish and human cerebellar cortex (Castejón and Caraballo 1980a,b; Castejón and Valero, 1980). Scheibel *et al.* (1981), by means of the creative tearing technique for SEM, exposed the outer surface of Purkinje cells, the surrounding basket cell axon collaterals, and segments of climbing fibers. Arnett and Low (1985), using ultrasonic microdissection, have shown at SEM level the Purkinje cells, the basket cell synapses, and the Purkinje dendritic spines.

Applying the freeze-fracture method for SEM to the study of mouse, rat, fish, and human cerebellum, the author has identified the parent climbing fibers and their crossing-over pattern of bifurcation, and the formation of tendril collaterals and climbing fiber glomeruli in the granule cell layer (Castejón 1983, 1986; Castejón, 1988; Castejón and Castejón, 1988). Later, Apergis *et al.* (1991) also described by using SEM that the climbing fibers form tendril collaterals, and climbing glomeruli in the cerebellar granular layer.

Castejón (1990, 1993), Castejón and Castejón (1991), and Castejón and Apkarian (1992) carried out a freeze-fracture SEM and comparative transmission electron microscopy (TEM) freeze-etching study of mossy fiber glomeruli, their proteoglycan content, and parallel fiber-Purkinje spine synapses of vertebrate cerebellar cortex. Takahashi-Iwanaga (1992) showed the reticular endings of Purkinje cell axons in the rat cerebellar nuclei by means of SEM and the use of an NaOH maceration technique. Castejón (1993) reported the sample preparation techniques for conventional and high-resolution scanning electron microscopy of the central nervous system. Castejón *et al.* (1994a, b, c), Castejón (1996), and Castejón and Castejón (1997) have described the three-dimensional morphology of cerebellar synaptic junctions, the field emission SEM features of parallel fiber-Purkinje spine synapses, and postulated the use of high-resolution scanning electron microscopy and cryofracture techniques as tools for tracing cerebellar short intracortical

Scanning Electron Microscopy for the Life Sciences, edited by H. Schatten. Published by Cambridge University Press

circuits. Hojo (1994, 1996) prepared specimens of human cerebellar cortex by means of a t-butyl alcohol freeze-drying device and examined by SEM the Purkinje cell somatic surface. Castejón and Castejón (1997) described in detail the three-dimensional morphology and synaptic connections of Purkinje cells of several vertebrates using conventional and high-resolution SEM combined with the freeze-fracture method for SEM. Castejón *et al.* (2000a, b; 2001a, b, c) and Castejón (2003) have made a correlative microscopic study of granule cells, mossy, and climbing fibers, and basket cells using confocal laser scanning microscopy (CLSM), TEM, and SEM. More recently, a comparative and correlative microscopic study of cerebellar intrinsic circuits using light and Golgi light microscopy, TEM, SEM, and CLSM using immunohistochemical techniques, such as Synapsin-I, PSD-95,CaMKII alfa, and N-cadherin have been reported (Castejón *et al.*, 2004b, Castejón and Dailey, 2009, and Castejón, 2010).

The present chapter describes the three-dimensional features of cerebellar intrinsic circuits as seen by conventional and field emission SEM, and outlines the contribution of SEM to cerebellar neurobiology. The topics covered are: characterization of afferent mossy and climbing fibers at cerebellar white matter, the mossy fiber-granule cell glomerular synapse, the climbing fiber intracortical course and their synaptic relationships, parallel fiber-Purkinje dendritic spine synapses, basket cell-Purkinje cell axosomatic junctions, and stellate neuron-Purkinje cell synapses.

11.2 SEM characterization of afferent mossy and climbing fibers at cerebellar white matter

At the center of each cerebellar folium lies a thin layer of white matter composed of myelinated afferent and efferent fibers connecting the cerebellar cortex with other central nervous system centers (Feirabend *et al.*, 1996; Castejón, 2003). By means of SEM low magnification, the mossy and climbing fibers can be identified according to differential caliber and branching pattern (Castejón, 2003). Exploration of teleost fish cerebellar white matter with the scanning electron probe at low magnification, in samples coated with gold-palladium, shows longitudinal bundles of thick mossy parent fibers intermingled with bundles of thin afferent climbing fiber. At higher magnification, both types of afferent fiber can be clearly distinguished by their different thicknesses. The mossy fibers are up to 2.5 μm in diameter and the climbing fibers up to 1 μm in diameter, measured in cross-sections of these fibers in conventional SEM fractographs. At the level of the entrance site to the granular layer, the afferent mossy and climbing fibers are additionally distinguished by their branching pattern. The mossy fibers exhibit a characteristic dichotomous pattern of bifurcation, whereas climbing fibers displayed a typical arborescence or crossing-over type of bifurcation (Castejón, 1988; Castejón and Sims, 2000; Castejón *et al.*, 2000b). The cross-over that follows climbing fiber branching was first described by Athias (1897), and later by O'Leary *et al.* (1971). The criteria for identification at SEM level are in agreement with previous Golgi light microscopic (Ramón y Cajal, 1911; Scheibel and Scheibel, 1954; Fox, 1962; Fox *et al.*, 1967) and TEM studies (Chan-Palay and Palay, 1971a, b; Chan-Palay, 1971; Mugnaini, 1972; Shinoda *et al.*, 2000).

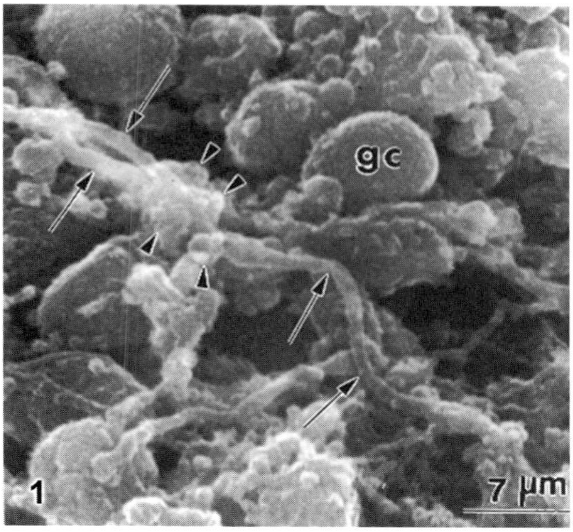

Figure 11.1 Teleost fish cerebellar cortex. Sagittally fractured mossy fiber glomerulus showing the longitudinal profile of a central mossy fiber (arrows) surrounded by the dendritic tips (arrowheads) of granule cells (gc). SEM freeze-fracture method. Gold-palladium coating. (Castejón *et al.*, 2000a.)

11.3 SEM of mossy fiber-granule cell dendrite glomerular synapses

With conventional SEM, the thick mossy fibers can be traced from the white matter entering into the granule cell layer establishing the mossy glomerular "en passant" synaptic regions (Castejón *et al.*, 2000a), and showing their typical topographic distribution around granule cell dendrites. With the use of the ethanol-cryofracturing technique for SEM (Humphreys *et al.*, 1975) applied to human cerebellum (Castejón and Caraballo, 1980a, b; Castejón and Valero, 1980), the afferent mossy fibers are seen entering the granular layer and penetrating into the granule cell groups. The cryofracture process exposes the outer surface of the glomerular region showing the mossy fiber rosettes surrounded by granule cell dendrites (Figure 11.1).

Hojo (1994, 1996), using a t-butyl alcohol freeze-drying device to prepare the human cerebellar cortex for SEM, also showed the synaptic relationship of mossy fiber with granule cell dendrites at the glomerular region.

With the freeze-fracture SEM method an "en face view" of the internal structure of mossy glomeruli is also obtained. The mossy fiber rosette appears in the center surrounded by up to 18 granule cells. The granule cell dendrites are traced in a radial orientation converging upon the outer surface of a mossy fiber rosette (Castejón *et al.*, 2000a) (Figure 11.2).

Figures 11.1 and 11.2 offer a real image of quantitative divergence of information of nerve impulse from one mossy fiber to numerous granule cells, and this opens new fields of investigation on information processing at the level of cerebellar glomeruli. These images support earlier quantitative studies of mossy glomeruli (Jakab and Hamori, 1988).

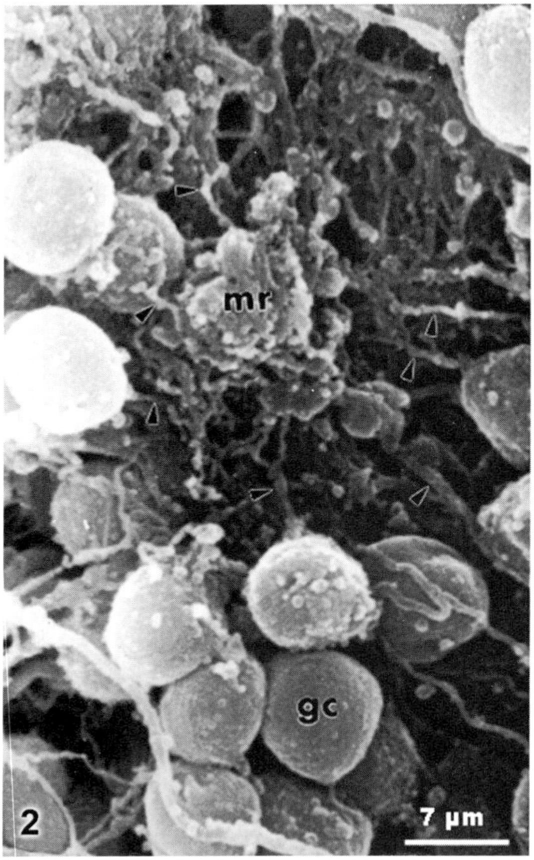

Figure 11.2 Teleost fish cerebellar cortex. Cross-fractured mossy fiber glomerulus showing the central mossy rosette fiber (mr) surrounded by the radially converging dendrites (arrowheads) upon the mossy rosette of up to 16 neighboring granule cells (gc). Note the inner radial configuration of mossy glomerulus. SEM freeze-fracture method. Gold-palladium coating. (Castejón *et al.*, 2000a.)

11.4 SEM climbing fiber intracortical course and synaptic relationships

At the level of the granular layer, the climbing fibers establish synaptic contacts with the Golgi and granule cell dendrites, and form small climbing fiber glomeruli (Chan-Palay and Palay, 1971b; Palay and Chan-Palay, 1974; Castejón and Caraballo, 1980a). Due to the depth of focus of scanning electron microscopes, the fine tendril collaterals of climbing fibers are clearly distinguished spreading throughout the granular layer Examination of the granular layer at low magnification shows the climbing fibers making axodendritic contacts with granule cell dendrites (Figure 11.3).

Castejón *et al.* (1994b), applying the ethanol cryofracturing technique to the human cerebellar molecular layer, described the cryo-dissected climbing fiber collaterals making axodendritic synaptic connections with Purkinje dendritic spines. The climbing fibers

Figure 11.3 Teleost fish cerebellar granular layer. The climbing fibers (CF) are observed making synaptic contacts (arrowheads) with granule cell (GC) dendrites. The arrows indicate the tendril collaterals of climbing fibers. The arrowheads label the synaptic relationship with granule cell dendrites. SEM slicing technique. Gold-palladium coating. (Castejón and Caraballo, 1980a.)

exhibit their typical crossing-over type of radial collateralization that characterizes the climbing fiber bifurcation pattern (Figure 11.4).

At the interface between the Purkinje cell layer and molecular layer, the climbing fiber collaterals are ascending vertically and giving off fine collateral processes (Castejón and Caraballo, 1980a,b; Castejón, 1988).

In samples of primate cerebellar molecular layer, coated with chromium and examined with the field emission SEM, the high mass density terminal endings of climbing fiber are found intimately applied to the outer surface of the less dense secondary and tertiary Purkinje cell dendritic branches (Castejón *et al.*, 1994c).

Fine terminal collaterals of cryodissected climbing fibers are observed in human cerebellum ending by means of round synaptic knobs upon the Purkinje dendritic spine bodies. In addition, some retrograde or Scheibel's collaterals were seen descending to the granular layer (Castejón and Valero, 1980; Castejón *et al.*, 1994c).

11.5 Purkinje cells and their synaptic contacts

The Purkinje cell layer shows the outer surface of Purkinje cell soma exposed by the selective removal of enveloping Bergmann glial cells by the cryo-fracture process. The lower pole of Purkinje cell soma shows the emergence of an axonal initial segment. The basket cell appears as a microneuron exhibiting a short axonal process directed to the

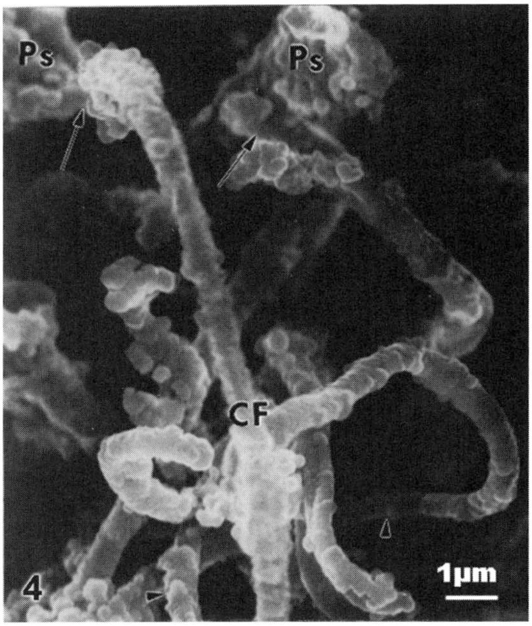

Figure 11.4 Human cerebellar cortex. Ethanol-cryofracturing technique. Climbing fiber (CF) in the molecular layer showing typical crossing-over type of bifurcation pattern. The arrows indicate the synaptic junctions with Purkinje dendritic spines (Ps). The arrowheads indicate the retrograde Scheibel's collaterals descending toward the granular layer. Gold-palladium coating. (Castejón *et al.*, 1994b.)

Purkinje cell soma. The climbing fibers are observed ascending to the Purkinje cell primary trunk, and toward the molecular layer (Figure 11.5).

11.6 Field emission SEM of parallel fiber-Purkinje dendritic spine synapses

Conventional SEM of a parallel fiber-Purkinje dendrite synaptic relationship demonstrated the cruciform arrangement of these synaptic contacts as formerly described by Cajal (1911) and Gray (1961) by TEM. Field emission scanning electron microscopy (FESEM) of a cryo-fractured outer third mouse cerebellar molecular layer shows the smooth outer surface of unattached mushroom shaped Purkinje dendritic spines (Pds) (Figure 11.6).

The spine neck ranges from about 0.68 to 1 µm in mushroom type-dendritic spines. The spine head has a maximum axial diameter ranging from 1.13 to 1.5 µm. The spine head transversal diameter is about 1 µm. The elongated spines exhibit an axial diameter up to 2.82 µm and a transversal diameter of 1.30 µm. The spine density is 18 dendritic spines per 8 μm^2. Some Pds exhibit elongated and lanceolate shapes, with a neck of 1 µm in length, and a body of up to 2.82 µm in axial diameter, and a transversal diameter of 1.30 µm. Close examination of the shaft of a tertiary Purkinje dendritic branch shows that the spines are separated by a distance ranging from 50 to 500 nm (Castejón *et al.*, 2004a).

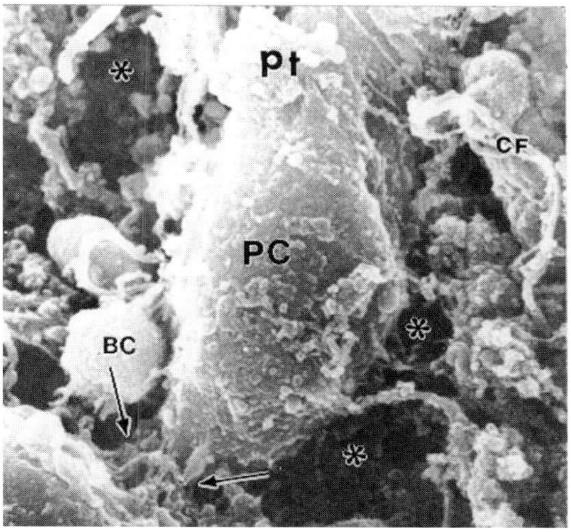

Figure 11.5 Sagittally cryo-fractured teleost fish cerebellar cortex showing the outer surface of Purkinje cell (PC) soma. The cryofracture process has removed the ensheathing Bergmann glial cell cytoplasm exposing the outer surface of the pear-shaped neuronal soma. Note the emergence of primary dendritic trunk (pt). Climbing fibers (CF) are seen approaching the primary dendritic trunk. The asterisks label the dark spaces previously occupied by Bergmann glial cell cytoplasm. A basket cell (BC) is observed sending its axon toward the Purkinje cell soma. The arrows indicate a partial view of the Purkinje cell infraganglionic plexus at the level of the Purkinje cell axon hillock region. Gold-palladium coating. (Castejón and Caraballo, 1980b.)

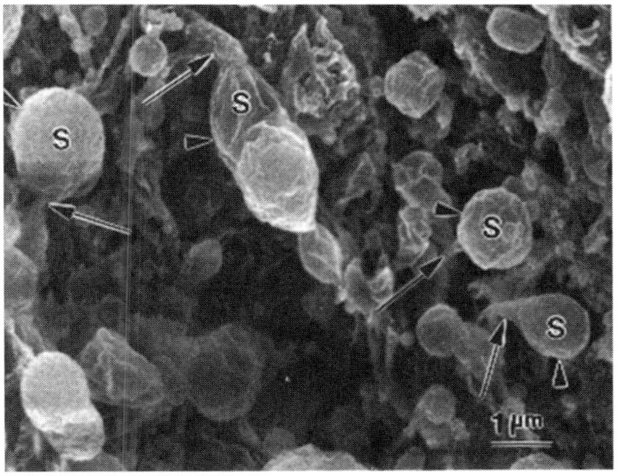

Figure 11.6 Field emission scanning electron micrograph of mouse cerebellar molecular layer showing, at high magnification, unattached Purkinje dendritic spines (S). The cryofracture process has removed the neighboring parallel fibers allowing the visualization of spine body (arrowheads) and neck (arrows). Chromium coating. (Castejón *et al.*, 2004a.)

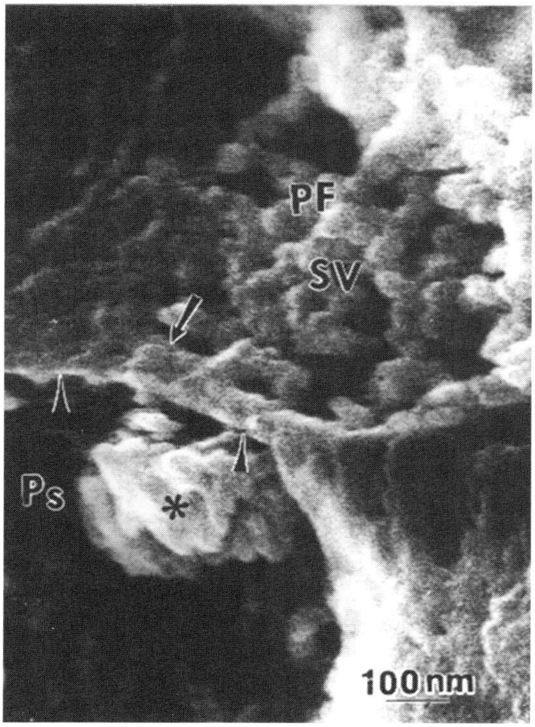

Figure 11.7 High-magnification field emission scanning electron micrograph of a parallel fiber (PF)-Purkinje spine (Ps) synaptic contact. The specialized synaptic membrane complex is indicated by arrowheads. Clustered spheroid synaptic vesicles (SV) are seen in the parallel fiber pre-synaptic ending. Some synaptic vesicles appear anchored to the pre-synaptic membrane (arrow). The asterisk labels the Purkinje spine post-synaptic density. Chromium coating.

High-magnification FESEM made it possible to resolve the synaptic membrane complex formed by parallel fiber-Purkinje spine synapses. The parallel fiber presynaptic vesicles, the pre- and post-synaptic synaptic membranes, and a partial view of post-synaptic density could be distinguished (Figure 11.7).

Higher magnification of a parallel fiber-Purkinje dendritic spine synapse using FESEM shows the substructure of the post-synaptic density of Pds, consisting of globular subunits, of 25–50 nm diameter, corresponding to the localization of post-synaptic proteins and/or post-synaptic receptors (Castejón, 1996, 2004a), (Figure 11.8).

11.7 Basket cell-Purkinje cell axo-somatic synaptic junctions

Scanning electron micrographs of human cerebellum processed by the ethanol-cryofracturing technique show the short axonal ramifications of basket cell embracing the Purkinje cell body (Figure 11.9A), as earlier described by Cajal by Golgi light microscopy (1911), and later by Bishop (1993) by means of TEM and horseradish

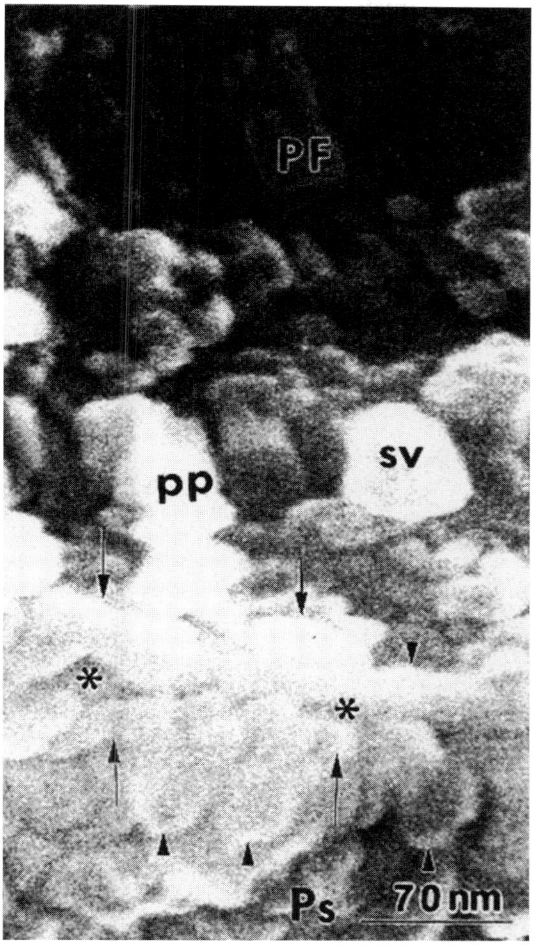

Figure 11.8 High-magnification field emission scanning electron micrograph of parallel fiber (PF)-Purkinje spine (Ps) synaptic membrane complex. The synaptic ending of parallel fiber shows a pre-synaptic dense projection (pp) and a synaptic vesicle (SV). The synaptic cleft (asterisks) is separating pre-and post-synaptic membranes (arrows). Round globular subunits (arrowheads) are observed associated to the post-synaptic membrane (arrow), corresponding to post-synaptic receptor and/or post-synaptic proteins. Chromium coating. (Castejón, 1996.)

peroxidase labeling. The ethanol-cryofracturing technique selectively removed the satellite Bergmann glial cell covering the Purkinje cell soma, and exposed the basket cell axonal collaterals applied to the Purkinje cell outer surface soma (Figure 11.9B). At higher magnification the short varicose twigs formed by the basket cell endings appear terminating as interconnected boutons. In the cerebellar cortex of teleost fish processed by the slicing technique, the axonal descending collaterals of basket cells are observed at the level of the Purkinje cell axonal initial segment contributing to the formation of infraganglionic plexus, and the formation of the pinceaux (Castejón *et al.*, 2001c).

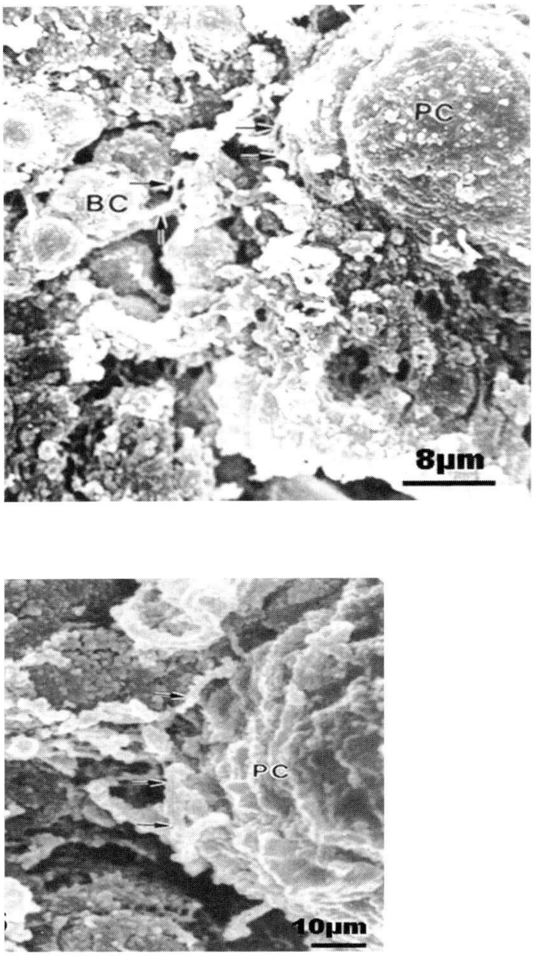

Figure 11.9 Scanning electron micrographs of human cerebellum. Ethanol-cryofracturing technique. (A) A basket cell (BC) located in the Purkinje cell layer is observed giving off its transverse axonal ramifications (arrows) toward a neighboring Purkinje cell (PC) (Castejón *et al.*, 2001c). (B) Higher magnification of the pericellular nest formed by basket cell axonal collaterals (arrows) around Purkinje cell body (PC). The ethanol-cryofracturing technique removed the satellite Bergmann glial cell cytoplasm covering the Purkinje cell soma, allowing visualization of Purkinje cell basket (Castejón, *et al.* 2001c).

11.8 Scanning electron microscopy of stellate neurons

Fish and human cerebellar cortex specimens conventionally processed for SEM show short-axon stellate neurons with round, elliptical, or fusiform somata in a parasagittal fracture of the outer third molecular layer. These stellate neurons are easy to recognize, since they are the only neurons in the upper molecular layer. Basket cells and fewer

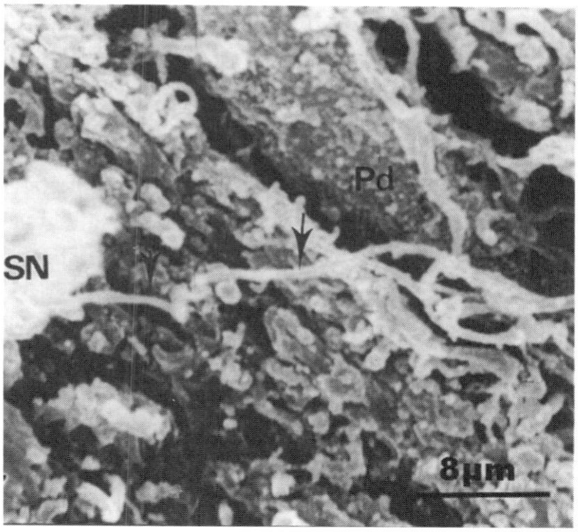

Figure 11.10 Scanning electron micrograph of human cerebellar molecular layer processed by the ethanol-cryofracturing technique showing a stellate neuron (SN) and its axonal process (arrows) directed to a neighboring Purkinje dendrite (Pd). Gold-palladium coating. (Castejón and Castejón, 1987.)

stellate cells are also found in the middle and inner thirds of the molecular layer. Numerous stellate cells are seen spreading throughout the outer third molecular layer. Three to five beaded dendrites radiate from the cell body toward the neighboring Purkinje dendrites or other stellate cells (Castejón and Castejón, 1987; Castejón, 2003). The axon originates by way of a typical triangular shaped axon hillock and, after a short initial segment, bifurcates into tenuous varicose collaterals. Short, ramified, and beaded dendrites emerge from the cell somata, directed toward the passing bundles of parallel fibers (Figure 11.10).

11.9 Contribution of SEM to cerebellar neurobiology

Characterization of afferent and efferent fibers in cerebellar white matter made by conventional SEM, using slicing and cryofracture methods, has allowed identification of afferent mossy and climbing fibers, and the efferent Purkinje cell axons in the white matter. As mentioned above, identification of these fibers was made taking into account the Golgi light microscopy criteria provided by previous published microscopic studies, mainly those concerned with the fiber thickness and branching pattern.

The SEM characterizations of extrinsic and intrinsic circuits fully support previous descriptions made by TEM (Gray, 1961; Larramendi and Victor, 1967; Larramendi, 1968; Mugnaini, 1972). The secondary electron images obtained by SEM can be used for morphometric studies of cerebellar afferent fibers, and the functional organization of these compartments (Castejón, 2003).

11.9.1 Identification of mossy fibers and their synaptic relationships

Conventional SEM and slicing technique (Castejón and Caraballo, 1980a; Castejón, 2003) show the "en passant" synaptic contacts of mossy fibers with various granule cell groups at the granular layer. The freeze-fracture method for SEM exposed the three-dimensional inner configuration of mossy glomeruli, and displayed the degree of divergence of nerve information processes at the level of each glomerular. This contribution emphasizes the unsurpassed potentiality of SEM and the freeze-fracture method for studying the three-dimensional morphology of multisynaptic complexes in the central nervous system. The SEM freeze-fracture method facilitated visualization of the outer surface of Purkinje cells and the supra- and infra-ganglionic plexuses, and therefore gave a partial view of the Purkinje pericellular nest. In addition, it made it possible to display the outer and inner surfaces of parallel fibers in the molecular layer, and their synaptic relationships with Purkinje dendritic spines (Castejón and Apkarian, 1992). The images of cross-fractured parallel fibers offer the future possibility to measure with high accuracy the parallel fiber diameter and to shed some light on their conduction velocity, a critical parameter for theory of timing in the cerebellar cortex (Sultan, 2000).

11.9.2 SEM proposed as a high-resolution tool for tracing cerebellar intracortical circuits

Conventional SEM and the sample preparation methods used have allowed us to trace the incoming mossy fibers in the granular layer, and the climbing fibers in the three-layered structure of the cerebellar cortex (Castejón and Caraballo, 1980 a,b; Castejón and Valero, 1980). The "en passant" nature of mossy fiber contacts with different granule cell groups demonstrates a higher degree of divergent information of mossy fibers in the granular layer than that provided by classical light and electron microscopic studies (Mugnaini, 1972; Ito, 1984), and support studies carried out with computer assisted microscopy (Hamori *et al.*, 1997), and confocal laser scanning microscopy (Castejón and Sims, 1999, 2000; Castejón *et al.*, 2000 a,b, 2001a). The SEM images bear the basic morphological information for future quantitative studies on mossy fiber-granule cell synaptic relationships using image analysis and quantitative morphometric methods.

The scanning electron micrographs of climbing fiber pathways in the three-layered structure of cerebellar cortex show a sagittal zonal organization, which adds further information on cerebellar zonal (Trott *et al.*, 1998), microzonal (Massion, 1993), sagittal (Garwicz, 1997), and parasagittal organization (Schweighofer, 1998).

An outstanding contribution of SEM was the finding of a high degree of lateral collateralization of climbing fibers in the granular and molecular layers, which offers new views on the participation of these fibers in information processing in the cerebellar cortex (De Schutter and Maex, 1996). Besides, it allows us to analyze in depth climbing fiber synaptic relationships with granule, Golgi, Purkinje, basket, and stellate neurons. These intracortical circuits deserve further investigation of new fractured hidden surfaces of cerebellar neurons using field emission SEM. Another conspicuous contribution of SEM has been the tracing of short cortical circuits, such as basket cell–Purkinje cell synaptic contacts, which gives additional evidence for a unitary relationship between both cells (O'Donohue *et al.*, 1989). SEM examination of granule cell–Purkinje spine synapses and of basket cell axonal collaterals embracing the Purkinje cell soma contributing to the pericellular nest reveals the potential contribution of SEM for tracing short intracortical circuits in the central nervous system.

Conventional and field emission SEM of cryofractured outer third mouse cerebellar molecular layer show the smooth outer surface, morphology, size, spine density, and interspine space of Purkinje dendritic spines, and the topographic and synaptic relationship of parallel fiber and Purkinje dendritic spines (Castejón and Apkarian, 1993; Castejón *et al.*, 2004a). These images reveal the SEM potentiality for future studies dealing with synaptic plasticity, motor learning, and cognitive processes in the cerebellar cortex (Castejón, 2003). High magnification of parallel fiber-Purkinje dendritic spine synapse using conventional and field emission SEM shows the substructure of the post-synaptic density of Purkinje dendritic spines, consisting of globular subunits of 25–50 nm diameter corresponding to the localization of post-synaptic proteins and/or post-synaptic receptors. This finding is basically important for research studies on receptor pathology. Scanning electron micrographs display the three-dimensional relief of a Purkinje pericellular nest formed by basket cell descending and transverse axonal collaterals.

The slicing technique for SEM, the ethanol-cryofracturing technique, and the freeze-fracture method for SEM have permitted us to demonstrate, with more accuracy than other microscopic techniques, the high degree of lateral collateralization of climbing fibers in the granular and molecular layers, and therefore their wide divergent field in the cerebellar cortex. In these layers the climbing fibers are not confined to a parasagittal and/or transverse plane, but dispersed in a wide radial fashion to a three-dimensional field (Castejón 1983, 1986; Castejón and Sims, 2000; Castejón, 2003, 2010). These findings should be considered for research on cerebellar information processing, especially from the theoretical point of view.

11.10 References

Apergis, G., Alexopoulos, T., Mpratakos, M., and Katsorchis, T. (1991). Scanning electron microscopy of the granular layer of rat cerebellar cortex. *Microsc. Electron. Biol. Cell.*, **15**, 119–129.

Arnett, C. E. and Low. F. N. (1985). Ultrasonic microdissection of rat cerebellum for scanning electron microscopy. *Scanning Electron Microscopy*, **I**, 274–255.

Athias, M. (1897). Recherches sur l'histogénèse de l'écorce du cervelet. *J. Anat. Physiol. Normal Pathol. de l'Homme et des Animaux*, **33**, 372–399.

Bishop, G. A. (1993). An analysis of HRP-filled basket cell axons in the cat's cerebellum. I Morphometry and configuration. *Anat. Embryol. (Berlin)*, **188**, 287–297.

Castejón, O. J. and Valero, C. J. (1980). Scanning electron microscopy of human cerebellar cortex. *Cell Tissue Research*, **212**, 263–374.

Castejón, O. J. and Caraballo, A. J. (1980a). Application of cryofracture and SEM to the study of human cerebellar cortex. *Scanning Electron Microscopy*, **IV**, 197–207.

Castejón, O. J. and Caraballo, A. J. (1980b). Light and scanning electron microscopic study of cerebellar cortex of teleost fishes. *Cell and Tissue Research*, **207**, 211–226.

Castejón, O. J. (1983). Scanning electron microscope recognition of intracortical climbing fiber pathways in the cerebellar cortex. *Scanning Electron Microscopy*, **III**, 1427–1434.

Castejón, O. J. (1986). Freeze-fracture of fish and mouse cerebellar climbing fibers. A SEM and TEM study. In: *Electron Microscopy*, Vol. IV, T. Imura, S. Maruse, and T. Susuki (ed.). Tokyo, Japan, Japanese Society of Electron Microscopy, pp. 3165–3166.

Castejón, O. J. (2010). Comparative and correlative microscopy of cerebellar cortex. Maracaibo, Venezuela, Zulia University, Astrodata, pp. 1–293.

Castejón, O. J. and Castejón, H. V. (1987). Electron microscopy and glycosaminoglycan histochemistry of cerebellar stellate neurons. *Scanning Microscopy*, **1**, 681–693.

Castejón, O. J. and Castejón, H. V. (1988). Scanning electron microscope, freeze etching and glycosaminoglycan cytochemical studies of the cerebellar climbing fiber system. *Scanning Microscopy* **2**, 2181–2193.

Castejón, O. J. (1988). Scanning electron microscopy of vertebrate cerebellar cortex. *Scanning Microscopy*, **2**, 569–597.

Castejón, O. J. (1990). Freeze-fracture scanning electron microscopy and comparative freeze-etching study of parallel fiber–Purkinje spine synapses of vertebrate cerebellar cortex. *J. Submicrosc. Cytol. Pathol.*, **22**, 281–295.

Castejón, O. J. (1993). Sample preparation technique for conventional and high resolution scanning electron microscopy of the central nervous system. The cerebellum as a model. *Scanning Microscopy*, **7**, 725–740.

Castejón, O. J. (1996). Conventional and high resolution scanning electron microscopy of cerebellar synaptic junctions. *Scanning Microscopy*, **10**, 177–186.

Castejón, O. J. (2003). The cerebellar white matter. In: *Scanning Electron Microscopy of Cerebellar Cortex*. New York, Kluwer Academic Publishing, pp. 25–28.

Castejón, O. J. and Castejón, H. V. (1991). Three-dimensional morphology of cerebellar protoplasmic islands and proteoglycan content of mossy fiber glomerulus: A scanning and transmission electron microscope study. *Scanning Microscopy*, **5**, 477–494.

Castejón, O. J. and Castejón, H. V. (1997). Conventional and high resolution scanning electron microscopy of cerebellar Purkinje cells. *Biocell*, **21**, 149–159.

Castejón, O. J. and Apkarian, R. P. (1992). Conventional and high resolution scanning electron microscopy of outer and inner surface features of cerebellar nerve cells. *J. Submicrosc. Cytol. Pathol.*, **24**, 549–562.

Castejón, O. J. and Apkarian, R. P. (1993). Conventional and high resolution field emission scanning electron microscopy of vertebrate cerebellar parallel fiber–Purkinje spine synapses. *Cell and Molecular Biology*, **39**, 863–873.

Castejón, O. J., Castejón, H. V., and Apkarian, R. P. (1994a). Proteoglycan ultracytochemistry and conventional and high resolution scanning electron microscopy of vertebrate cerebellar parallel fiber pre-synaptic endings. *Cell and Molecular Biology*, **40**, 795–801.

Castejón, O. J., Apkarian, R. P., and Valero, C. (1994b). Conventional and high resolution scanning electron microscopy and cryofracture techniques as tools for tracing cerebellar short intracortical circuits. *Scanning Microscopy*, **8**, 315–324.

Castejón, O. J., Castejón, H. V., and Apkarian, R. P. (1994c). High resolution scanning electron microscopy features of primate cerebellar cortex. *Cell and Molecular Biology*, **40**, 1173–1181.

Castejón, O. J. and Sims, P. (1999). Confocal laser scanning microscopy of hamster cerebellum using FM4–64 as an intracellular staining. *Scanning*, **21**, 15–21.

Castejón, O. J. and Sims, P. (2000). Three-dimensional morphology of cerebellar climbing fibers. A study by means of confocal laser scanning microscopy and scanning electron microscopy. *Scanning*, **22**, 211–217.

Castejón, O. J., Castejón, H. V., and Sims, P. (2000a). Confocal, scanning and transmission electron microscopic study of cerebellar mossy fiber glomeruli. *J. Submicrosc. Cytol. Pathol.*, **32**, 247–260.

Castejón, O. J., Castejón, H. V., and Alvarado, M. V (2000b). Further observations on cerebellar climbing fibers. A study by means of light microscopy, confocal laser scanning microscopy, and scanning and transmission electron microscopy. *Biocell*, **24**, 197–212.

Castejón, O. J., Castejón, H. V., and Apkarian, R. P. (2001a). Confocal laser scanning and scanning and transmission electron microscopy of vertebrate cerebellar granule cells. *Biocell*, **25**, 235–255.

Castejón, O. J., Apkarian, R. P., Castejón, H. V., and Alvarado, M. V. (2001b). Field emission scanning electron microscopy and freeze-fracture transmission electron microscopy of mouse cerebellar synaptic contacts. *J. Submicrosc. Cytol. Pathol.*, **33**, 289–300.

Castejón, O. J., Castejón, H. V., and Sims, P. (2001c). Light microscopy, confocal laser scanning microscopy, scanning and transmission electron microscopy of cerebellar basket cells. *J. Submicrosc. Cytol. Pathol.*, **33**, 23–32.

Castejón, O. J., Castellano, A., Arismendi, G., and Apkarian, R. P. (2004a). Correlative microscopy of Purkinje dendritic spines: a field emission scanning and transmission electron microscopic study. *J. Submicrosc. Cytol. Pathol.*, **36**, 29–36.

Castejón, O. J., Leah, F., and Dailey, M. (2004b). Localization of synapsin and PSD-95 in developing postnatal rat cerebellar cortex. *Develop. Brain Res.*, **151**, 25–32.

Castejón, O. J. and Dailey, M. (2009). Immunochemistry of GluR1 subunits of AMPA receptors of rat cerebellar nerve cells. *Biocell*, **33**, 71–80.

Chan-Palay, V. (1971). The recurrent collaterals of Purkinje cell axons. A correlated study of the rat's cerebellar cortex with electron microscopy and the Golgi method. *Zeitschrift Anat. Enwicklungs Geschichte*, **134**, 200–234.

Chan-Palay, V. and Palay, S. L. (1971a). The synapse "en marron" between Golgi II neurons and mossy fiber in the rat's cerebellar cortex. *Zeitschrift Anat. Enwicklungs Geschichte*, **133**, 274–287.

Chan-Palay, V. and Palay, S. L. (1971b). Tendril and glomerular collaterals of climbing fibers in the granular layer of the rat's cerebellar cortex. *Zeitschrift Anat. Enwicklungs Geschichte*, **133**, 247–273.

De Schutter, E. and Maex, R. (1996). The cerebellum: cortical processing and theory. *Current Opinion Neurobiol.*, **6**, 759–764.

Eccles, J. C., Ito., M., and Szentagothai, J. (1967). *The Cerebellum as a Neuronal Machine*. New York, Springer-Verlag, pp. 58–61.

Feirabend, H. K., Choufoer, H., and Voogd, J. (1996). White matter of the cerebellum of the chicken (*Gallus domesticus*): a quantitative light and electron microscopic analysis of myelinated fibers and fiber compartments. *J. Comp. Neurol.*, **369**, 236–251.

Fox, C. A. (1962). Fine structure of the cerebellar cortex. In: *Correlative Anatomy of the Nervous System*, E. C. Crosby, T. Humphreys, and E. W. Lauer (ed.). New York, MacMillan Co., pp. 192–198.

Fox, C. A., Hillman, D. E., Siegesmund, K. A, and Dutta, C. R. (1967). The primate cerebellar cortex. A Golgi and electron microscopic study. *Prog. Brain Res.*, **25**, 174–225.

Gray, E. G. (1961). The granule cells, mossy synapses and Purkinje spine synapses of the cerebellum: Light and electron microscope observations. *J. Anat.*, **95**, 345–356.

Hamori, J., Jakab, R. L., and Takacs, J. (1997). Morphogenetic plasticity of neuronal elements in cerebellar glomeruli during differentiation-induced synaptic reorganization. *J. Neural. Transplant. Plast.*, **6**, 11–20.

Hojo, T. (1994). An experimental scanning electron microscopic study of human cerebellar cortex using t-butyl alcohol freeze-drying device. *Scanning Microscopy*, **8**, 303–313.

Hojo, T. (1996). Specimen preparation of the human cerebellar cortex for scanning electron microscopy using a t-butyl alcohol freeze-drying device. *Scanning Microscopy (Suppl)*, **10**, 345–348.

Garwicz, M. (1997). Sagittal organization of climbing fibre input to the cerebellar anterior lobe of the ferret. *Experimental Brain Research*, **117**, 389–398.

Humphreys, W. J., Spurlock, B. O., and Johnson, J. S. (1975). Transmission electron microscopy of tissue prepared for scanning electron microscopy by ethanol cryofracturing. *Stain Tech.*, **50**, 119–125.

Ito, M. (1984). Mossy fibers. In: *The Cerebellum and Neural Control*. New York, Raven Press, pp. 74–85.

Jakab, R. L. and Hamori, J. (1988). Quantitative morphology and synaptology of cerebellar glomeruli in the rat. *Anat. Embryology*, **179**, 81–88.

Larramendi, L. M. H. and Victor, L. M. (1967). Synapses on the Purkinje cell spines in the mouse. An electron microscopic study. *Brain Research*, **5**, 15–30.

Larramendi, L. M. H. (1968). Morphological characteristics of extrinsic and intrinsic nerve terminals and their synapses in the cerebellar cortex of the mouse. In: *The Cerebellum in Health and Diseases*. W. S. Field and W. D. Willes (ed.). St. Louis, Warren H. Green, Inc., pp. 63–110.

Massion, J. (1993). Major anatomical functional relations in the cerebellum. *Rev. Neurol. (Paris)*, **149**, 600–606.

Mugnaini, E. (1972). The histology and cytology of the cerebellar cortex. In: *The Comparative Anatomy and Histology of the Cerebellum. The Human Cerebellum, Cerebellar Connections and Cerebellar Cortex*, O. Larsell and J. Jansen (ed.). Minneapolis, The University of Minnesota Press, pp. 201–264.

O'Donoghue, D. L, King, J. S., and Bishop, G. A. (1989). Physiological and anatomical studies of the interactions between Purkinje cells and basket cells in the cat's cerebellar cortex: evidence for a unitary relationship. *Journal of Neuroscience*, **9**, 2141–2150.

O'Leary, J. L., Inukai, F., and Smith. M. B. (1971). Histogenesis of the cerebellar climbing fiber in the rat. *J. Comp. Neurol.*, **I42**, 377–392.

Palay, S. L. and Chan-Palay, V. (1974). *Cerebellar Cortex. Cytology and Organization*. Berlin, Springer-Verlag, pp. 1–348.

Ramón y Cajal, S. (1911). *Histologie du Système Nerveux de L'Homme et Des Vertébrés*. Vol. I, pp. 73–105. Translated by L Azoulay, Maloine, París.

Scheibel, M. E. and Scheibel, A. B. (1954). Observations on the intracortical relations of the climbing fibers of the cerebellum. *J. Comp. Neurol.*, **101**, 733–763.

Schweighofer, N. (1998). A model of activity-dependent formation of cerebellar microzones. *Biol. Cybern.*, **79**, 97–107.

Shinoda, M., Sughihara, I., Wu, H. S., and Sugiuchi Y. (2000). The entire trajectory of single climbing and mossy fibers in the cerebellar nuclei and cortex. In: *Cerebellar Modules: Molecules, Morphology and Function*. N. M. Gerrits, T. J. H. Ruigrok, and C I. De Zeeuw (ed.). Elsevier, Europe, *Prog. Brain Research*, **124**, 173–186.

Sultan, F. (2000). Exploring a critical parameter of timing in the mouse cerebellar microcircuitry: the parallel fiber diameter. *Neuroscience Letters*, **280**, 41– 44.

Scheibel, A. B., Paul, L., and Fried. I. (1981). Scanning electron microscopy of the central nervous system. I: The cerebellum. *Brain Res. Rev.*, **3**, 207–228.

Takahashi-Iwanaga, H. (1992). Reticular endings of Purkinje cell axons in the rat cerebellar nuclei: scanning electron microscopic observations of the pericellular plexus of Cajal. *Arch. Histol. Cytol.*, **55**, 307–314.

Trott, J. R., Apps, R., and Armstrong, R. M. (1998). Zonal organization of cortico-nuclear and nucleo-cortical projections of the paramedian lobule of the cat cerebellum. I: The C1 zone. *Experimental Brain Research*, **118**, 298–315.

12 Application of *in vivo* cryotechnique to living animal organs examined by scanning electron microscopy

Shinichi Ohno, Nobuo Terada, Nobuhiko Ohno, and Yasuhisa Fujii

12.1 Introduction

For the past several decades, morphological study by electron microscopy has been one of the major approaches to understanding physiological and pathological features of living animals, including humans, in the medical and biological fields. In particular, transmission (TEM) or scanning electron microscopy (SEM) has greatly facilitated enormous progress in ultrastructural analyses of cells and tissues, including many applications for clinical medicine and molecular biology. In such cases, they must reflect some functional aspects of living animal organs. For morphological observation, both chemical fixation and alcohol dehydration have commonly been used as established preparation procedures, but they bring about many morphological artifacts in dynamically changing cells and tissues at an electron microscopic level (Figure 12.1 steps a,b) (Furukawa *et al.*, 1991; Yoshimura *et al.*, 1991; Ohno *et al.*, 1992).

On the contrary, conventional cryotechniques, by which animal tissues are quickly frozen for physical fixation, have greatly contributed to reduction of such morphological artifacts, but they have to be resected from living animal organs for quick-freezing and high-pressure freezing (Figure 12.1 steps c,d) (Yu *et al.*, 1997, 1998). These animal specimens are inevitably exposed to biological stresses of ischemia and anoxia, exhibiting only dead morphological states of their cells and tissues without normal blood circulation. Therefore, we have developed the "*in vivo* cryotechnique" (IVCT) to clarify morphofunctional significance of cells and tissues in living animal organs (Figure 12.1 step f), and have reported dynamically changing morphology *in vivo* and also immunolocalization of functional proteins in cells and tissues at both light and electron microscopic levels (Ohno *et al.*, 1996a, 2004; Ohno *et al.*, 1996b; Ohno *et al.*, 2006; Terada *et al.*, 2006; Ohno *et al.*, 2007).

The IVCT, by which living animal organs are directly cryofixed *in vivo*, can prevent the common morphological artifacts of resected tissues caused by ischemia and anoxia. The purpose of this chapter is to describe the technical details and significance of IVCT, and also review some morphofunctional findings of living animal organs examined by SEM.

Scanning Electron Microscopy for the Life Sciences, edited by H. Schatten. Published by Cambridge University Press

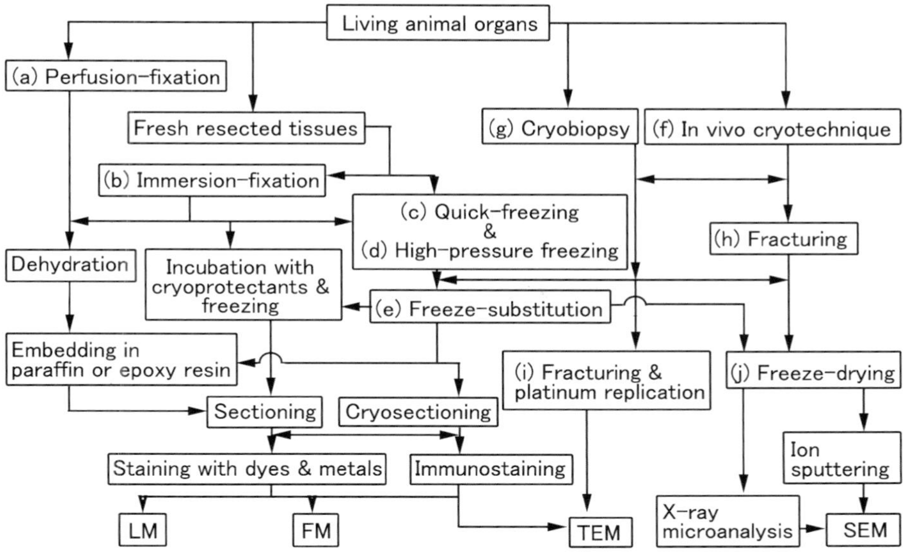

Figure 12.1 A flowchart of various preparation steps for light microscopy (LM), fluorescence microscopy (FM), transmission electron microscopy (TEM), scanning electron microscopy (SEM), and X-ray microanalysis. The perfusion (a) or immersion fixation (b) and alcohol dehydration steps, quick-freezing (c) and high-pressure freezing (d) methods, and cryobiopsy (g) or "*in vivo* cryotechnique" (IVCT) (f), are shown in connection with their following preparation steps. Note that all the preparation steps, following the quick-freezing and high-pressure freezing methods, are also available after the IVCT preparation. To apply IVCT to examination for a series of living animal tissues or human organs, a new cryotechnique of biopsy, termed as "cryobiopsy" (g), would be necessary.

12.2 How to perform IVCT

The IVCT was originally performed with a handmade "*in vivo* cryoapparatus," which has been described before (Ohno *et al.*, 1996a). In the original IVCT method, prepared isopentane–propane cryogen (−193 °C) was manually poured over exposed animal organs under anesthesia after simultaneously cutting them with a precooled cryoknife in liquid nitrogen (Ohno *et al.*, 1996a). After the cryogen had been poured for several seconds, specimens were additionally treated with liquid nitrogen (−196 °C) to maintain the low temperature. The *in vivo* frozen organs were then cracked off from the animal body in the liquid nitrogen and plunged into it for preservation. Thereafter, some tissue pieces were taken out with a dental electric drill in liquid nitrogen. However, it was sometimes difficult for an operator to manually perform all such preparation procedures by himself and constantly obtain well-frozen tissue specimens. To perform IVCT more easily, a new "*in vivo* cryoapparatus" (VIO-10, Eiko Corporation, Hitachinaka, Ibaraki, Japan) is now commercially available all over the world (Ohno *et al.*, 2004; Zea-Aragon *et al.*, 2004) (Figure 12.2). The operation manual for the "*in vivo* cryoapparatus" is briefly summarized below.

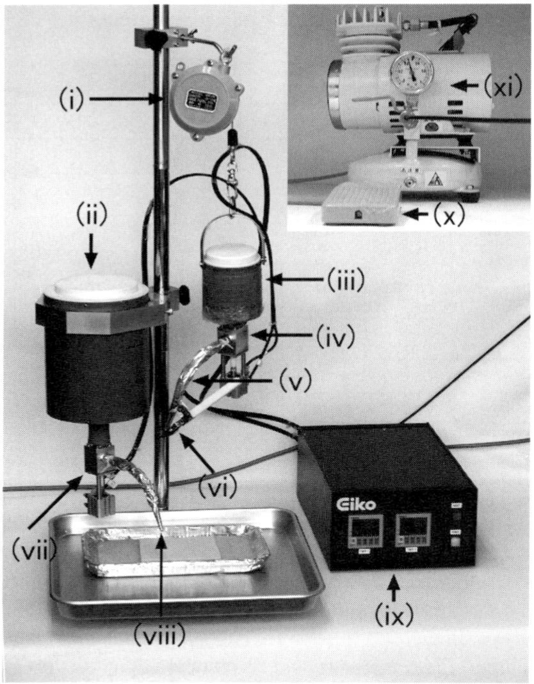

Figure 12.2 An overview of an "*in vivo* cryoapparatus," which is composed of the main part mounted on a table and a compressor (upper inset, xi) and a foot switch (upper inset, x) under the table. IVCT can be performed easily with the "*in vivo* cryoapparatus": (i) balancer, (ii) reservoir for liquid nitrogen, (iii) reservoir for isopentane–propane cryogen, (iv) valve for the isopentane–propane cryogen, (v) nozzle, (vi) cryoknife, (vii) valve for liquid nitrogen, (viii) nozzle for liquid nitrogen, (ix) electrical controller, (x) foot switch, and (xi) air compressor for opening or closing the valves.

Step 1: Pour some liquid nitrogen into two reservoirs [Figure 12.2(ii), (iii)] to be cooled down. After cooling of the reservoirs starts, they should be continuously cooled with liquid nitrogen. If they are temporarily warmed up, water produced by melting of attached frost covers all parts of the valves. When the reservoirs are cooled again below zero degrees, the valves are completely frozen and immobilized.

Step 2: Set timers of the controller [Figure 12.2(ix)] and press the foot switch [Figure 12.2, inset (x)] to check if the liquid nitrogen in the reservoirs correctly passes through the nozzles [Figure 12.2(v), (viii)]. This trial is important to prevent machinery accidents.

Step 3: Pour the isopentane–propane cryogen into the reservoir [Figure 12.2(iii)]. The cryogen is prepared beforehand by bubbling propane gas in liquid isopentane precooled in liquid nitrogen under constant agitation with a magnet stirrer. The ratio of isopentane to propane should be 1:2 to 1:3 to achieve the maximal cooling ability at −193 °C (Jehl *et al.*, 1981).

Step 4: Expose a target organ of living animals under anesthesia.

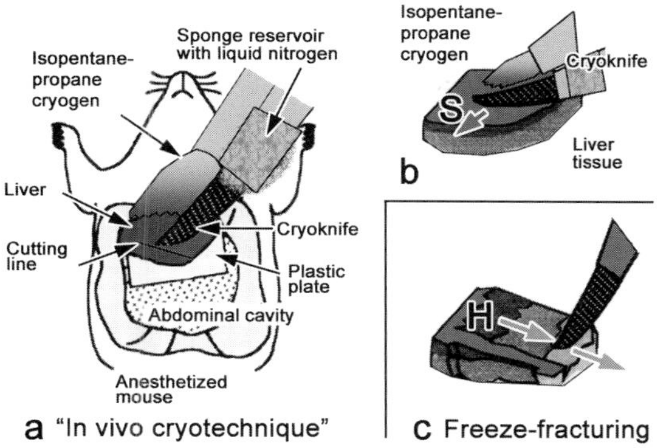

Figure 12.3 A representation of IVCT applied to anesthetized mouse livers. Note that the tissue surface is cryocut on a plastic plate by the IVCT (a, b), and the well-frozen tissue is additionally freeze-fractured by a cryoknife in liquid nitrogen (c). S: Cryo-cutting plane of liver tissues in contact with the cryoknife. H: Horizontally freeze-fractured liver tissues (with another cryoknife) in liquid nitrogen.

Step 5: Set timers of the controller [Figure 12.2(ix)], and precool the cryoknife [Figure 12.2(vi)] in liquid nitrogen pooled in another container. A piece of sponge (Fig. 12.3a) is sometimes attached to the cryoknife [Figure 12.2(vi)], to absorb some liquid nitrogen and keep the cryoknife cooled over the exposed organ during transfer.

Step 6: For this step, see Figure 12.3. Bring the cooled cryoknife [Figure 12.2(vi), 12.3a] onto the exposed target organ, and press the foot switch [Figure 12.2(x)]. Immediately after pressing, cut the animal organ manually with the cryoknife (Figure 12.3b), followed by the isopentane–propane cryogen being applied through the nozzle [Figure 12.2(v), 12.3a], which has already been initiated by pressing the foot switch and automatically regulated by the controller [Figure 12.2(ix)]. In several seconds after pouring the isopentane–propane cryogen, liquid nitrogen is also automatically poured onto the frozen organ through another nozzle [Figure 12.2(viii)].

Step 7: Put the frozen organ as a whole in liquid nitrogen, and preserve it until removal.

Step 8: Get necessary tissue parts from the frozen organ in the liquid nitrogen using a dental electric drill. After performing IVCT and then removing the frozen tissues, various subsequent preparation steps are chosen for each morphological analysis (Figure 12.1, i, j).

12.3 Technical processes and merits of IVCT

The main purpose of IVCT is to render all biological components of cells and tissues promptly frozen in living animal organs (Ohno *et al.*, 1996a, 1996b). However, the necessary freezing times are always different at each depth from the frozen tissue surface, because thermal conductance of cooling within cells and tissues is due to the continuous movement of thermal energy. So, only surface tissue layers within certain depths, such as about 10μm, are frozen enough to prevent formation of visible ice crystals in ultrathin sections at an electron microscopic level (Ohno *et al.*, 1996a, 1996b). The layered areas at the same depth from the frozen tissue surface are expected to be quickly frozen nearly at the same time. Therefore, by freeze-fracturing almost horizontally to the frozen tissue surface (Figure 12.3c), we can obtain similarly well-frozen tissue morphology over wide areas, which can be examined later by SEM.

To examine areas deeper below the frozen tissue surface, it is necessary to cryocut target organs of living animals under anesthesia (Figure 12.3a, b). When a cryoknife precooled in liquid nitrogen passes through the anesthetized animal organ, the exposed tissue surface in direct contact with the cryoknife is first frozen in the same way as with the metal contact freezing method (Ohno *et al.*, 1996a). Mechanical damage of cryocut tissues is rarely detected, if the target organs can be quickly cut through with a cryoknife. Then, the widely cryocut tissues are additionally frozen by another application of liquid cryogen, isopentane–propane mixture, which is simultaneously poured over them. In this process, physical speeds of the moving cryoknife and freezing intensity of the cryogen are critical for the necessary time to achieve *in vivo* freezing of living animal organs. Practically, at an electron microscopic level, well-frozen tissue areas appear to occupy a narrow band in ultrathin sections, less than 10 μm deep from the contact tissue surface (Ohno *et al.*, 1996a).

12.4 Further preparation after IVCT

12.4.1 Freeze-substitution fixation

All biological components, including proteins, lipids, carbohydrates, and electrolytes, of living animal organs are immediately embedded in tiny ice crystals *in situ* with IVCT. In past years, the freeze-substitution (FS) fixation method has often been used to cross-link molecules in cells and tissues at low temperatures by substituting ice crystals with organic solvents containing fixatives (Figure 12.1e). Then, specimens can often be transferred to organic solvents, such as t-butyl alcohol, for freeze-drying (Figure 12.1j). We have already reported the dynamic morphological changes of renal glomeruli and flowing erythrocytes in living mouse organs, which were revealed by TEM and SEM (Ohno *et al.*, 1996a, 1996b; Terada *et al.*, 1998a; Xue *et al.*, 1998; Ohno *et al.*, 2001; Xue *et al.*, 2001). As described in the previous papers, well-frozen tissue areas without visible ice crystals are obtained to examine the dynamically changing morphology of living animal organs *in vivo* at an electron microscopic level.

12.4.2 Freeze-drying for X-ray microanalysis

To analyze soluble electrolytes in cells and tissues at an electron microscopic level, they must be retained by the common freeze-drying (FD) method (Figure 12.1j), resulting in dehydrated electrolyte deposits *in vivo*. For example, X-ray microanalysis of elements in freeze-dried erythrocytes, as already observed with SEM after cryofixation of erythrocytes, showed a spectrum with different energy peaks of several atomic elements (Terada *et al.*, 2006). Thus, a dynamically changing morphology of flowing erythrocytes and also detection of certain electrolyte elements can be obtained with the FD method, following IVCT (Terada *et al.*, 1998, 2006; Terada and Ohno, 1998), as described below.

12.5 Application of IVCT to organs

12.5.1 Renal glomeruli under hemodynamic conditions

The renal glomerulus consists of intricate networks of blood capillaries, through which the blood always passes influenced by the blood pressure. It is well known that hemodynamic factors, including blood flow, exert an important influence on glomerular functions and structures (Griffith *et al.*, 1967; Kanwar, 1984; Kriz *et al.*, 1994). The alteration in glomerular hemodynamics probably affects the driving force that modulates permeability properties of the glomerular filtration barrier (Ryan *et al.*, 1976; Bohrer *et al.*, 1977; Brenner *et al.*, 1977). The distribution of albumin proteins in rat glomeruli has also been reported to be changed due to different hemodynamic conditions (Ryan and Karnovsky, 1976; Olivetti *et al.*, 1981). The functional role of normal blood circulation is definitely important in maintenance of the glomerular barrier function. In the present section, we describe three-dimensional ultrastructures of functioning kidneys *in vivo* under various hemodynamic conditions prepared by IVCT for SEM.

Left kidneys of anesthetized mice under normal blood circulation were prepared by IVCT, as described before (Ohno *et al.*, 1996a, 2001). Other mouse kidneys were prepared in the same way after heart arrest with an overdose of the anesthetic to stop blood supply, or ligation of the lower abdominal aorta to acutely increase blood supply into kidneys. The frozen specimens were routinely freeze-substituted in absolute acetone containing 2% osmium tetroxide, transferred into t-butyl alcohol and freeze-dried at −5 °C in the Hitachi ES-2030 apparatus. They were then mounted on aluminum stages, evaporated with platinum/palladium (10–15 nm) in the Hitachi E-1030 apparatus and observed in a Hitachi S-4500 scanning electron microscope at an accelerating voltage of 5 kV.

Many interdigitating foot processes under normal blood circulation, which are covering the outer capillary surface (Figure 12.4a), reflect their *in vivo* arrangement, as prepared by the *in vivo* cryotechnique. The foot processes are located almost in parallel with each other, and filtration slits are irregularly opened in some parts (Figure 12.4a, inset). Many erythrocytes are flowing in the capillary loops (Figure 12.4b), and foot processes with slit spaces are slightly extended (Figure 12.4b, inset), as seen in freeze-fractured capillary loops. To the contrary, glomerular foot processes are tightly attached to each other after heart arrest (Figure 12.5a, inset), and urinary spaces become collapsed (Figure 12.5a). At higher magnification, the foot processes are tall after heart arrest (Figure 12.5b, c), as seen in freeze-fractured capillary loops. In addition, another different morphology of stretched capillary loops is revealed after ligation of the aorta

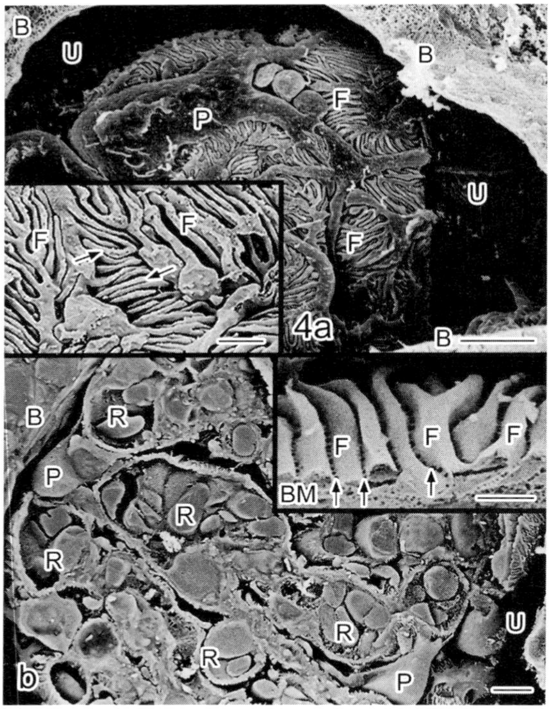

Figure 12.4 Scanning electron micrographs of freeze-fractured renal corpuscles of mice under normal blood flow condition, as prepared by IVCT: (a) Interdigitating foot processes are covering the outer capillary loop surface, P: podocytes, F: foot processes, U: urinary space, B: Bowman's capsule, bar: 5 μm. Inset: At higher magnification, surface contours of foot processes (F) with various slit spaces are arranged almost in parallel with each other, bar: 1 μm. (b) A freeze-fractured glomerulus is seen to have many flowing erythrocytes (R) with various shapes in blood capillary networks, bar: 5 μm. Inset: Filtration slits (arrows) are also seen to be open between foot processes (F) under the normal blood flow condition, BM: basement membrane, bar: 0.5 μm.

(Figure 12.6). As compared with the morphology of capillary loops after heart arrest (Figure 12.5), interdigitating foot processes with wide filtration slits are more loosely arranged with each other after ligation of the aorta (Figure 12.6a, inset). In a freeze-fractured renal corpuscle, urinary spaces and lumens of capillary loops are more widely dilated (Figure 12.6b), showing flowing blood cells in the capillary lumens.

The glomerular capillary loops of living mouse kidneys are perpetually stretched by hydraulic pressures in blood capillaries, whereas those in excised kidneys receive no such stretching forces. The filtration slits between foot processes of podocytes are widely open to facilitate passage of the glomerular filtrate *in vivo* (Kriz *et al.*, 1994). It has been reported that the foot processes are rich in contractile proteins, such as actin, myosin, and α-actinin, and capable of such actin-mediated movement (Andrews, 1988). Through this active movement, the podocytes may regulate the glomerular filtration rate, thereby influencing hydraulic pressures across the glomerular basement membrane, as reported previously (Kriz *et al.*, 1994). The blood pressure must be maintained steadily before

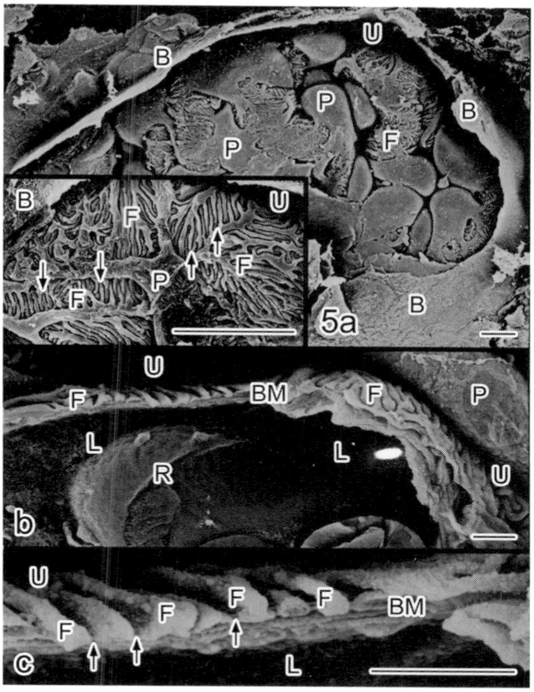

Figure 12.5 Scanning electron micrographs of freeze-fractured renal corpuscles of mice under heart arrest condition, as prepared by IVCT: (a) The urinary space (U) is almost collapsed, which is also covered by outer Bowman's capsule (B), bar: 5 μm. Inset: At higher magnification, filtration spaces between foot processes (F) are compactly interdigitated by each other (arrows), P: podocyte, bar: 5 μm. (b) and (c) A freeze-fractured capillary loop is seen, and foot processes (F) with narrow filtration slits (arrows) are attached to the basement membrane (BM), bars: 1 μm. (c) Higher magnification, L: capillary lumen, U: urinary space, R: erythrocyte.

exposure to freezing with IVCT, and the filtration slits are found to be wide under normal blood circulation. The glomerular hydraulic pressure might be regulated by the relative width of the filtration slits. The total width of the filtration slits plays an important role in providing a porous theory that controls hydraulic conductivity and water flow (Drumond and Deen, 1994; Kriz *et al.*, 1994). The present findings provide the morphological confirmation of glomerular capillary loops in living states, indicating that the passage of solutes is always affected by glomerular hemodynamics under different blood flow conditions.

12.5.2 Changing shapes of erythrocytes in mouse organs

Mammalian erythrocytes usually keep their biconcave discoid shapes *in vitro*, but they change their form under complicated dynamic conditions in blood vessels (Maeda, 1996). The deformability of erythrocytes is an important factor in such a situation. However, the morphological appearance of erythrocytes had been unknown in various

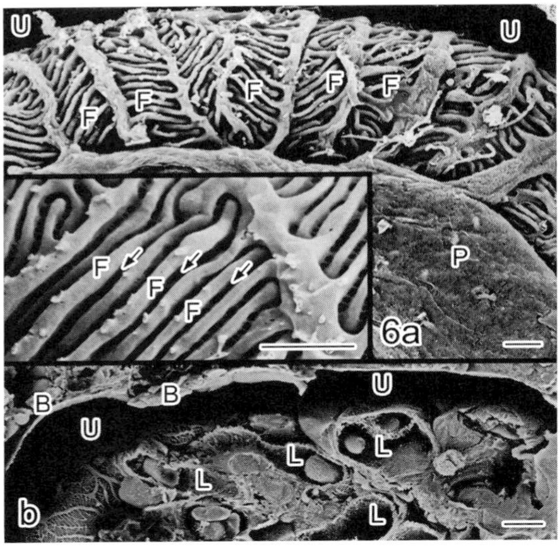

Figure 12.6 Scanning electron micrographs of freeze-fractured renal corpuscles of mice under aorta ligation condition, as prepared by IVCT: (a) Interdigitating foot processes (F) are extended to become slender, covering the outer surface of blood capillary loops, P: podocyte, U: urinary space, bars: 1 μm. Inset: Higher magnification view of foot processes (F), filtration slits between thin foot processes are more widely open (arrows). (b) The lumens of freeze-fractured capillary loops (L) are also more widely open with dilatation of the urinary space (U), B: Bowman's capsule, bar: 5 μm.

blood vessels of living animal organs, because of limitations in preparative techniques for electron microscopy. It was also difficult to observe behavior of erythrocytes in large blood vessels directly with their thick walls and solid organs. Therefore, conventional morphological studies with the immersion- or perfusion-fixation method have not revealed the true morphology of circulating erythrocytes under various blood flow conditions. In the present section, the authors have used IVCT to examine the behavior of flowing erythrocytes in living mouse organs, such as large blood vessels and hepatic sinusoids (Terada *et al.*, 1998; Xue *et al.*, 1998). The abdominal aorta, inferior vena cava, and livers of anesthetized mice were prepared by IVCT, as described above. The frozen organs were submitted to the routine freeze-substitution fixation method.

12.5.2.1 Flowing erythrocytes in aorta and inferior vena cava

Three-dimensional shapes of flowing erythrocytes *in vivo* appear to be various in the abdominal aorta, and some ellipsoidal or curved erythrocytes are always observed (Figure 12.7a). On the contrary, most erythrocytes flowing in the inferior vena cava form almost biconcave discoid shapes (Figure 12.7b). In past years, erythrocytes in stored blood have been known to have the typical biconcave discoid shape. However, in large blood vessels *in vivo*, rheology should also be considered, in relation to their bulk flow (Stuart and Nash, 1990). Arterial flow is reported to be laminar, especially in the abdominal aorta (Robinson, 1978). A high shear rate in a large artery is determined by a rapid flow velocity, ~100 cm/s (Klug *et al.*, 1974). It thus provides the possibility for

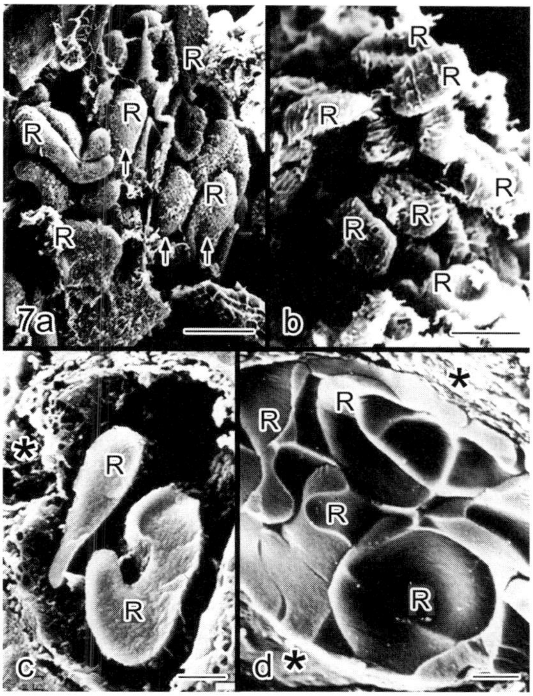

Figure 12.7 (a) and (b) Scanning electron micrographs of flowing erythrocytes in abdominal aorta (a) and inferior vena cava (b) of mice under normal blood flow condition, as prepared by IVCT, bars: 5 µm. (a) Flowing erythrocytes in the abdominal aorta appear to have various shapes (R), and some of them are stretched into ellipsoidal shapes (arrows). (b) Flowing erythrocytes in the inferior vena cava mostly resemble the typical biconcave discoid shapes (R). (c) and (d) Scanning electron micrographs of flowing erythrocytes in hepatic sinusoids of mice under normal blood flow (c) or heart arrest (d) condition, as prepared by IVCT, bars: 1 µm. (c) In a freeze-fractured sinusoid, flowing erythrocytes are seen to be various shapes (R), asterisks: open space of Disse. (d) Under the heart arrest condition, erythrocytes are aggregated in the collapsed sinusoid to form the typical biconcave discoid shapes (R), asterisks: collapsed space of Disse.

erythrocyte deformation as a result of external force stress. In the present study, erythrocyte shapes in the aorta are seen to be varied, some being stretched along the direction of blood flow. Conversely, a low shear rate in venous blood flow of the inferior vena cava, probably at a velocity of 30 cm/s (Klug *et al.*, 1974), results in erythrocyte shapes being approximately the biconcave discoid.

In response to fluid shear forces, erythrocytes are changed from the resting biconcave discoid shape into an ellipsoid form and align with their long axes parallel to the fluid stream. Such temporary erythrocyte deformability in large blood vessels can be examined by IVCT. It is tempting to conclude that the erythrocyte deformability differs between the abdominal aorta and the inferior vena cava, because haematocrit, plasma viscosity, and erythrocyte aggregation are significantly higher in venous than in arterial blood (Mokken *et al.*, 1996). Erythrocyte deformability in blood vessels is adapted to blood flow conditions and in relation to their function in oxygen delivery. It has been apparent that the shape and elasticity of human erythrocytes are important for explaining the etiology of certain pathological conditions (Athanassiou *et al.*, 1992; Cynober *et al.*, 1996). Some hemolytic anemias, for example, are related to increased mechanical fragility of erythrocyte membranes (Rybicki *et al.*, 1993), which can be examined *in vivo* by IVCT.

12.5.2.2 Flowing erythrocytes in hepatic sinusoid

The authors have also applied IVCT to examining flowing erythrocyte shapes in hepatic sinusoids of living mouse livers (Figure 12.7c). Their morphological features are different from those flowing in the aorta and inferior vena cava. Some spaces between flowing erythrocytes are observed in the hepatic sinusoids with open spaces of Disse (Figure 12.7c, asterisk). After artificial heart arrest, they are congested in the sinusoidal lumen (Figure 12.7d). Such erythrocyte shapes are completely different from those in the sinusoids of living mouse livers, and most of them appear to be typical biconcave discoid shapes. Moreover, spaces of Disse between hepatocytes and endothelial cells are completely collapsed (Figure 12.7d, asterisk).

The flowing erythrocytes had various shapes in hepatic sinusoids, other than biconcave discoid shapes, responding to hemodynamic stresses. In isotonic physiological solution, erythrocyte shapes are generally known to be regular and uniform to form biconcave discoid shapes. Some distance between the erythrocyte surface and the endothelium has a physiological significance for the erythrocyte-capillary relationship, which results in reduction of the diffusion distance and increased shear stresses. Moreover, the erythrocyte deformation to allow its passage through networks of narrow blood capillaries also increases the contact surface of the endothelium and erythrocytes, and effectively broadens their surface area for gas diffusion. With the erythrocyte deformation, the surface structure of erythrocytes, including skeletal proteins of erythrocyte membranes, must be responsible for flow through the narrow blood vessels (Shiga *et al.*, 1990; Terada *et al.*, 1997).

12.5.2.3 X-ray microanalysis of erythrocytes in hepatic sinusoids

The IVCT approach has already been developed for examining the erythrocyte shapes flowing in large blood vessels and hepatic sinusoids of living mice. However, there have been no reports about electrolyte concentrations of erythrocytes *in vivo* under blood flowing conditions. The variable-pressure SEM has been used to observe hydrated and uncoated biological specimens. In the present section, by using IVCT combined with the common freeze-drying method, the authors have examined the uncoated erythrocyte morphology and electrolyte elements of sinusoidal erythrocytes under the normal blood flow or heart-arrest condition, using a variable-pressure SEM equipped with an X-ray microanalysis system.

The mice were anesthetized with sodium pentobarbital, and their livers were routinely prepared by IVCT, as described before (Figure 12.3a, b). The frozen liver tissue surface was freeze-fractured with another cryoknife in liquid nitrogen (Figure 12.3c). The specimens were then freeze-dried at −95 °C in a freeze-etching apparatus (10^{-5} Pa) for 24 hr, as reported previously (Yoshimura *et al.*, 1991; Ohno *et al.*, 1992). The freeze-dried specimens were gradually warmed up to room temperature. They were attached to carbon plates by using graphite containing resin, and some were coated with carbon alone. They were finally analyzed with a Hitachi S-4300SE/N or S-3000N variable-pressure SEM equipped with an X-ray microanalysis system (accelerating voltage: 10 kV, vacuum condition: 30 Pa, illumination current: 50 μA, analytical time: 950 s), or an S-4500 conventional SEM (Hitachi High-Technologies Corp., Tokyo, Japan) at an accelerating voltage of 10 kV. The analyzed elements were Na(sodium), P(phosphorus), S(sulfur), Cl(chloride), and K(potassium).

Figure 12.8a shows freeze-fractured liver tissues of living mice under normal blood flow condition, as observed by the variable-pressure SEM. Natural images of live hepatocytes and various shapes of flowing erythrocytes in hepatic sinusoids are detected without any metal coating. Small oval-shaped cell organelles are seen in the cytoplasm of

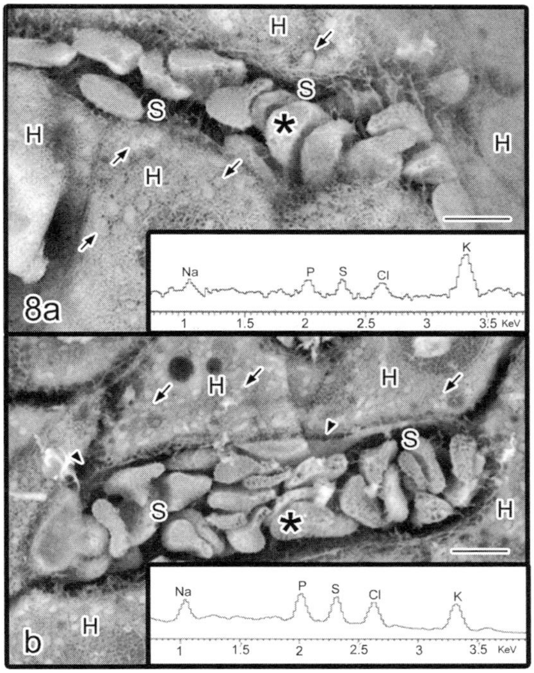

Figure 12.8 Scanning electron micrographs of freeze-fractured hepatic sinusoids of mice, examined by variable-pressure SEM. (a) Normal blood flow condition: images of liver tissues without any metal coating are obtained with back-scattered electron beams. Flowing erythrocytes with various shapes are observed in the hepatic sinusoids (S). Freeze-fractured hepatocytes (H) are also observed and some cell organelles can be detected in their cytoplasm (arrows), bars: 5 μm. Inset: X-ray microanalysis data of flowing erythrocytes (asterisk) under the normal blood flow condition. The potassium peak (K) is the highest, and the left sodium peak (Na) is the lowest. (b) In hepatic sinusoids (S) of the heart-arrested mouse, erythrocytes are aggregated to each other, but a space of Disse is still observed (arrowheads). Inset: X-ray microanalysis data of erythrocytes (asterisk) in the heart-arrested mouse liver. The potassium peak (K) is decreased, but the sodium peak (Na) is relatively increased, H: hepatocyte, arrows: cell organelle.

hepatocytes (Figure 12.8a, arrows). The X-ray microanalysis data of flowing erythrocytes were obtained under normal blood flow condition (Figure 12.8a, inset). The highest potassium peak is seen at the right side, and the lower sodium peak is observed at the left side. Figure 12.8b shows other freeze-fractured liver tissues of mice under a heart-arrest condition. Many erythrocytes are congested in the sinusoidal spaces. The space of Disse can be detected slightly (Figure 12.8b, arrowheads). The X-ray microanalysis data of erythrocytes were also obtained under heart-arrest condition. The potassium peak is relatively decreased, but the sodium peak is increased, probably because of anoxia. Other phosphorus, sulfur, and chloride are relatively almost at a similar peak level.

Elemental changes of flowing erythrocytes in hepatic sinusoids of mouse liver tissues have been clarified by using variable-pressure SEM with X-ray microanalysis. The IVCT in combination with the variable-pressure SEM is a powerful tool for analyzing biological samples with less technical artifacts for morphology and atomic elements of cells and tissues. Some microenvironmental conditions against flowing erythrocytes in hepatic sinusoids may cause functional changes in ion transport systems of erythrocyte membranes, which are also related to their morphological changes, as revealed in the present study.

12.6 Concluding remarks

Many biological components and structures of living animal organs are dynamically changing to accommodate functions of numerous extracellular or intracellular body fluids. Our ultimate goal is to obtain the real histology of animal organs. The IVCT is a first step for keeping normal blood circulation into various organs of living animals (Figure 12.1f), which can be followed by various electron microscopic procedures (Figure 12.1). So, by IVCT, it is now possible to perform morphological analyses of living animal organs without the major stresses of ischemia and anoxia. Some findings regarding cells and tissues, which had never been demonstrated by other conventional preparation methods for SEM, have been reviewed.

12.7 References

Andrews, P. (1988). Morphological alterations of the glomerular (visceral) epithelium in response to pathological and experimental situations. *Journal of Electron Microscopy Technique*, **9**, 115–44.

Athanassiou, G., Symeonidis, A., Kourakli, A., Missirlis, Y. F., and Zoumbos, N. C. (1992). Deformability of the erythrocyte membrane in patients with myelodysplastic syndromes. *Acta Haematologica*, **87**, 169–72.

Bohrer, M. P., Deen, W. M., Robertson, C. R., and Brenner, B. M. (1977). Mechanism of angiotensin II-induced proteinuria in the rat. *American Journal of Physiology*, **233**, F13–F21.

Brenner, B. M., Bohner, M. P., Baylis, C., and Deen, W. M. (1977). Determinants of glomerular permselectivity: Insights derived from observations *in vivo*. *Kidney International*, **12**, 229–37.

Cynober, T., Mohandas, N., and Tchernia, G. (1996). Red cell abnormalities in hereditary spherocytosis: relevance to diagnosis and understanding of the variable expression of clinical severity. *Journal of Laboratory and Clinical Medicine*, **128**, 259–69.

Drumond, M. C. and Deen, W. M. (1994). Structural determinants of glomerular hydraulic permeability. *American Journal of Physiology*, **266**, F1–F12.

Furukawa, T., Ohno, S., Oguchi, H., *et al.* (1991). Morphometric study of glomerular slit diaphragms fixed by rapid-freezing and freeze-substitution. *Kidney International*, **40**, 621–4.

Griffith, L. D., Bulger, R. E., and Trump, B. F. (1967). The ultrastructure of the functioning kidney. *Laboratory Investigation*, **16**, 220–46.

Jehl, B., Bauer, R., Dorge, A., and Rick, R. (1981). The use of propane/isopentane mixtures for rapid freezing of biological specimens. *Journal of Microscopy*, **123**, 307–9.

Kanwar, Y. S. (1984). Biophysiology of glomerular filtration and proteinuria. *Laboratory Investigation*, **51**, 7–21.

Klug, P. P., Lessin, L. S., and Radice, P. (1974). Rheological aspects of sickle cell disease. *Archives Internal Medicine*, **133**, 577–90.

Kriz, W., Hackenthal, E., Nobiling, R., Sakai, T., and Elger, M. (1994). A role for podocytes to counteract capillary wall distension. *Kidney International*, **45**, 369–76.

Maeda, N. (1996). Erythrocyte rheology in microcirculation. *Japanese Journal of Physiology*, **46**, 1–14.

Mokken, F. C., Waart, F. J., Henny, C. P., Goedhart, P. T., and Gelb, A. W. (1996). Differences in peripheral arterial and venous hemorheologic parameters. *Annals of Hematology*, **73**, 135–7.

Ohno, N., Terada, N., Saitoh, S., *et al.* (2007). Recent development of *in vivo* cryotechnique to cryobiopsy for living animals. *Histology and Histopathology*, **22**, 1281–90.

Ohno, N., Terada, N., Fujii, Y., Baba, T., and Ohno, S. (2004). "*In vivo* cryotechnique" for paradigm shift to "living morphology" of animal organs. *Biomedical Review*, **15**, 1–19.

Ohno, S., Terada, N., Ohno, N., Fujii, Y., and Baba, T. (2006). "*In vivo* cryotechnique" for examination of living animal organs, further developing to "cryobiopsy" for humans. *Recent Research Development Molecular Cell Biology*, **6**, 65–90.

Ohno, S., Kato, Y., Xiang, T., *et al.* (2001). Ultrastructural study of mouse renal glomeruli under various hemodynamic conditions by an "*in vivo* cryotechnique." *Italian Journal of Anatomy and Embryology*, **106**, 431–8.

Ohno, S., Terada, N., Fujii, Y., Ueda, H., and Takayama, I. (1996a). Dynamic structure of glomerular capillary loop as revealed by an *in vivo* cryotechnique. *Virchows Archiv*, **427**, 519–27.

Ohno, S., Baba, T., Terada, N., Fujii, Y., and Ueda, H. (1996b). Cell biology of kidney glomerulus. *International Review of Cytology*, **166**, 181–230.

Ohno, S., Hora, K., Furukawa, T., and Oguchi, H. (1992). Ultrastructural study of the glomerular slit diaphragm in fresh unfixed kidneys by a quick-freezing method. *Virchows Archiv B Cell Pathology*, **61**, 351–8.

Olivetti, G., Kithier, K., Giacomelli, F., and Wiener, J. (1981). Glomerular permeability to endogenous proteins in the rat. *Laboratory Investigation*, **44**, 127–37.

Robinson, K. (1978). Abdominal aorta. In: *Circulation of the Blood* (ed. James, D. G.). Tunbridge Wells, England, Pitman Medical, pp. 173–175.

Ryan, G. B. and Karnovsky, M. J. (1976). Distribution of endogenous albumin in the rat glomerulus: Role of hemodynamic factors in glomerular barrier function. *Kidney International*, **9**, 36–45.

Ryan, G. B., Hein, S. J., and Karnovsky, M. J. (1976). Glomerular permeability to proteins. Effects of hemodynamic factors on the distribution of endogenous immunoglobulin G and exogenous catalase in the rat glomerulus. *Laboratory Investigation*, **34**, 415–27.

Rybicki, A. C., Qiu, J. J., Musto, S., *et al.* (1993). Human erythrocyte protein 4.2 deficiency associated with hemolytic anemia and a homozygous 40 glutamic acid-lysine substitution in the cytoplasmic domain of band 3. *Blood*, **81**, 2155–65.

Shiga, T., Maeda, N., and Kon, K. (1990). Erythrocyte rheology. *Critical Reviews in Oncology and Hematology*, **10**, 9–48.

Stuart, J. and Nash, G. B. (1990). Red cell deformability and haematological disorders. *Blood Reviews*, **4**, 141–7.

Terada, N., Ohno, N., Li, Z., *et al.* (2006). Application of *in vivo* cryotechnique to the examination of cells and tissues in living animal organs. *Histology and Histopathology*, **21**, 265–72.

Terada, N., Kato, Y., Fuji, Y. *et al.* (1998). Scanning electron microscopic study of flowing erythrocytes in hepatic sinusoids as revealed by "in vivo cryotechnique." *Journal of Electron Microscopy*, **47**, 67–72.

Terada, N. and Ohno, S. (1998). Dynamic morphology of erythrocytes revealed by cryofixation technique. *Acta Anatomica Nipponica*, **73**, 587–93.

Terada, N., Fujii, Y., Ueda, H., and Ohno, S. (1997). Immunocytochemical study of human erythrocyte membrane skeletons under stretching conditions by quick-freezing and deep-etching method. *Journal of Anatomy*, **190**, 397–404.

Xue, M., Baba, T., Terada, N., *et al.* (2001). Morphological study of erythrocyte shapes in red pulp of mouse spleens revealed by an *in vivo* cryotechnique. *Histology and Histopathology*, **16**, 123–9.

Xue, M., Kato, Y., Terada, N. *et al.* (1998). Morphological study by an "*in vivo* cryotechnique" of the shape of erythrocytes circulating in large blood vessels. *Journal of Anatomy*, **193**, 73–9.

Yoshimura, A., Ohno, S., Nakano, K., *et al.* (1991). Three-dimensional ultrastructure of anionic sites of the glomerular basement membrane by a quick-freezing and deep-etching method using a cationic tracer. *Histochemistry*, **96**, 107–13.

Yu, Y., Leng, C. G., Kato, Y., *et al.* (1998). Ultrastructural study of anionic sites in glomerular basement membranes at different perfusion pressures by quick-freezing and deep-etching method. *Nephron*, **78**, 88–95.

Yu, Y., Leng, C. G., Kato, Y., and Ohno, S. (1997). Ultrastructural study of glomerular capillary loops at different perfusion pressures as revealed by quick-freezing, freeze-substitution and conventional fixation methods. *Nephron*, **76**, 452–459.

Zea-Aragon, A., Terada, N., Ohno, N., *et al.* (2004). Effects of anoxia on serum immunoglobulin and albumin leakage through blood-brain barrier in mouse cerebellum as revealed by cryotechniques. *Journal of Neuroscience Methods*, **138**, 89–95.

13 SEM in dental research

Vladimir Dusevich, Jennifer R. Melander, and J. David Eick

13.1 Introduction

The versatility of scanning electron microscopy (SEM) makes it an extremely useful tool in the exciting and remarkably wide field of dental research. Dental research ranges from the study of soft and hard tissues (such as dentin, enamel, and bone) to cell culture (i.e. mineralizing bone cells) to dental materials (such as adhesives and composites, metals, and ceramics). Characterization of these materials relies to a large extent on SEM. For example, in 2009, 33% of papers in the journal of *Dental Materials* and 47% in *Dental Materials Journal* (Japan) contained information obtained with SEM. It is difficult to overestimate the importance of SEM in dental research. It is equally difficult to summarize in more than general terms the dental investigations that employ SEM in one book chapter. This chapter, therefore, is focused on certain practical aspects of the utilization of SEM in dental research with some examples based on current literature.

Teeth are covered with a layer of enamel protecting the dentin which surrounds the pulp (Figure 13.1). The pulp is a living viable tissue composed of nerves and blood vessels which connect directly to the supporting bone. Biological fluid is transported from the blood supply in the bone through the dentin tubules to the dentin and creates a dynamic circulatory system between the living bone and tooth structure. A carious lesion involves breakdown of the hard tooth structure by acid-producing bacteria, many times leading to infection of the pulp and possible death of the tooth and supporting bone. A large proportion of dental research is devoted to restoration of carious lesions with dental composite. Accordingly, a number of examples in this chapter involve composite restorations, so a basic explanation of this restorative technique is presented. Preparation of a carious lesion (i.e. cavity) with dental cutting instruments creates a smear layer on top of the exposed enamel and dentin (Eick *et al.*, 1970). The smear layer consists of collagen debris and mineral (hydroxyapatite). For proper adhesion of dental bonding agents, the smear layer and the underlying dentin should be etched with acids to remove the mineral and expose the dentin collagen network. Adhesives penetrate the network creating a hybrid layer (effectively a collagen-resin composite) and form a bond to dentin (Nakabayashi, 1982; Eick *et al.*, 1997). Subsequently, dental composite, a mixture of resin and filler particles, is filled over the adhesive layer. The properties of etched dentin, hybrid layers, adhesives, and composites determine the longevity of the restoration and are extensively studied with a wide application of SEM techniques.

Scanning Electron Microscopy for the Life Sciences, edited by H. Schatten. Published by Cambridge University Press

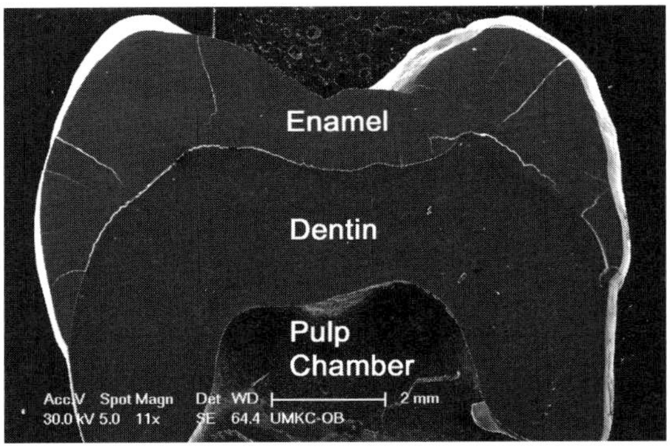

Figure 13.1 Tooth structure. The three main components of a tooth are the enamel, dentin, and pulp. Courtesy Dr. Changqi Xu, unpublished.

13.2 Specimen preparation

Preparation of biological specimens for conventional SEM is described elsewhere (Bozzola and Russell, 1999; Dykstra and Reuss, 2003) and is only discussed briefly in this chapter. Aldehyde fixation is frequently adequate for soft tissues (Dykstra and Reuss, 2003), but in some cases post-fixation with osmium tetroxide can help stabilize beam sensitive specimens. The so-called non-coating techniques involve deposition of a thin layer of a conductive metal on a specimen by reducing the metal (usually osmium) from an aqueous salt solution (Bozzola and Russell, 1999; Jongebloed *et al.*, 1999). It must be noted that in some instances this procedure may cause severe tissue damage (Goldstein *et al.*, 2003).

Specimens must withstand a vacuum environment for conventional SEM. Therefore, specimens are generally dehydrated in such a way as to prevent the disruptive forces of water surface tension from altering specimen morphology. The dehydration procedure usually begins by replacing water in the specimen with ethanol, methanol, or acetone by passing the specimen through increasing concentrations of these dehydrating liquids. The surface tension of these liquids is still high enough to damage specimens during air drying. So, these liquids are substituted with a subsequent liquid, often liquid CO_2 in a special unit called a critical point drier (CPD). After this substitution, it is possible to remove the CO_2 without crossing a phase boundary. The specimen is dehydrated with minimal damage by bypassing the critical point where the distinction between gas and liquid ceases to apply and the specimen is not subject to surface tension forces. Instead of using a CPD, the dehydrating liquid could be substituted with a highly volatile liquid with low surface tension, such as hexamethyldisilazane (HMDS), and then air dried. In dental research HMDS is widely used and gives satisfactory results (Perdigao *et al.*, 1995). Another method of dehydration, freeze-drying, has been used occasionally in dental research (Rojas-Sanchez *et al.*, 2007; Aida *et al.*, 2009), but its advantages for mineralized tissues (such as decreased specimen shrinkage) are questionable, and the strict requirements of small specimen dimensions and the rather long drying times prevent its wide use.

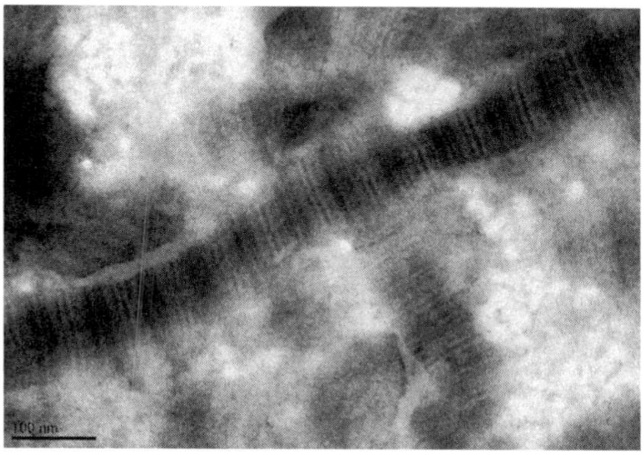

Figure 13.2 TEM micrograph of hybrid layer of not fixed, air dried tooth. Ultrathin sections were cut without embedding (hybrid layer performed sufficient embedding function). Good preservation of collagen fibers. Bar 100 nm.

While soft tissues such as cell cultures and the etched layer of dentin should always be fixed and dehydrated in a proper way, mineralized tissues can often be observed without fixation in an air dried state. When tooth or bone specimens are embedded in resin and polished, fixation does not improve the quality of the results when observing mineral content. In many instances of dentin–adhesive interface investigations, when the only biological organic component of interest is collagen, neither fixation nor dehydration is essential and in many cases neither is performed (Takahashi *et al.*, 2002; Osorio *et al.*, 2005; Purk *et al.*, 2009). Collagen in the hybrid layer has its structure preserved even after air drying without fixation, as can be seen from a TEM micrograph of the hybrid layer of a tooth that was neither fixed nor embedded, but simply air dried (Figure 13.2).

The two most common specimen types for the study of dentin–adhesive interfaces are cross-sections and cross fractures. Again, fixation and dehydration are not essential in this case and often are not performed. Cross-sections are prepared by cutting the specimen in a plane orthogonal to the interface, which optionally can be polished. Polished cross-sections are sometimes ion etched (Aida *et al.*, 2009). Planes orthogonal to the interface can also be created by fracturing; pre-made grooves in the specimen can be helpful to direct fracture propagation. Both cross-sections and cross fractures can be acid etched (5–60 s using 37% phosphoric acid) with subsequent deproteinization with sodium hypochlorite (NaOCl) for 5–30 min (Guo *et al.*, 2007; Tay *et al.*, 2007). Since sodium hypochlorite is the main component of household bleach, such deproteinization is often called "bleaching." While etching is essential for cross-sections to remove the smear layer created by cutting, cross fractures can be observed without any additional treatments. Both etched and unetched specimens offer a good view of the hybrid layer (Figure 13.3). Etched specimens provide a full view of the resin tags, but do not show dentin (Figure 13.3a). Unetched cross fractures illustrate the dentin morphology and its interface with the hybrid layer and resin tags (Figure 13.3b). Common artifacts of these preparations, both cut and fractured, are separations (cracks) along the adhesive–hybrid layer and hybrid layer–dentin interfaces. Resin tags presented on Figure 13.3b are polymerized dental adhesive resin, which has

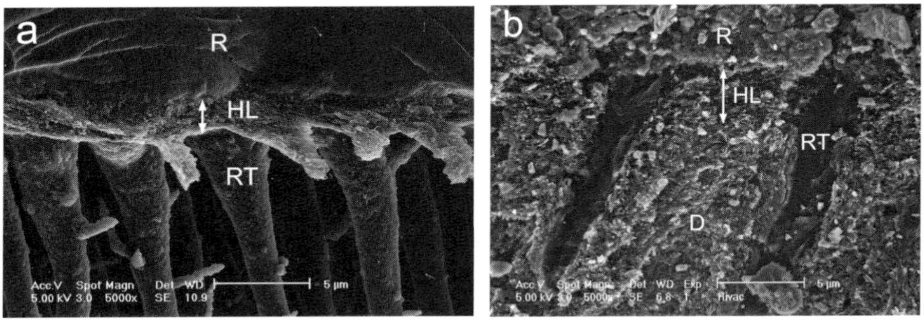

Figure 13.3 Cross fractures of bonding system–dentin interface; R – adhesive resin, D – dentin, HL – hybrid layer, RT – resin tag in dental tubule. (a) Scotchbond Multi-Purpose Plus adhesive. Etched for 30 s with 37% phosphoric acid and deproteinized with sodium hypochlorite for 15 min. Bar 5 µm. (b) PQ1 adhesive. No etching, no deproteinization. Bar 5 µm.

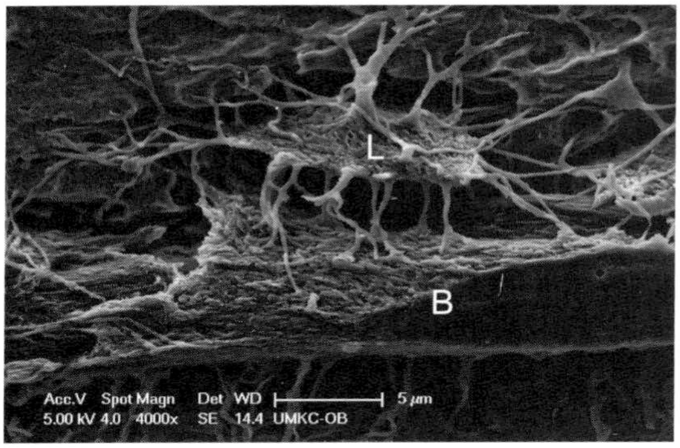

Figure 13.4 Resin casts of osteocyte lacunae (L), connected by its canaliculi to blood vessel (B). Bar 5 µm. Courtesy of Dr. Bonewald, unpublished.

filled into dentin tubules. Dentin tubules are one of the most prominent features of dentin and represent canals that radiate from the pulp through the dentin.

To visualize the non-mineralized volumes in dentin (tubules) and bone (lacuno-canalicular system) a special method of resin casting can be employed. Specimens embedded in methyl methacrylate resin are cut, ground (to create a flat surface), etched with acid (e.g. with 37% phosphoric acid for 10 s), deproteinized in sodium hypochlorite (10–20 min), and air dried (Martin *et al.*, 1978; Feng *et al.*, 2006; Lu *et al.*, 2007; Gorustovich, 2010). Dentin casts resemble those produced by dental adhesives, seen in Figure 13.3a. For bone, the method provides an image of lacunae, canaliculi, and blood vessels, illustrated in Figure 13.4.

13.3 Conventional SEM, secondary electrons

Among the signals generated by the interaction of the electron beam with a specimen in an SEM, the secondary electron (SE) signal is the most widely used. Other signals used for imaging include back-scattered electrons (BSE), X-rays, and cathodoluminescence. Cathodoluminescence (emitting of light by a specimen under the electron beam) is the least utilized signal in dental research, and while there has been some research that relied on application of cathodoluminescence (Boyde and Reid, 1983; Boyde and Jones, 1996; Piattelli *et al.*, 1999), it has not found a wide application in dental research and is not discussed further here.

In choosing accelerating voltages some microscopists are advocates of the "old school" approach and always work at 15–25 kV; old microscopes simply could not provide a suitable resolution at lower voltages. Working at these voltages can result in significant disadvantages: loss of fine surface details and a drastically increased edge effect (Boyde and Jones, 1996). However, there are circumstances when higher voltages are preferable. The SE signal consists of secondary electrons generated by both the incoming electron beam and outgoing back-scattered electrons (BSE). Therefore, the SE signal may reflect, at least to some degree, the composition of the specimen (discussed later in more detail in the BSE section). With increasing accelerating voltage, a greater number of beam electrons penetrate the conductive coating, creating more back-scattered electrons from the specimen itself, which increases the fraction of compositional contrast in the SE signal/image. Figure 13.5 shows the effect of increasing acceleration voltage on MLO-A5 bone cell culture images. The micrograph of the cell culture obtained at 1 kV accelerating voltage (Figure 13.5a) looks fine, and only a comparison with the micrograph obtained at 4 kV (Figure 13.5b) shows the drawbacks of the lower voltage image. The most striking difference is that the 1 kV image could not make a distinction between the cell surface and the substrate surface, displaying them at the same brightness level. When the voltage is increased to 4 kV, the difference between the cells and the substrate becomes clearly visible. The cell attachments (the fine details on cell edges) become far more noticeable and the micrograph is overall better suited for examining the cell culture. A further increase in accelerating voltage (15.0 kV, Figure 13.5c) did not yield additional improvements. In fact, edge effects are more pronounced at higher voltages, which make some image features disproportionately bright, thus decreasing the overall image quality. It is thus necessary to find the optimal accelerating voltage value for specimen observation. Sometimes the use of a higher voltage (i.e. 15 kV) can be justified, for instance to help

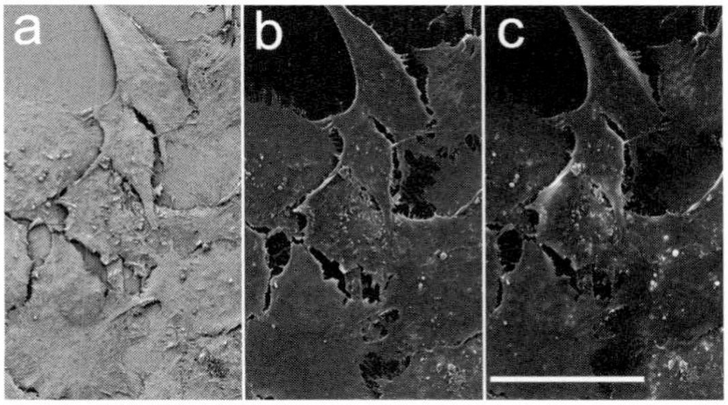

Figure 13.5 MLO-A5 cell culture observed at: (a) 1 kV, (b) 4 kV, and (c) 15 kV. Bar 50 μm. Reprinted with permission (Dusevich *et al.*, 2010).

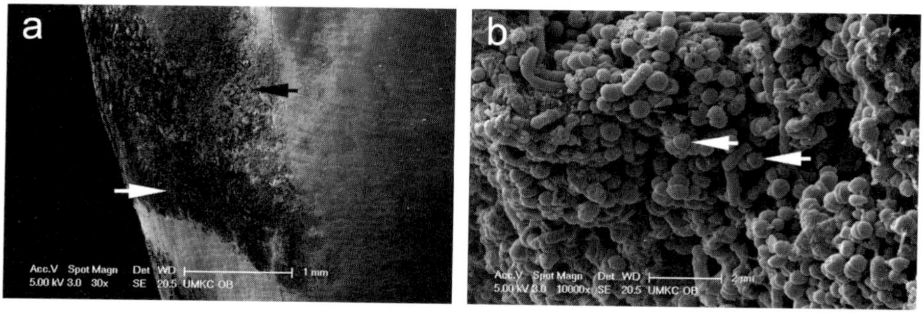

Figure 13.6 (a) Low-magnification view of tooth enamel with adhering microbial biofilm in various stages of maturation, relatively thin microbial layers (white arrow) to those of greater thickness (black arrow). Perikymata are representative of incremental enamel formation and typically show a transverse orientation relative to the long axis of the tooth. Bar 1 mm. (b) Maturing supragingival biofilm on enamel surface. Based on dominan bacterial morphotypes this biofilm appears to be approximately 2–3 days old. The biofilm is dominated by cocci (black arrow), many of which are in various stages of replication, i.e. binary fission (white arrow). Other bacterial morphotypes noted are short rods and medium length rods (black arrow), which typically follow cocci in the process of successional colonization of a developing biofilm. Courtesy of Dr. C. M. Cobb, unpublished.

distinguish composite from adhesive in the case of fractographical examinations at low and intermediate magnifications.

Low-voltage imaging can be performed without using a conductive coating on specimens when incoming and outgoing electrons are balanced. However, this technique is not often employed in dental research. Biological specimens contain low atomic number elements resulting in a low secondary electron yield. This leads to a low signal-to-noise ratio, problems with image quality, and limitations in high-resolution imaging. Therefore, thin conductive coatings with heavy metals are often beneficial with biological specimens to improve image quality/resolution as well as specimen conductivity (Goldstein *et al.*, 2003).

SE imaging provides morphological details for biofilm research (Li *et al.*, 2009; Van Essche *et al.*, 2009) (Figure 13.6). It can also be used for estimating the surface quality of polished dental composites (Ozel *et al.*, 2008) and the efficacy of dentin etching agents (Blomlof *et al.*, 1997). SE images are widely used for studies of the adhesive–dentin interfaces (Tay *et al.*, 2000; Purk *et al.*, 2004; Wang and Spencer, 2005; Tay *et al.*, 2007; Kim *et al.*, 2010). SE imaging is also sufficient for bone marrow studies (Bi *et al.*, 2000).

An interesting feature of SEM is its ability to produce 3-D images by making "stereo pairs." With this technique, two micrographs of the specimen are taken at different tilt angles. A full 3-D effect is produced by viewing them through a stereo viewer or with two-color glasses after conversion of these two images to an anaglyph (a combination of two superimposed images color coded for the left and right eye). So far stereo SEM has had limited application in dental research (Lester and Boyde, 1987; Dusevich and Eick, 2002). However, as stereo capabilities are provided with some new scanning electron microscopes as a built in feature and new commercial software packages are capable of creating anaglyphs, profiles, and 3-D models, a wider application of 3-D SEM in dental

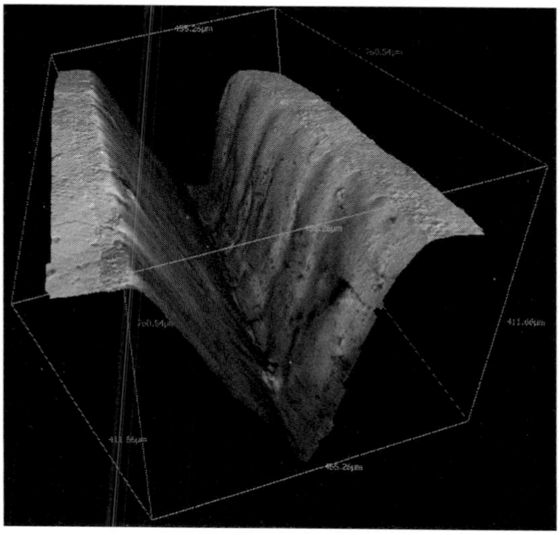

Figure 13.7 3-D model of a laser generated groove in an enamel. Courtesy of Drs C. Xu and Y. Wang, unpublished. (See plate section for color version.)

research can be expected. Figure 13.7 (and color plate) shows how a 3-D model can be used to analyze surface topography.

13.4 Back-scattered electrons

Back-scattered electrons (BSE) are the electrons of a beam reflected from the specimen after a number of elastic scattering events. They are collected with a special BSE detector. Since their number depends on the mean atomic number of the specimen (Z), they provide information about the composition of the specimen (Goldstein *et al.*, 2003). The number of BSE depends also on the surface tilt angle (i.e. topography) and is subject to edge effects. If the specimen consists of phases with significant differences in atomic number, such as dental composites with low Z resin and much higher Z filler particles, or a combination of dentin or bone (high Z) and resin (low Z), then compositional contrast will prevail over topography. In these cases, BSE images, even of rough surfaces such as fractures, provide useful compositional information. Figure 13.8 presents a BSE image of the same field of view as Figure 13.3b. The BSE signal highlighted filler particles in PQ1 adhesive and non-demineralized dentin, in contrast to the easily identified dark hybrid layer.

The atomic number of gold (79) is much higher than the atomic numbers of all tissues and embedding media used in dental research. Therefore, the BSE contrast between gold and other materials is so high that even small gold particles can be detected on specimens with developed topography. This makes the application of immunogold staining possible in SEM studies (Erlandsen *et al.*, 1999; Breschi *et al.*, 2002; Kaneko *et al.*, 2008).

Flat specimen surfaces are needed to obtain the best results for detecting variations in mineralization, especially when BSE contrast is weak. While polishing can be used to

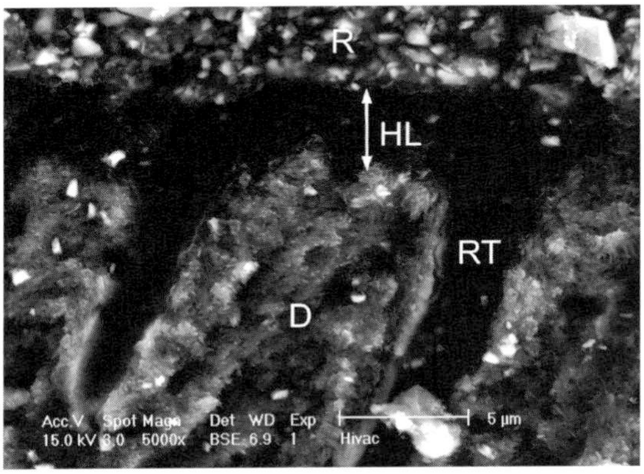

Figure 13.8 Cross fractures of PQ1 bonding system–dentin interface; the same field of view as Figure 13.3b, R – adhesive resin, D – dentin, HL – hybrid layer, RT – resin tag in dental tubule. Bar 5 μm.

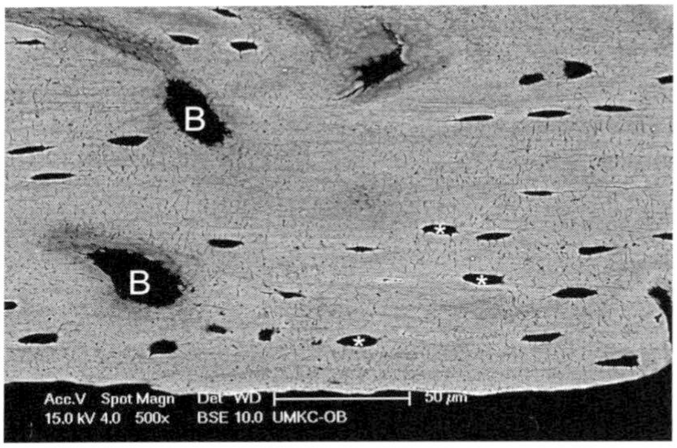

Figure 13.9 BSE micrograph of embedded mouse cortical bone, B - blood vessels, * - osteocyte lacunae. Bar 50 μm. Courtesy of Dr. Bonewald, unpublished.

create flat specimens, its application to dentin or bone alters their surface by creating a smear layer with smear plugs in dentin tubules and osteocyte lacunae. Careful resin embedding of mineralized tissue with good infiltration of tubules and lacunae is therefore desirable for polishing. Figure 13.9 is a BSE micrograph of an embedded and polished cortical mouse bone. The osteocyte lacunae and blood vessels are surrounded by areas with a lower degree of mineralization (as evidenced by slightly decreased brightness around lacunae). A high contrast between resin and bone means that a BSE image could be easily converted to a binary image and used to quantify the morphological parameters of bone (Kingsmill *et al.*, 2007). Non-mineralized features of bone and teeth, such as lacunae and dentin tubules, can be analyzed by quantitative methods.

Cutting the specimen with a diamond knife, similar to creating thin sections for transmission electron microscopy (TEM), gives better surface quality compared to polishing. Figure 13.10a is a micrograph of a specimen (mouse trabecular bone) that has been

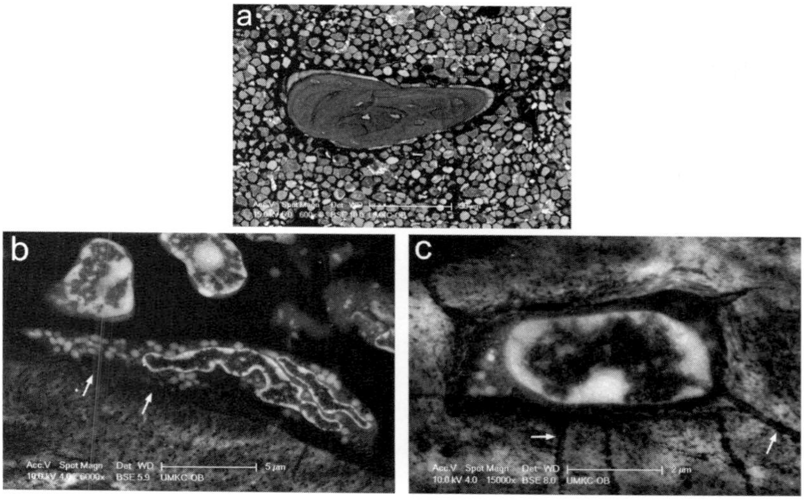

Figure 13.10 BSE micrographs of specimen post-fixed with OsO4 and surfaced with diamond knife: (a) Mouse trabecular bone, bar 50 μm, (b) after additional staining with uranil acetate and lead citrate; osteoclast with ruffled border (arrows), bar 5 μm, (c) after additional staining with uranil acetate and lead citrate; osteocyte with cell processes in canaliculi (arrows), bar 2 μm. Courtesy of Dr. Bonewald, unpublished.

demineralized, post-fixed with OsO_4, embedded, and cut with a diamond knife. Bone, osteocytes, lining cells, and bone marrow cells are clearly visible, but they lack detail. Further improvement is possible by staining the specimen in a similar way to the staining of specimens for TEM observation (Boyde and Jones, 1996; Wierzchos *et al.*, 2008). Figures 13.10b and c represent micrographs of a specimen that was additionally stained in water solutions of uranyl acetate and lead citrate for 15 min each and then carbon coated. Two nuclei and a ruffled border of an osteoclast are visible in Figure 13.10b. An osteocyte with well-defined cell processes in canaliculi is shown in Figure 13.10c. These micrographs are not subject to additional dimensional distortions, which can arise in TEM ultrathin sections; the entire surface of the specimens is visible, not just in the openings of TEM grids. This allows for convenient inspection at low magnifications.

There have been a number of publications dedicated to quantification of BSE signals (gray levels) in order to determine mineral density in bone and dental tissues. These were based: 1) on the fact that the dependence of BSE signal on Z (in a Z interval corresponding to an average Z of mineralized tissues) could be approximated by a linear function; and 2) on the assumption that all water in mineralized tissues is replaced by the equivalent volume of resin during resin embedding. Therefore, in these publications all of the analyzed specimens were embedded in resin. Carbon and aluminum (Roschger *et al.*, 1998) or monobrominated and monoiodinated dimethacrylate (Kingsmill and Boyde, 1998) were used as standards for quantifying BSE gray levels. A very strong correlation was found between the bone gray levels and mineral content, measured by the ash method (Bloebaum *et al.*, 1997), or by EDS (Roschger *et al.*, 1995). At the same time there is some evidence that both EDS data and quantified BSE measurements tend to underestimate the mineral content of bone when compared with traditional ash measurements (Vajda *et al.*, 1998). Possibly, reporting relative changes in mineralization or just intensity (gray levels) of BSE signal (Rawlinson *et al.*, 2009), instead of the calculated absolute amount of mineral, is a better approach.

13.5 X-ray microanalysis

Utilization of the X-ray signal produced from the interaction of the electron beam with the specimen for the purpose of elemental analysis is called X-ray microanalysis. The most frequently used acronym for X-ray microanalysis is EDS (Energy Dispersive Spectroscopy), which coexists with others, such as EDX and EDXS (Energy Dispersive X-ray Spectroscopy), and even EDAX (the name of the company that pioneered the field of EDS). It is possible that among all of the fields of life sciences, X-ray microanalysis is most often used in dental research, where its use centers on mineralized tissues (i.e. containing relatively "heavy" elements such as Ca and P) and on dental materials of different compositions. The weakness of X-ray microanalysis when working with light elements, such as carbon and oxygen, is not as critical in dental research as when working with non-mineralized tissue.

In addition to EDS, wavelength dispersive spectrometers (WDS) can be used for X-ray microanalysis. Ideally both types of spectrometer should be used together since their strengths are complementary. EDS has a much higher X-ray collection efficiency than WDS and thus needs lower probe currents and can usually work at the same setting as for SEM imaging. It is also faster and less affected by specimen topography. WDS has a much better energy (or wavelength) resolution and thus has a much better peak separation and peak to background ratio which leads to a better minimum detection limit. This characteristic makes it superior for quantification and trace element analysis. While microprobe (or EPMA, which is essentially an SEM optimized for X-ray microanalysis) ordinarily has spectrometers of both types, there are many more SEMs than microprobes and SEMs are overwhelmingly equipped with EDS. WDS is certainly a useful tool in dental research (Eick *et al.*, 1970; Tantbirojn *et al.*, 1998; Chang *et al.*, 2006), but in the rest of this chapter it is mostly EDS, as the most common method, that is considered.

X-ray microanalysis is routinely used in many fields of dental research, such as the study of implants (Otulakowska and Nicholson, 2006; Zinelis *et al.*, 2010); analysis of composition of mineralized tissue (Eick *et al.*, 1970; Shore *et al.*, 2010); study of dental materials (Chakmakchi *et al.*, 2009; Scougall-Vilchis *et al.*, 2009); and study of the dentin–adhesive interface (Yuan *et al.*, 2007; Hosoya *et al.*, 2010).

X-ray data can be acquired as either a spectrum, a line scan, or an X-ray map. Information from the spectrum is restricted to just one spot analysis on the specimen (usually a volume 1–2 microns in diameter), but it contains the most complete information about a specimen's composition. Examples of spectra obtained from a dental implant (Biomet 3i The Certain® Internal Connection Implant System) at accelerating voltages of 15 kV and 7 kV are

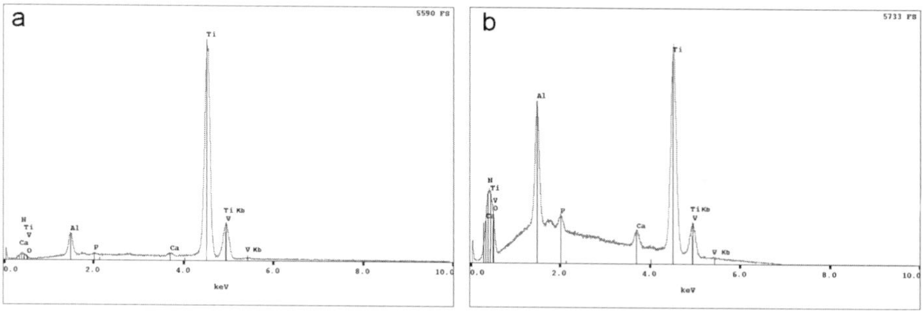

Figure 13.11 Spectra obtained from dental implant (Biomet 3i The Certain® Internal Connection Implant System) at accelerating voltages: (a) 15 kV, and (b) 7 kV. Courtesy of Dr. Liu, unpublished.

presented in Figures 13.11a and b. The spectrum obtained at 15 kV shows the presence of titanium, aluminum, and some calcium and phosphorus (the low-voltage end of the spectrum has overlapping peaks of titanium, oxygen, and nitrogen and is not discussed here in detail). The peak to the right of the strong titanium peak, marked as "Ti Kb; V" is higher than it should be for pure titanium and consists of the overlapping peaks of Ti Kb and V Ka. The small peak of V Kb confirms the presence of vanadium in the implants. Decreasing the accelerating voltage to 7 kV (Figure 13.11b) leads to a relative increase of the heights of Ca and P peaks compared to the Ti and Al peaks. Lower accelerating voltage produces a smaller range of generated X-rays and consequently a greater proportion of radiation comes from the surface layer of the specimen. Therefore, in this example, EDS results suggest the implant has a layer of Ca and P on its surface. This corroborates the manufacturer's information that the implants are coated with calcium phosphate (CaP).

X-ray spectra are best for qualitative analysis (identification of elements present in a specimen) and are the only source of data available for quantitative analysis. When a spatial distribution of elements in the specimen is of interest, line scans and X-ray maps can provide better information. Figure 13.12 is an example of a line scan and shows the distribution of Ca and P along a line marked on the picture as "scan line." The specimen is a cross-section of dentin treated with the all-in-one adhesive system Prompt L-Pop. Prompt L-Pop is both an etchant and adhesive, thus all of the dissolved Ca remains in the adhesive. This leads to an elevated concentration of Ca in the adhesive, as is shown in the picture. Prompt L-Pop contains phosphoric esters in addition to dissolved P, which is reflected in the P line scan, where the intensity of P in the adhesive is close to that in dentin. The hybrid layer contains less resin as it is "diluted" by collagen, and thus contains less Ca and P than can be found in the pure adhesive. This appears as a dip in the concentration profiles of the hybrid layer region. An electron beam voltage of 15 kV was used for this analysis.

X-ray maps have the power of visually presenting the spatial distribution of elements, and in many cases deliver essential information about the specimen. Figure 13.13 (and color plate) is a map of bone healing with the help of a bioactive glass scaffold. It presents the spatial distribution of Ca, P, and Si and shows glass particles in different stages of conversion to hydroxyapatite (particles closest to the bone are fully converted and have Ca and P in their composition, but no Si).

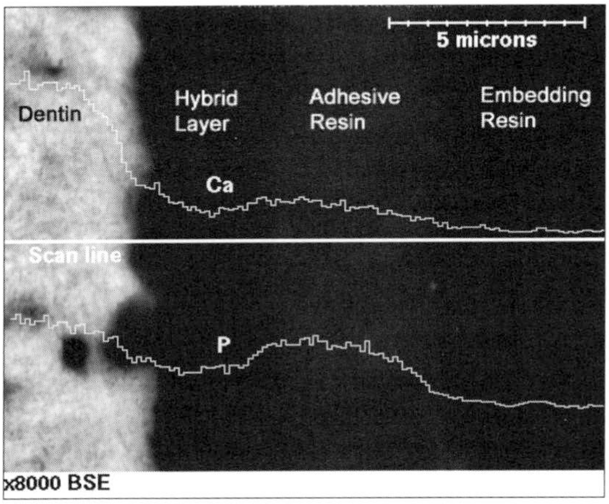

Figure 13.12 Profiles of Ca and P of a bulk cross-section of dentin, treated with all-in-one adhesive system Prompt L-Pop. Bar 5 μm.

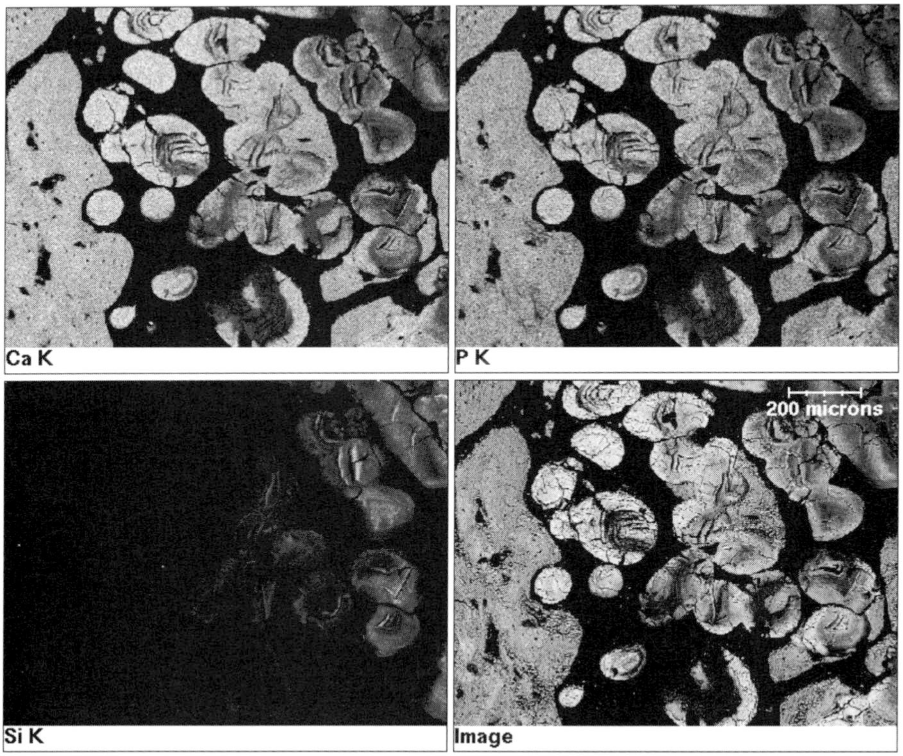

Figure 13.13 X-ray map (for Ca, P, and Si) and BSE image of bone healing with the help of scaffold of bioactive glass. Round glass particles are in different stages of conversion to hydroxyapatite (particles closest to bone are fully converted and have Ca and P in their composition, but no Si). Bar 200 μm. Courtesy of Dr. Bi, unpublished. (See plate section for color version.)

Table 13.1 Dependence of Ca/P and Ca/O X-ray peak count ratios on SEM accelerating voltage (instrument dependent)

kV	Ca/P	Ca/O
5.0	0.23	0.18
10.0	1.27	4.87
14.1	**1.67**	13.54
15.0	1.78	16.20
20.0	2.19	35.63
30.0	2.89	82.08

The calcium to phosphorus content ratio has an important role in mineralized tissue research. It indicates how close the mineral in a specimen is to stoichiometric hydroxyapatite $Ca_{10}(PO_4)_6(OH)_2$. Hydroxyapatite has ten atoms of Ca and six atoms of P, so its atomic (molar) Ca/P ratio is 10/6 = 1.67. Since the atomic weight of Ca is 40.078 and the atomic weight of P is 30.974, the Ca/P weight ratio is 10*40.078/6*30.974 = 2.16. Determined by ash content in the pre-EDS days, the Ca/P weight ratio in normal dentin was equal to 2.03 (Little *et al.*, 1965). With proper quantitative EDS analysis, Ca/P ratio calculation is pretty straightforward. However, it is tempting to look for an easier way to calculate directly the ratio of characteristic peak heights. Unfortunately, the peak ratio changes with variation of accelerating voltage (Table 13.1). Additionally, the peak ratio depends not only on accelerating voltage, but also on the instrument and its geometry, design, and settings. At 14.1 kV, the Ca/P peak ratio is equal to stoichiometric molar ratio of 1.67; and we can see that at 20 kV, the peak ratio is 2.19 and is close to the stoichiometric weight ratio of 2.16. Therefore, it is possible to obtain results that look correct even when applying a completely wrong procedure. While calculation of the Ca/P ratio has been reported in many papers, all too often there has been no indication of whether quantitative analysis was used and whether the reported result is the molar or the weight ratio. If, for example, the reported Ca/P ratio were equal to 1.61 (Shibata *et al.*, 2008), it could be assumed that the molar ratio was calculated. However, when the following results are analyzed, 2.14, 2.11, 2.02 (close to weight ratios), and 1.66 (close to a molar ratio) (Sakoolnamarka *et al.*, 2005), it is unclear as to how the measurements were obtained.

An important artifact of EDS, WDS maps, and line scans is the contrast due to changes in the background. The intensity of the background (or X-ray continuum) is proportional to the average atomic number Z, based upon the weight fractions of the elemental constituents of the specimen (Goldstein *et al.*, 2003). If, for example, an X-ray map (or line scan) is made from a cross-section containing enamel and embedding resin, the intensity of the background of the low Z resin will be considerably lower than the intensity of the background of the higher Z mineralized tissue. Figure 13.14a (and color plate) is a map with the superimposed line scans (Ca, Al, Ar) of the cross-section of a resin embedded tooth. The analyzed area includes three major components: resin, enamel, and dentin. Calcium, of course, is a real component, but neither aluminum nor argon is present in the specimen. Nevertheless, maps as well as line scans show a similar "distribution" for all three elements: low in resin, high in enamel, and a bit lower but still high in dentin. All of the line maps follow the distribution of mineral within the

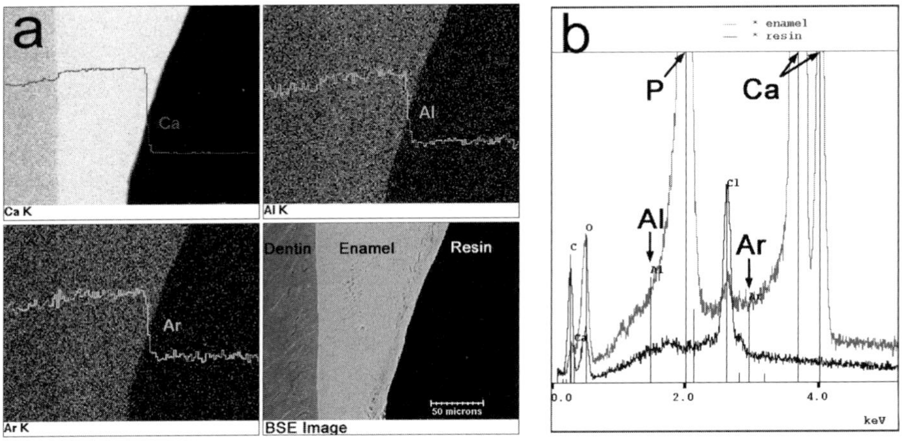

Figure 13.14 (a) Maps and scan lines of Ca, Al, and Ar obtained from polished cross-section of resin embedded tooth. There was no Al nor Ar in the specimen; registered distribution of X-ray intensities was actually a distribution of background intensities. Bar 50 μm. (b) Superimposed spectra of enamel and of embedding resin; substantial difference in background intensities recorded; no characteristic X-ray peaks for Al and Ar were detected. (See plate section for color version.)

specimen. The calcium map represents its presence as a constituent of the mineral, however, the other two (Al and Ar) reflect intensities of the background, which essentially depend on overall mineral density. Only spectrum analysis, but neither a map nor a line scan, can definitively prove the presence of an element. As can be seen from the enamel and resin spectra (Figure 13.14b), the difference in background intensities is clearly visible and no peaks of Al and Ar are detected.

Auto scaling of X-ray maps can cause skewing of image contrast due to the background effect causing absent elements to appear at levels comparable to legitimate elements (Goldstein *et al.*, 2003). Map comparison between specimens is questionable when auto scaling is used. For example, auto scaling can make the background based maps of magnesium fairly bright in normal dentin where magnesium is not present. On the other hand, magnesium maps of caries affected dentin show bright spots for the Mg-containing mineral in tubules and much darker dentin. Comparison of such maps leads to the incorrect conclusion that caries affected dentin is depleted in magnesium (Nakajima *et al.*, 2005). Thus care should be taken when interpreting results of X-ray line scans or maps used to study minor or trace elements. With proper precautions, meaningful comparisons to the background or control specimen can be made (Harnirattisai *et al.*, 2007). The best way to avoid background artifacts is to use quantitative maps. However, this method may be time-prohibitive due to the need to acquire spectra suitable for the quantification of each pixel. Newer silicon drift detectors (SDD) can achieve a much better count rate and are more suitable for obtaining quantitative maps. Another method, background subtraction, is less time consuming, but still rarely used.

So-called edge resolution is a practical approach to measuring resolution (Vladar *et al.*, 2009). When an electron beam is scanned across a specimen with a sharp vertical edge,

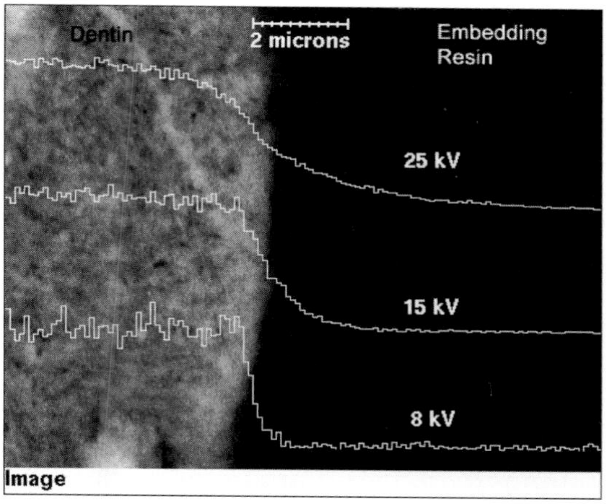

Figure 13.15 Ca profiles obtained at different accelerating voltages from the same scan line on enamel–embedding resin boundary. Bar 2 μm.

the signal generated by the beam changes gradually as the signal production volume crosses the edge. Often the edge resolution is measured as "25–75% resolution," i.e. the distance between the two positions of the beam whose signal intensity changes from 25% to 75% of full intensity. Figure 13.15 illustrates the effect of accelerating voltage on EDS spatial resolution, showing a striking difference in the width of the transition region from enamel to embedding resin on Ca profiles at 25, 15, and 8 kV. An estimate of 25–75 resolutions, made with these profiles, gave the following values: the resolution for 25 kV is 2.0 micron, for 15 kV it is 1.0 micron, and for 8 kV it is 0.3 micron.

The dramatic difference in resolution (2–3 orders of magnitude) between the SE and X-ray signal can lead to misunderstanding and incorrect interpretation of the results. While the SE image is in perfect focus, the EDS image of the same area at the same magnification can be completely out of focus. The profiles (scan lines) of the sharp changes of concentration could be displayed as those in Figure 13.15. Such profiles could be mistakenly reported as diffusion zones or partially demineralized zones, and then the width of these "zones" would be measured at about 5–95% of the signal value. The width of 5–95% signal transitionary zones, measured for profiles in Figure 13.13, corresponds to 5.2 micron at 25 kV, 2.2 micron at 15 kV, and 1.0 micron at 8 kV. Neglect of spatial resolution limitations often leads to wrong conclusions, especially when relatively small features, such as dentin–adhesive hybrid layers, are considered. Often in the dental literature, profiles similar to the ones in Figure 13.15 are incorrectly presented as a "proof" of a "gradual increase of Ca and P concentrations" in the interface (hybrid layer) of adhesive with dentin or enamel (Ferracane *et al.*, 1998; Weerasinghe *et al.*, 2007).

A further improvement of spatial resolution is possible by utilizing ultrathin (usually less than 100 nm) specimens prepared for TEM. High resolution for EDS in TEM/STEM is achieved by thin specimens due to the smaller region of electron scattering and X-ray generation. There is simply not enough space for electrons to scatter. Since the physics of electron beam interaction with a specimen is the same for STEM and SEM, EDS spatial

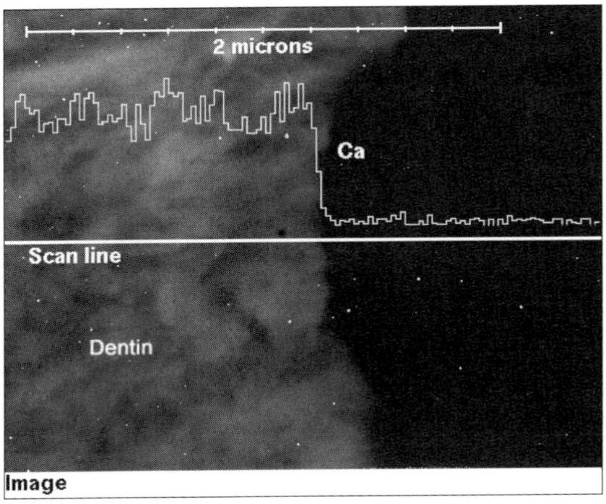

Figure 13.16 Ca profile of the edge (tear) in an ultrathin section of dentin (TEM specimen on copper grid, cut with diamond knife and placed on carbon stub). Bar 2 μm.

resolution depends not only on the instrument, but mostly on the specimen used. Analyzing ultrathin specimen sections with SEM is gaining popularity as special adapters for such observation in transmitted electrons for SEM are now marketed. For mineralized tissue, these adapters are not necessary, since the mineral is nicely visible in SEM with a standard BSE detector.

Figure 13.16 represents the edge of an ultrathin section of dentin (cut with a diamond knife) on a TEM copper grid, which was placed on a carbon stub. This dentin edge (actually, a tear in the section) was used as a test specimen for measuring resolution. A rough estimate of 25–75% resolution gives 40 nm. This extraordinary improvement in resolution comes at the price of a sharp drop in X-ray intensity, making this method unsuitable for trace elements. The element should be a major component of at least a microscopical particle, like silicon in Figure 13.17, where Si is a major component of the particle and a trace element of the specimen as a whole.

Figure 13.17 represents high-resolution line scans from an ultrathin section of a non-demineralized cross-section of dentin treated with the Scotchbond Multi-Purpose Plus bonding system. The line scans of Ca, P, and Si show the presence of particles (of less than 200 nm in size) on the surface of the hybrid layer, consisting of hydroxyapatite (that did not dissolve during etching) and of Si from the etching gel. All three of the line scans represent element distribution along the same line, shown as a white straight line in the middle of the micrograph. The sharp drop in the concentration of Ca and P on the edge of the dentin demonstrates that there was almost no partially demineralized layer of dentin on the micro level.

However, it does not mean that bulk specimens cannot be used for the study of hybrid layers with size close to the EDS spatial resolution. A bulk cross-section of dentin, which was shown in Figure 13.12, provided useful results as long as no assumptions were made about gradual (and not sharp) changes in intensities.

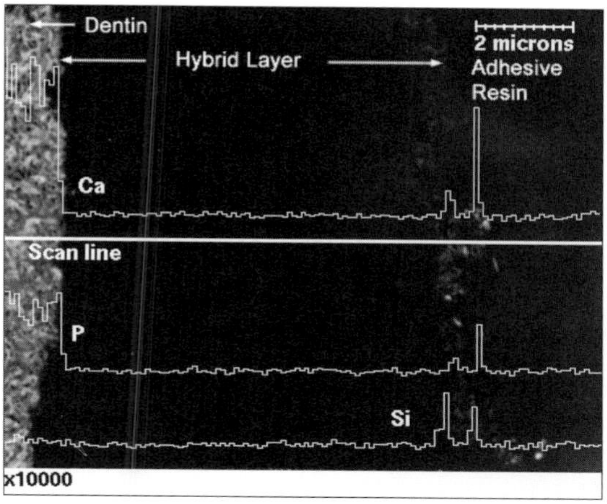

Figure 13.17 Line scans of Ca, P, and Si of non-demineralized ultrathin section of dentin treated with Scotchbond Multi-Purpose Plus bonding system. Bar 2 μm.

13.6 Fractography

Fractography is the interpretation of features observed on fracture surfaces. Fractographic studies often require simultaneous utilization of SE, BSE, and X-ray signals for a proper description of observed features.

Cross-fractured specimens of caries affected dentin often contain dentin tubules filled with the acid resistant mineral, whitlockite (Frank *et al.*, 1964; Ceballos *et al.*, 2003). Whitlockite can contain about 5% magnesium by weight (Tay and Pashley, 2004). Such mineral casts of tubules and small branches are seen in Figure 13.18, after etching and bleaching removed a layer of hydroxyapatite, but not whitlockite. On the dentin–adhesive interface of a caries affected tooth (Figure 13.19a), mineral casts could be misinterpreted as adhesive resin tags. BSE imaging (Figure 13.19b) shows relatively high Z-contrast for casts, which can help with the interpretation. Additional data obtained from an EDS spectrum (Figure 13.19c) shows the presence of mineral with some magnesium. This spectrum was acquired at a lower accelerating voltage of 8 kV to reduce the range of X-ray signal generation. Thus the areas of interest were proven to be mineral casts and not adhesive resin tags.

Fractographic methods are routinely used to determine the cause of failure of materials. The fracture surface of a Z250 composite specimen fractured during a 4-point bend test (Figure 13.20a) contains hackle lines radiating in the direction of crack propagation and pointing at the fracture origin. Observation of the critical flaw at the fracture origin at higher magnifications in SE (Figure 13.20b) and especially in BSE (Figure 13.20c) shows an inclusion of about the same structure as the main body of the composite. EDS analysis also showed a similar composition for composite and inclusion, which was possibly an inclusion of prepolymerized composite.

A wealth of information can be obtained by fractographic analysis of fractures generated by mechanical testing of dentin–adhesive interfaces, such as the microtensile

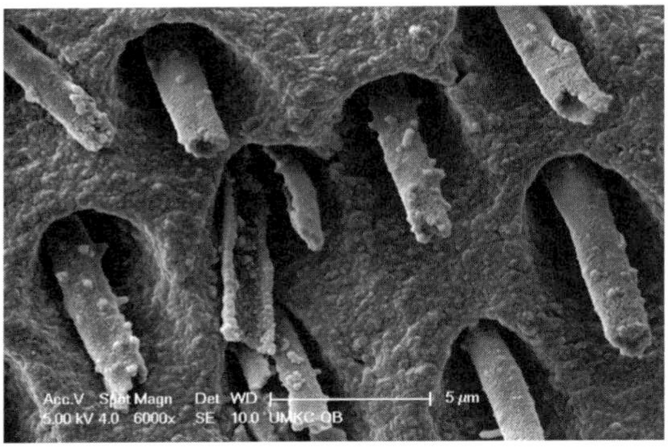

Figure 13.18 Etched and bleached cross section of caries affected dentin with acid resistant mineral casts. Bar 5 µm.

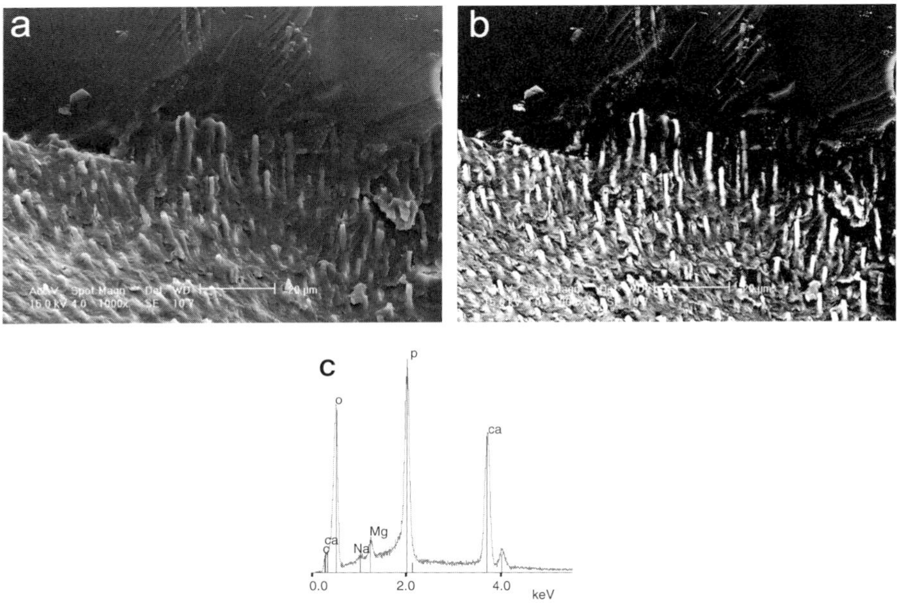

Figure 13.19 Cross fracture of adhesive–caries affected dentine interface: (a) SE signal, (b) BSE signal, bar 20 µm, (c) EDS spectra acquired from tubule cast at 8 kV. Courtesy of Dr. Bohaty, unpublished.

test. The crack trajectory, displayed by the fracture surface, reflects the properties of various regions of the tested specimens. Examination of both sides of fractured specimens is often necessary for a proper explanation of observed phenomena (Schreiner *et al.*, 1998; Carrilho *et al.*, 2007). Figures 13.21a and b represent micrographs in SE and BSE respectively of the enamel side of the fracture surface created during a microtensile test of a Clearfil SE bonding system to enamel. It could have been concluded that the fracture occurred through the adhesive (appearing dark on the BSE image) and enamel (bright) if it were not for micrographs taken from the same place on the other half of the

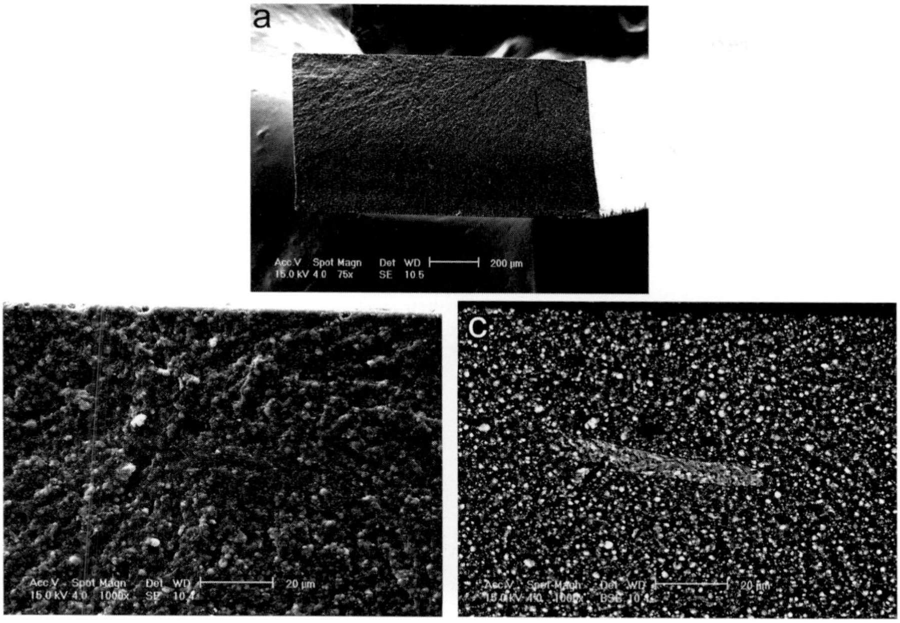

Figure 13.20 Fracture surface of Z250 composite, fractured during 4-point bend test: (a) Hackle lines radiating from fracture origin, arrows. SE, bar 200 μm. b) Fracture origin, SE, bar 20 μm. c) Fracture origin, BSE, bar 20 μm. Courtesy of Dr. M. Walker, unpublished.

specimen (Figure 13.21c and d). Highlighted enamel and dark adhesive in Figure 13.21d demonstrated that the crack propagated mostly along the enamel–adhesive interface, with pieces of enamel pulled out of the bulk enamel on the other side of the specimen.

Quantification can expose a correlation between fracture modes and mechanical properties. Voids were detected due to over-wetness on the fracture surfaces of specimens prepared from class II composite restorations under *in vivo* class II conditions. According to statistical analysis, the fraction (percent) of the fracture surface occupied by voids was highly predictive of bond strength (Purk *et al.*, 2007).

13.7 Environmental scanning electron microscope (ESEM)

There is no conventional definition of the term *ESEM*. Sometimes it is used to define an SEM capable of working with wet specimens, keeping them hydrated in a specimen chamber. In other cases it specifies an SEM that permits a gaseous environment in the specimen chamber. Historically (due to marketing issues) SEMs capable of working with a gaseous environment, but not with wet specimens, were called VP SEMs (for variable pressure SEM), and those that could work with wet specimens in addition to gaseous environment were called ESEMs.

The vacuum system of the ESEM pumps atmospheric gases out of the specimen chamber and replaces them with low-pressure water vapor. To keep the specimen wet,

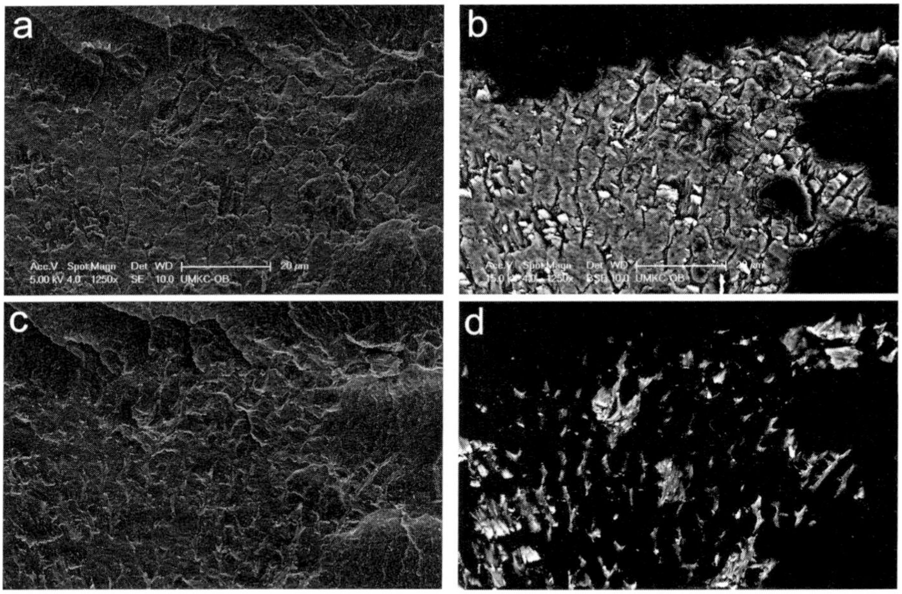

Figure 13.21 Fracture surfaces of enamel specimen treated with Clearfil SE bonding system and tested for microtensile strength, bar 20 μm: (a,b) Enamel side of the specimen in SE and BSE respectively, (c, d) Exactly the same place as on previous micrographs on composite side of the specimen, images were flipped around their vertical axes, SE and BSE respectively. Courtesy of Dr. J. Purk, unpublished.

i.e. at dew point, water vapor pressure should be 4.6 torr at 0 °C and 17.5 torr at 20 °C. Working at pressures lower than 17 torr requires a cooling stage to keep the specimen at dew point. Often the ESEM is operated at temperatures of about 5 °C and pressures of about 6 torr. A rather common misconception is that a VP SEM is capable of keeping the specimens wet. For example, in an attempt to keep carious dentin wet, work was performed at a pressure of 1.5 torr, using ambient air (not water vapor) as the introduced gas, and without the cooling stage (Angker *et al.*, 2004). Under these conditions humidity in the specimen chamber was low and the specimen would have dried out quickly. In dental research ESEMs are used for various purposes, such as investigations of cements (Gandolfi *et al.*, 2010), hybrid layers (Yiu *et al.*, 2004; Eick *et al.*, 2003; Franz *et al.*, 2006), etched enamel (Cowan *et al.*, 1996), and etched dentin (Gilbert and Doherty, 1993; Gwinnett, 1994; de Wet *et al.*, 2000; Dusevich and Eick, 2002). ESEMs are well suited for conducting dynamic experiments, such as observation of the surface structure of cement paste under wet conditions and in real time during setting (Gandolfi *et al.*, 2010), or measurements of the dimensional changes of a layer of demineralized dentin while drying in a specimen chamber (Dusevich and Eick, 2002). Figure 13.22 represents two superimposed micrographs of etched dentin, immediately after pumping down the specimen chamber and after controlled drying.

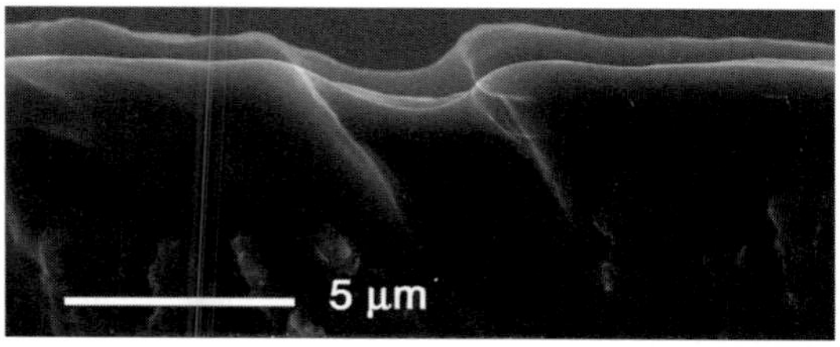

Figure 13.22 Superimposed micrographs of wet layer of etched dentin before and after controlled drying in ESEM specimen chamber; decrease of the thickness of the layer is clearly observable. Bar 5 μm. Reprinted with permission (Dusevich and Eick, 2002).

13.8 Conclusion

There are many exciting developments in the field of scanning electron microscopy and its implementation in dental research. In-lens field-emission scanning electron microscopes have greatly improved resolution. Silicon drift detectors (SDD) offer much higher count rates than traditional Si(Li) detectors at the same beam current, a feature which is extremely useful in most EDS applications. SEM can now be equipped with attachments that bring a wide array of analytical techniques in a SEM specimen chamber: X-ray microscopy with 3-D capabilities; X-ray fluorescence spectroscopy (XRF) with better detection limits than EDS (but with spatial resolution); and Raman spectroscopy.

SEM continues to be used extensively with success in dental research applications. This chapter has attempted to summarize the uses of SEM in dental research and has pointed out some advantages and disadvantages of the various methodologies available with the SEM.

13.9 References

Aida, M., Odaki, M., Fujita, K., *et al.* (2009). Degradation-stage effect of self-etching primer on dentin bond durability. *J. Dent. Research*, **88** (*5*), 443–448.

Angker, L., Nockolds, C., Swain, M. V., and Kilpatrick, N. (2004). Quantitative analysis of the mineral content of sound and carious primary dentine using BSE imaging. *Arch. Oral Biology*, **49** (*2*), 99–107.

Bi, L. X., Simmons, D. J., Hawkins, H. K., Cox, R. A., and Mainous, E. G. (2000). Comparative morphology of the marrow sac. *Anat. Rec.*, **260** (*4*), 410–415.

Bloebaum, R. D., Skedros, J. G., Vajda, E. G., Bachus, K. N., and Constantz, B. R. (1997). Determining mineral content variations in bone using backscattered electron imaging. *Bone*, **20** (*5*), 485–490.

Blomlof, J., Blomlof, L., and Lindskog, S. (1997). Effect of different concentrations of EDTA on smear removal and collagen exposure in periodontitis-affected root surfaces. *Journal of Clinical Periodontology*, **24**, (*8*), 534–537.

Boyde, A. and Reid, S. A. (1983). Tetracycline cathodoluminescence in bone, dentine and enamel. *Histochemistry*, **77** (*4*), 525–533.

Boyde, A. and Jones, S. J. (1996). Scanning electron microscopy of bone: instrument, specimen, and issues. *Microscopy Research and Technique*, **33**, (*2*), 92–120.

Bozzola, J. J. and Russell, L. D. (1999). *Electron Microscopy: Principles and Techniques for Biologists*. Sudbury, MA, Jones and Bartlett Pub.

Breschi, L., Lopes, M., Gobbi, P., *et al.* (2002). Dentin proteoglycans: an immunocytochemical FEISEM study. *J. Biomed. Mater. Res.*, **61** (*1*), 40–46.

Carrilho, M. R., Carvalho, R. M., de Goes, M. F., *et al.* (2007). Chlorhexidine preserves dentin bond *in vitro*. *J. Dent. Research*, **86** (*1*), 90–94.

Ceballos, L., Camejo, D. G., Victoria Fuentes, M., *et al.* (2003). Microtensile bond strength of total-etch and self-etching adhesives to caries-affected dentine. *J. Dent.*, **31** (*7*), 469–477.

Chakmakchi, M., Eliades, G., and Zinelis, S. (2009). Bonding agents of low fusing cpTi porcelains: elemental and morphological characterization. *J. Prosthodont. Research*, **53** (*4*), 166–171.

Chang, J., Platt, J. A., Yi, K., and Cochran, M. A. (2006). Quantitative comparison of the water permeable zone among four types of dental adhesives used with a dual-cured composite. *Oper. Dent.*, **31** (*3*), 346–353.

Cowan, A. J., Wilson, N. H., Wilson, M. A., and Watts, D. C. (1996). The application of ESEM in dental materials research. *J. Dent.*, **24** (*5*), 375–377.

de Wet, F. A., van der Vyver, P. J., Eick, J. D., and Dusevich, V. M. (2000). Environmental scanning electron microscopy of hydrated conditioned/etched dentine. *SADJ*, **55** (*11*), 603–609.

Dusevich, V. M. and Eick, J. D. (2002). Evaluation of demineralized dentin contraction by stereo measurements using environmental and conventional scanning electron microscopy. *Scanning*, **24** (*2*), 101–105.

Dusevich, V. M., Purk, J. H., and Eick, J. D. (2010). Choosing the right accelerating voltage for SEM (An introduction for beginners). *Microscopy Today*, **18**, 48–52.

Dykstra, M. J. and Reuss, L. E. (2003). *Biological Electron Microscopy: Theory, Techniques, and Troubleshooting*. New York, Kluwer Academic/Plenum Publishers, pp. 534.

Eick, J. D., Wilko, R. A., Anderson, C. H., and Sorensen, S. E. (1970). Scanning electron microscopy of cut tooth surfaces and identification of debris by use of the electron microprobe. *J. Dental Research*, **49** (*6*), *Suppl*, 1359–1368.

Eick, J. D., Gwinnett, A. J., Pashley, D. H., and Robinson, S. J. (1997). Current concepts on adhesion to dentin. *Crit. Rev. Oral. Biol. Med.*, **8** (*3*), 306–335.

Eick, J. D., de Wet, F. A., van der Vyver, P. J., and Dusevich, V. M. (2003). Application of the ESEM in dental research of the dentin–resin interface. In: *Science, Technology and Education of Microscopy: an Overview*, Badajoz, Spain, Formatex, pp. 363–370.

Erlandsen, S. L., Macechko, P. T., and Frethem, C. (1999). High resolution backscatter electron (BSE) imaging of immunogold with in-lens and below-the-lens field emission scanning electron microscopes. *Scanning Microscopy*, **13** (*1*), 43–54.

Feng, J. Q., Ward, L. M., Liu, S., *et al.* (2006). Loss of DMP1 causes rickets and osteomalacia and identifies a role for osteocytes in mineral metabolism. *Nature Genetics*, **38** (*11*), 1310–1315.

Ferracane, J. L., Mitchem, J. C., and Adey, J. D. (1998). Fluoride penetration into the hybrid layer from a dentin adhesive. *Am. Journal Dent.*, **11** (*1*), 23–28.

Frank, R. M., Wolff, F., and Gutmann, B. (1964). Electron microscopy of caries at the level of human dentine. *Arch. Oral Biol.*, **9**, 163–179.

Franz, N., Ahlers, M. O., Abdullah, A., and Hohenberg, H. (2006). Material-specific contrast in the ESEM and its application in dentistry. *J. Mater. Science*, **41** (*14*), 4561–4567.

Gandolfi, M. G., Van Landuyt, K., Taddei, P., *et al.* (2010). Environmental scanning electron microscopy connected with energy dispersive X-ray analysis and Raman techniques to study

ProRoot mineral trioxide aggregate and calcium silicate cements in wet conditions and in real time. *J. Endod.*, **36** (*5*), 851–857.

Gilbert, L. C. and Doherty, R. E. (1993). Using ESEM and SEM to compare the performance of dentin conditioners. *Microscopy Research and Technique*, **25** (*5–6*), 419–423.

Goldstein, J., Newbury, D., Joy, D., *et al.* (2003). *Scanning Electron Microscopy and X-ray Microanalysis*. New York, Kluwer Academic Plenum Publishers, pp. 699.

Gorustovich, A A. (2010). Imaging resin-cast osteocyte lacuno-canalicular system at bone–bioactive glass interface by scanning electron microscopy. *Microscopy and Microanalysis*, **16** (*2*), 132–136.

Guo, X., Spencer, P., Wang, Y., *et al.* (2007). Effects of a solubility enhancer on penetration of hydrophobic component in model adhesives into wet demineralized dentin. *Dent. Mater.*, **23** (*12*), 1473–1481.

Gwinnett, A. J. (1994). Chemically conditioned dentin: a comparison of conventional and environmental scanning electron microscopy findings. *Dent. Mater.*, **10** (*3*), 150–155.

Harnirattisai, C., Senawongse, P., and Tagami, J. (2007). Microtensile bond strengths of two adhesive resins to discolored dentin after amalgam removal. *J. Dent. Research*, **86** (*3*), 232–236.

Hosoya, Y., Ando, S., Yamaguchi, K., *et al.* (2010). Quality of the interface of primary tooth dentin bonded with antibacterial fluoride-releasing adhesive. *J. Dent.*, **38** (*5*), 423–430.

Jongebloed, W. L., Stokroos, I., Kalicharan, D., and van der Want, J. J. L. (1999). Is cryopreservation superior over tannic acid/arginine/osmium tetroxide non-coating preparation in field emission scanning electron microscopy? *Scanning Microscopy*, **13** (*1*), 93–109.

Kaneko, T., Okiji, T., Kaneko, R., Nor, J. E., and Suda, H. (2008). Antigen-presenting cells in human radicular granulomas. *J. Dent. Research*, **87** (*6*), 553–557.

Kingsmill, V. J. and Boyde, A. (1998). Mineralisation density of human mandibular bone: quantitative backscattered electron image analysis. *J. Anat.*, **192** (*Pt 2*), 245–256.

Kim, J., Mai, S., Carrilho, M. R., *et al.* (2010). An all-in-one adhesive does not etch beyond hybrid layers. *J. Dent. Research*, **89** (*5*), 482–487.

Kingsmill, V. J, Gray, C. M., Moles, D. R., and Boyde, A. (2007). Cortical vascular canals in human mandible and other bones. *J. Dent. Research*, **86** (*4*), 368–372.

Lester, K. S. and Boyde, A. (1987). Relating developing surface to adult ultrastructure in Chiropteran enamel by SEM. *Adv. Dent. Res.*, **1** (*2*), 181–190.

Li, F., Chai, Z. G., Sun, M. N., *et al.* (2009). Anti-biofilm effect of dental adhesive with cationic monomer. *J. Dent. Research*, **88** (*4*), 372–376.

Little, M. F., Dirksen, T. R., and Schlueter, G. (1965). The Ca, P, Na, and ash content at different depths in caries. *J. Dent. Research*, **44**, 362–365.

Lu, Y., Xie, Y., Zhang, S., *et al.* (2007). DMP1-targeted Cre expression in odontoblasts and osteocytes. *J. Dent. Research*, **86** (*4*), 320–325.

Martin, D. M., Hallsworth, A. S., and Buckley, T. A. (1978). Method for the study of internal spaces in hard tissue matrices by SEM, with special reference to dentine. *Journal of Microscopy*, **112** (*3*), 345–352.

Nakabayashi, N. (1982). Resin reinforced dentin due to microinfiltration of monomers into the dentin at the adhesive interface. *J. Jpn Dent. Mater.*, **1**, 78–81.

Nakajima, M., Kitasako, Y., Okuda, M., Foxton, R. M., and Tagami, J. (2005). Elemental distributions and microtensile bond strength of the adhesive interface to normal and caries-affected dentin. *J. Biomed. Mater. Res. B Appl. Biomater.*, **72** (*2*), 268–275.

Osorio, R., Erhardt, M. C., Pimenta, L. A., Osorio, E., and Toledano, M. (2005). EDTA treatment improves resin-dentin bonds' resistance to degradation. *J. Dent. Research*, **84** (*8*), 736–740.

Otulakowska, J. and Nicholson, J. W. (2006). Scanning electron microscopy and energy dispersive X-ray study of a recovered dental implant. *J. Mater. Sci. Mater. Med.*, **17** (*3*), 277–9.

Ozel, E., Korkmaz, Y., Attar, N., and Karabulut, E. (2008). Effect of one-step polishing systems on surface roughness of different flowable restorative materials. *Dent. Mater. J.*, **27** (*6*), 755–764.

Perdigao, J., Lambrechts, P., Van Meerbeek, B., Vanherle, G., and Lopes, A. L. (1995). Field emission SEM comparison of four postfixation drying techniques for human dentin. *J. Biomed. Mater. Res.*, **29** (*9*), 1111–1120.

Piattelli, A., Scarano, A., Di Alberti, L., and Piattelli, M. (1999). Bone–hydroxyapatite interface in retrieved hydroxyapatite-coated titanium implants: a clinical and histologic report. *International Journal of Oral Maxillofac Implants*, **14** (*2*), 233–238.

Purk, J. H., Dusevich, V., Glaros, A., Spencer, P., and Eick, J. D. (2004). *In vivo* versus *in vitro* microtensile bond strength of axial versus gingival cavity preparation walls in Class II resin-based composite restorations. *J. Am. Dent. Assoc.*, **135** (*2*), 185–193; quiz 228.

Purk, J. H., Dusevich, V., Glaros, A., and Eick, J. D. (2007). Adhesive analysis of voids in Class II composite resin restorations at the axial and gingival cavity walls restored under *in vivo* versus *in vitro* conditions. *Dent. Mater.*, **23** (*7*), 871–877.

Purk, J. H., Dusevich, V., Atwood, J., *et al.* (2009). Microtensile dentin adhesive bond strength under different positive pulpal pressures. *Am. J. Dent.*, **22** (*6*), 357–360.

Rawlinson, S. C., Boyde, A., Davis, G. R., *et al.* (2009). Ovariectomy vs. hypofunction: their effects on rat mandibular bone. *J. Dent. Research*, **88** (*7*), 615–620.

Rojas-Sanchez, F., Alaminos, M., Campos, A., Rivera, H., and Sanchez-Quevedo, M. C. (2007). Dentin in severe fluorosis: a quantitative histochemical study. *J. Dent. Research*, **86** (*9*), 857–861.

Roschger, P., Plenk H., Jr., Klaushofer, K., and Eschberger, J. (1995). A new scanning electron microscopy approach to the quantification of bone mineral distribution: backscattered electron image grey-levels correlated to calcium K alpha-line intensities. *Scanning Microscopy*, **9** (*1*), 75–86; discussion 86–88.

Roschger, P., Fratzl, P., Eschberger, J., and Klaushofer, K. (1998). Validation of quantitative backscattered electron imaging for the measurement of mineral density distribution in human bone biopsies. *Bone*, **23** (*4*), 319–326.

Sakoolnamarka, R., Burrow, M. F., Swain, M., and Tyas, M. J. (2005). Microhardness and Ca:P ratio of carious and Carisolv treated caries-affected dentine using an ultra-micro-indentation system and energy dispersive analysis of X-rays – a pilot study. *Aust. Dent. J.*, **50** (*4*), 246–250.

Schreiner, R. F., Chappell, R. P., Glaros, A. G., and Eick, J. D. (1998). Microtensile testing of dentin adhesives. *Dent. Mater.*, **14** (*3*), 194–201.

Scougall-Vilchis, R. J., Hotta, Y., Hotta, M., Idono, T., and Yamamoto, K. (2009). Examination of composite resins with electron microscopy, microhardness tester and energy dispersive X-ray microanalyzer. *Dent. Mater. J.*, **28** (*1*), 102–112.

Shibata, Y., He, L. H., Kataoka, Y., Miyazaki, T., and Swain, M. V. (2008). Micromechanical property recovery of human carious dentin achieved with colloidal nano-beta-tricalcium phosphate. *J. Dent. Research*, **87** (*3*), 233–237.

Shore, R. C., Backman, B., Elcock, C., *et al.* (2010). The structure and composition of deciduous enamel affected by local hypoplastic autosomal dominant amelogenesis imperfecta resulting from an ENAM mutation. *Cells Tissues Organs*, **191** (*4*), 301–306.

Takahashi, A., Sato, Y., Uno, S., Pereira, P. N., and Sano, H. (2002). Effects of mechanical properties of adhesive resins on bond strength to dentin. *Dent. Mater.*, **18** (*3*), 263–268.

Tantbirojn, D., Douglas, W. H., Ko, C. C., and McSwiggen, P. L. (1998). Spatial chemical analysis of dental stain using wavelength dispersive spectrometry. *Eur. J. Oral. Sci.*, **106** (*5*), 971–976.

Tay, F. R. and Pashley, D. H. (2004). Resin bonding to cervical sclerotic dentin: a review. *J. Dent.*, **32** (*3*), 173–196.

Tay, F. R., Carvalho, R. M., Yiu, C. K., *et al.* (2000). Mechanical disruption of dentin collagen fibrils during resin–dentin bond testing. *J. Adhes. Dent.*, **2** (*3*), 175–192.

Tay, F. R., Pashley, D. H., Kapur, R. R., *et al.* (2007). Bonding BisGMA to dentin – a proof of concept for hydrophobic dentin bonding. *J. Dent. Research*, **86** (*11*), 1034–1039.

Vajda, E. G., Skedros, J. G., and Bloebaum, R. D. (1998). Errors in quantitative backscattered electron analysis of bone standardized by energy-dispersive X-ray spectrometry. *Scanning*, **20** (*7*), 527–535.

Van Essche, M., Quirynen, M., Sliepen, I. Van Eldere, J., and Teughels, W. (2009). Bdellovibrio bacteriovorus attacks Aggregatibacter actinomycetemcomitans. *J. Dent. Research*, **88** (*2*), 182–186.

Vladar, A. E., Postek, M. T., and Ming, B. (2009). On the sub-nanometer resolution of scanning electron and helium ion microscopes. *Microscopy Today*, **17** (*2*), 6–13.

Wang, Y. and Spencer, P. (2005). Continuing etching of an all-in-one adhesive in wet dentin tubules. *J. Dent. Research*, **84** (*4*), 350–354.

Weerasinghe, D. D., Nikaido, T., Ichinose, S., Waidyasekara, K. G., and Tagami, J. (2007). Scanning electron microscopy and energy-dispersive X-ray analysis of self-etching adhesive systems to ground and unground enamel. *J. Mater. Sci. Mater. Med.*, **18** (*6*), 1111–1116.

Wierzchos, J., Falcioni, T., Kiciak, A., *et al.* (2008). Advances in the ultrastructural study of the implant–bone interface by backscattered electron imaging. *Micron*, **39** (*8*), 1363–1370.

Yiu, C. K., Tay, F. R., King, N. M., *et al.* (2004). Interaction of glass-ionomer cements with moist dentin. *J. Dent. Research*, **283** (*4*), 283–289.

Yuan, Y., Shimada, Y., Ichinose, S., and Tagami, J. (2007). Qualitative analysis of adhesive interface nanoleakage using FE-SEM/EDS. *Dent. Mater.*, **23** (*5*), 561–569.

Zinelis, S., Thomas, A., Syres, K., Silikas, N., and Eliades, G. (2010). Surface characterization of zirconia dental implants. *Dent. Mater.*, **26** (*4*), 295–305.

14 SEM, teeth, and palaeoanthropology: the secret of ancient human diets

Alejandro Romero and Joaquín De Juan

14.1 Introduction

In palaeoanthropological research, the relationship between diet, dental morphology, and tooth use is biologically very significant. Teeth are the body's most enduring component; hard tooth enamel retains its structure during taphonomic and fossilization processes, which is why teeth are so prevalent in the fossil record of vertebrate organisms and provide such a valuable research resource. Advances in scanning electron microscopy (Romero and De Juan, 2003, 2005; Ungar *et al.*, 2003), topographic and 3-D analysis (Ungar and M'Kirera, 2003; Kullmer *et al.*, 2009), and micro-CT imaging techniques (Smith and Tafforeau, 2008) have provided new ways for teeth to be studied (Irish and Nelson, 2008; Koppe *et al.*, 2009). Analysis of dental morphometrical characteristics and tooth wear is of great interest when reconstructing dietary behavior in fossil hominins and past human populations from an evolutionary perspective (Ungar, 1998; Teaford and Ungar, 2000). Changes in subsistence and socioeconomic system economies around 10 000 years ago produced significant biological and cultural drifts in our species (Kaifu *et al.*, 2003; Larsen, 2006). Specifically, historically recent changes in our diets imply adaptability to a new lifestyle, and direct consequences of many contemporary health problems can be found (Cordain *et al.*, 2005). In this context, scanning electron microscopy (SEM) technologies in tooth enamel exploration are an important tool for documenting dietary habits in extant and extinct primate species (Pérez-Pérez, 2004; Teaford, 2007). The microscopic damage or microwear found on enamel surfaces can be used to record patterns of consistency in foods consumed and related dietary implications. As well as examining various morphological and structural dental characteristics, this paper reviews existing SEM literature relating to chewing mechanisms and the physical properties of food in the context of dental microwear studies from a palaeoanthropological perspective. SEM technology applied to *in vivo* samples raises new questions and areas of discussion about our understanding of the importance of nutrition during human evolution and adaptation.

Scanning Electron Microscopy for the Life Sciences, edited by H. Schatten. Published by Cambridge University Press

14.2 Tooth shape and function

The form and function of hominoid teeth and dentition are related to dietary adaptations (Fleagle, 1999; Teaford and Ungar, 2007). These phenotypic features within and among primates reflect specific biological structures for processing food types (Teaford and Ungar, 2000; Ungar and Bunn, 2008). However, molars present the most complicated morphology and topography of all tooth types (Henke, 1998). Among primates, for example, folivorous species adapted to a vegetarian diet tend to have high cusps and long molar shearing crests, which are perpendicular to the tooth row and capable of shearing through tough leaves, whereas frugivores, especially those that eat hard objects, tend to have blunt cusp teeth that are better suited to crushing (Kay and Hiiemae, 1974; Lucas, 2004) and processing food items that are not very tough and have a high elastic modulus (Teaford, 2000).

Nevertheless, functional tooth morphology in extinct hominin species can show the limits of a direct analogy. The interspecific variability of tooth shape does not indicate the amount of a particular food item eaten by extant mammals, and analysis of enamel loss during lifetime is an important way to approach several questions linked to diet (Teaford and Glander, 1996; Teaford, 2000; Galbany *et al.*, 2011).

Recently, three-dimensional measurements of teeth and molar wear pattern analysis have provided new tools in the study of functional tooth morphology and diets in living and extinct hominin species (Ungar, 2004; Ungar and Bunn, 2008). Digital models of tooth crowns show that changes in crown shapes occur when teeth wear down during an individual's life. For example, folivorous species tend to have more sloping surfaces and relief than frugivorous species (Ungar and Bunn, 2008). The relationships between tooth shape and wear have also been used to infer feeding adaptations in fossil forms. Results indicate that *Australopithecus africanus* (~3.0–2.4 Myr) and early *Homo* (~2.4–1.5 Myr) relied on more elastic foods than *Paranthropus robustus* (~2.0–1.5 Myr), which were better adapted to harder, more brittle items as foods, as was the case for *Australopithecus afarensis* (~4.0–3.0 Myr) (Ungar, 2004, 2007).

The molar cusps in primates have functional significance in mastication, whilst in hominin and Homo species the crown is more rounded. Fossil hominin species probably differed as regards fracture and type of food consumed, given the enamel structure and morphological trends of the teeth (Lucas and Peters, 2000; Lucas *et al.*, 2008). Early members of the genus *Homo* had relatively larger anterior teeth but much smaller posterior teeth than modern humans, and this reduction in tooth size in what is a large-bodied omnivore constitutes one of the crucial aspects that is still poorly understood in human evolutionary adaptation. However, it is probable that dental reduction in *Homo sapiens* was due to new dietary habits. There have been periods of marked dental reduction during the evolution of our genus since ~2 Myr (Hillson, 1996), events that have been related to shifts in dietary trends (Calcagno and Gibson, 1991; Larsen, 2003), and new food processing technologies for reducing food toughness (Wrangham *et al.*, 1999; Lucas and Peters, 2000; Lucas, 2004).

Due to the potential significance of the relationship between human diet and tooth-wear, this phenomenon still requires further investigation. Measurements of dental wear are difficult, if not impossible, to obtain as the nature of the diets in question and longitudinal data samples are unattainable for fossils and difficult to collect from dental patients. One of the most important questions in the study of enamel loss that researchers have tried to answer is how molar cusp-tip wear facets change over the course of a lifetime. However, a further important question is how food abrasiveness produces changes in teeth.

14.3 SEM and teeth research

Since the late 1960s, scanning electron microscopy (SEM) technology has offered many advantages over optical microscopy in the analysis of archaeological and palaeoanthropological samples (Brothwell, 1969). The specific biodynamic problems of a wide variety of paleobiological samples can be studied using SEM, and it also provides a greater depth of focus and more detailed resolution than light microscopes (Olsen, 1988). In recent decades, considerable effort has been invested in reconstruction of primate diets using quantifiable approaches, together with studies of dental and cranial gross morphology. In this chapter, the authors focus particularly on the conceptual changes affecting classical hypotheses of feeding behavior in fossil hominins and ancient human groups resulting from the application of SEM analysis. The use of SEM techniques to analyze microwear patterns (dental microwear) on the enamel surface of occlusal and non-occlusal molars produced by abrasive food particles has proved to be an effective method for resolving several questions regarding food processing, type of food eaten, and jaw movements of modern and extinct species (Teaford 1994, 2007; Ungar, 1998; Pérez-Pérez, 2004; Romero and De Juan, 2005; Ungar *et al.*, 2008; Galbany *et al.*, 2009), and represents a useful approach to study of dietary adaptations during a hominoid's lifetime.

14.4 Dental microwear analyses

Dental microwear refers to microscopic-scale tooth wear caused by hard objects in food, and indicates how teeth were used during mastication. The most commonly employed method for dental microwear analysis is SEM, and microwear analysis of tooth enamel surfaces is one of the most effective methodologies available for determining ecological and dietary factors in ancient and modern species. Furthermore, microwear analysis of tooth surfaces is of particular interest, since differences in diet may have a relative correlation with enamel loss. However, interpretation of dental microwear data of fossil teeth is based on extant species from museum collections with uncertain life histories. Thus, many new key questions still need to be addressed. For instance, we still do not know what *in vivo* microwear patterns indicate or how enamel microwear changes with specific diets over time. Again, much more research is necessary.

14.4.1 Methods of analysis

Teeth form an important part of the fossil record, and SEM analyses of tooth surfaces can be carried out using either original surfaces or high-resolution replicas (Romero *et al.*, 2004; Galbany *et al.*, 2004, 2006; Fiorenza *et al.*, 2009). However, tooth observation under SEM requires sample metallization and a vacuum chamber, whilst environmental scanning electron microscopy (ESEM), where available, is only suitable for isolated samples (Romero *et al.*, 2004). Therefore, SEM research generally uses casts. Microscopic features of the tooth are then analyzed using digital images or SEM micrographs, which show electron scattering on the enamel surface. These represent an electronic map with a different intensity of grey levels on the tooth surface.

14.4.1.1 Moulding

High-resolution moulding techniques for replacing original teeth are invaluable in microwear studies because they have been shown to reproduce features to a fraction of a micron for both *in vitro* (Galbany *et al.*, 2004, 2006) and *in vivo* (Teaford and Oyen, 1989a; Romero and De Juan, 2007) specimens (Figure 14.1). Replicas used for SEM analysis are currently obtained with negative moulds. However, cast quality remains optimal for SEM analysis at high magnification for four consecutive replicas (Galbany *et al.*, 2006). Moulds can be made using hydrophobic polyvinylsiloxane silicones employed for dentistry (Galbany *et al.*, 2006; Fiorenza *et al.*, 2009), and various products present excellent

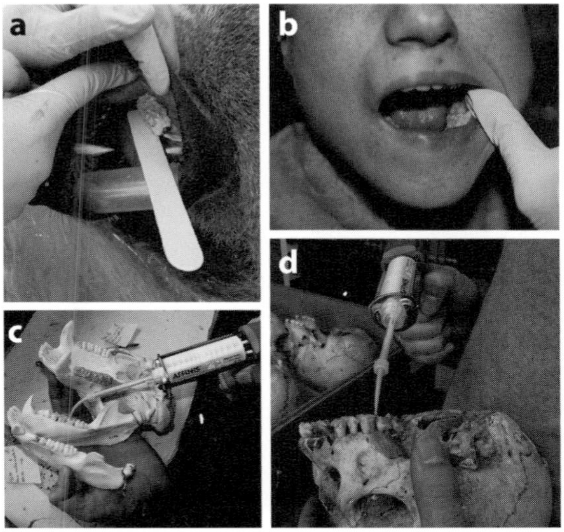

Figure 14.1 Moulding technique for dental morphology and microwear studies. Examples of application of dental impression material, using a regular body polyvinylsiloxane silicone: (a) Living baboon (*Papio cynocephalus*) from Amboseli (Kenya) in June 2007 (photo courtesy of Dr. Jordi Galbany, Universitat de Barcelona); (b) with adult volunteers in February 2004, Departamento de Biotecnología (Universidad de Alicante). Moulding by the authors of the postcanine dentition of original specimens from primatological (c), and human (d), osteological European collections.

dimensional stability and reproduction detail. Replicas are obtained by pouring epoxy or polyurethane resin and hardener into the moulds (Galbany *et al.*, 2004, 2006). Since observation with SEM requires conductive samples, resin replicas are coated under vacuum with gold-palladium. They are then mounted on stubs and a silver contact is placed at the point where the replica and the stub meet to further improve electrical conductivity. By contrast, variable pressure SEM does not require pre-coating of samples, and thus wet, oily, and non-conductive samples can be analyzed in their natural state (Romero *et al.*, 2004; Romero and De Juan, 2005). Lastly, SEM micrographs are then recorded at the magnification levels and within the methodological parameters previously established.

14.4.1.2 Tooth surfaces, facets and microwear

Microscopic wear patterns on tooth enamel surfaces are highly correlated with abrasion (tooth-food-tooth contact), but are also determined by the relative hardness and physical properties of tooth surfaces, with some enamel chips being caused by tooth-on-tooth contact, or attrition (Newesely, 1993; Romero and De Juan, 2003, 2005; Teaford, 2007). During food breakdown, masticatory movement includes both puncture crushing and chewing cycles. The first stage involves food-tooth contact and then, once the food has been softened, tooth-tooth interaction is produced. In this oral food processing scenario based on abrasion and attrition, teeth facets show different wear patterns depending on the shearing facets (Phase I) and grinding facets (Phase II), on occlusal molar surfaces (Kay and Hiiemae, 1974; Gordon, 1984; Teaford, 1988a; Hillson, 1996; Kullmer *et al.*, 2009) (Figure 14.2). The type of facet examined and the microwear pattern observed

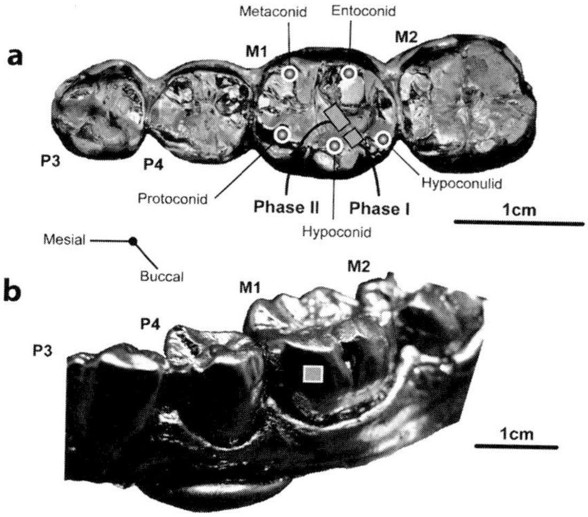

Figure 14.2 Occlusal (a) and occlusal-buccal (b) views of a human adult postcanine (P3-M2) dental complex (high-resolution replicas sputter-coated with gold palladium). Teeth and cusp-facet terms correspond to the conventional paleological and odontological terminology (see Krueger *et al.*, 2008; Kullmer *et al.*, 2009). Grey squares mark the preferable areas for examination under SEM in occlusal (Phase I–II facets 4–9 respectively) and buccal microwear (below), in the medial third of the surface under cusp tip.

reflect the biomechanisms of mastication in humans and non-human primates (Gordon, 1982; Mahoney, 2006a; Krueger *et al.*, 2008).

Microscopic wear features are categorized as pits (length:width ratio of 4:1) and scratches or striations (length:width >4:1) (Teaford, 1994; Pérez-Pérez *et al.*, 1999). Striations exhibit the appearance of scratches, but are thinner and less pronounced than scratches (Gordon, 1988; Pérez-Pérez *et al.*, 1994). Because the type of occlusal surface facet can reveal a wide variation in features of microwear fabrics due to the biomechanisms of chewing and compression loads (Gordon, 1984; Kullmer *et al.*, 2009), Phase II-facet 9 is usually the standard area used for diet-microwear studies (Teaford, 1994). In theory, an abrasive particle produces an impact pit at the start of a scratch or striation. Scratches and pits can be formed by similar particles but vary in the angle of compression load and attack on the enamel. Therefore, occlusal and non-occlusal tooth surfaces present different micro-feature fabrics, yielding different information concerning the angle of attrition and abrasion (Pérez-Pérez *et al.*, 2003; Romero *et al.*, 2003a; Teaford, 2007) (Figure 14.3). Non-occlusal surfaces do not have a direct effect on attrition during the process of chewing, and dietary information is barely affected by biomechanical factors (Pérez-Pérez *et al.*, 1994, 1999; Romero *et al.*, 2003b). We can assume that the buccal microwear pattern resulting from abrasion, and higher densities of microwear, will be located on occlusal-buccal contact areas rather than on mid-buccal or cervical areas (Lalueza *et al.*, 1996; Pérez-Pérez *et al.*, 2003).

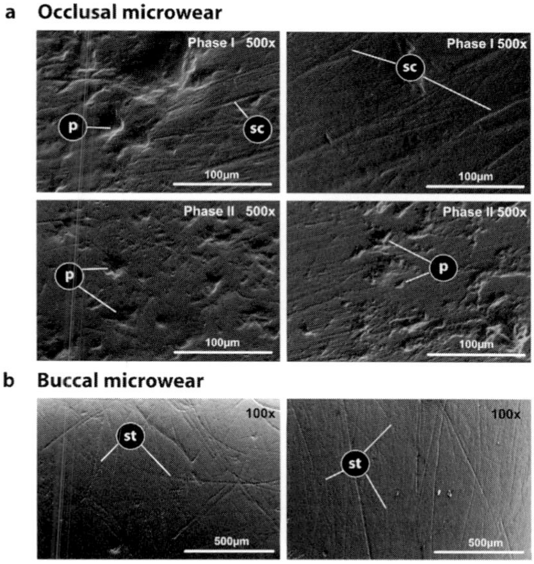

Figure 14.3 SEM micrographs showing *ante-mortem* microwear on occlusal (a) and non-occlusal (buccal) (b) surfaces in permanent lower first human molars (see Figure 14.2). Occlusal microwear facets show patterns with differences in predominance of pits (p) and scratches (sc). Buccal microwear of different feature densities is characterized by a less complex pattern of preferably occlusal to cervical striation (st) distribution.

14.4.1.3 Data collection and image processing

In dental microwear research, density and micro-feature frequencies (pits and scratches or striations), their length or width in micrometers (μm), and orientation (0°–180°) are computed through semi-automated methods identifying individual features marked using a mouse-driven pointer (Grine *et al.*, 2002; Galbany *et al.*, 2004, 2005a). Dental microwear parameters are used to assess the dietary differences of interspecific specimens. Researchers usually study areas at different magnification levels, generally at 500× for the occlusal surfaces (Teaford, 1988a), and 100× for the buccal (Pérez-Pérez, 2004). Differences in microwear between tooth and enamel occlusal facets may be significant (Gordon, 1982, 1984; Mahoney, 2006a). Despite these factors, which are not found in buccal microwear studies (Pérez-Pérez *et al.*, 1994; Romero and De Juan, 2007), wherever possible it is necessary to record SEM micrographs from similar facets on the same teeth at the same magnification level (Gordon, 1988; Pérez-Pérez, 2004). Differences within and between species are evaluated using statistical analysis (Pérez-Pérez *et al.*, 1994, 2003; Grine *et al.*, 2002; Galbany *et al.*, 2005a). Dental microwear is subject to error rates (Grine *et al.*, 2002; Galbany *et al.*, 2005a); however, researchers tend to show low intraobserver measurement error. Recently, a quantitative approach using confocal microscopy and scale sensitive fractal analysis (Ungar *et al.*, 2003) has offered the possibility of rapid surface characterization in microwear studies. This new tool has made it possible to re-examine baseline microwear patterns obtained in living and extinct species using texture analysis (Ungar *et al.*, 2008, 2010).

14.5 Microwear formation processes

Dietary adaptations influence cranial, mandibular, and dental anatomical structures on an evolutionary timescale (Teaford and Ungar, 2000, 2007; Larsen, 2006; Lucas *et al.*, 2008). The relationship between teeth and the mechanical properties of food items during mastication may be a principal factor in explaining variations in dental morphology and the biomechanisms of mastication among primates, including humans (Lucas, 2004; Strait, 1997). However, tooth microwear is a complex result of chewing processes, tooth and enamel morphological characteristics, and food properties.

14.5.1 Food, enamel, and microwear

Teeth are subject to physiological wear processes that affect the occlusal surfaces. During mastication, the postcanine dentition acts to reduce food toughness. Tooth abrasion occurs as the enamel, the hardest calcified tissue in mammals, comes into contact with a food item that is harder than enamel. However, food texture depends both on its nature and on cooking modes. Foods present heterogeneous structures with variable physical and chemical characteristics: hard or elastic foods may produce mechanical differences during mastication rather than abrasive effects on the enamel (Romero and De Juan, 2005). In fact, many foods are not hard enough to scratch tooth surfaces (Newesely 1993;

Gügel *et al.*, 2001; Romero and De Juan, 2005, 2007; Sanson *et al.*, 2007). Only those foods that include intrinsic or extraneous abrasive particles harder than enamel on the Mohs scale of mineral hardness (range 1–10), such as siliceous opal phytoliths in monocotyledone plants (i.e. *Poaceae* and *Cyperaceae*) or exogenous grit ranging between 5.5 and 6.5 (quartz is 7.0), or than dental apatite crystals (from 4.5 to 5.0) (Newesely, 1993; Romero and De Juan, 2005), can scratch tooth enamel surfaces (Teaford and Lytle, 1996; Gügel *et al.*, 2001; Romero *et al.*, 2007). As a result, abrasives in food and masticatory action leave a clear enamel imprint. In fact, research has established microwear formation processes based on specific steps in food breakdown. The physical consistency of food (physical nature of the food's texture) and the abrasives it contains are the main variables influencing wear.

14.5.2 *In vitro* and *in vivo* experimental microwear studies

Pioneering experimental *in vitro* microwear studies suggested that abrasive particles in different dietary food types (i.e. exogenous grit or grass phytoliths) leave similar microwear morphology (Peters, 1982). Later, enamel microwear patterns were found to be indistinguishable in species that were given different diets including insect chitin and plant fibers (Covert and Kay, 1981), or grit and plant opals (Kay and Covert, 1983). Nevertheless, these results were questioned on the grounds of dietary medium, additives included in the controlled diets, or insufficient duration of the experiment (Gordon and Walker, 1983). However, this research did establish an experimental basis for future studies, since investigators can never be sure of the diet of the museum specimens analyzed. More recent studies have produced microscopic features on human dental enamel produced by several cereal species (Gügel *et al.*, 2001), and results have revealed the importance of silica bodies (opal phytoliths) in the abrasion process and enamel loss (Figure 14.4a).

Experimental *in vivo* analysis of tooth wear and microwear linked to laboratory or wild-caught non-human primates as well as dental patients raised on known diets (controlled abrasiveness) have produced surprising results. For example, studies with living primates show that harder foods caused more pits than scratches, which are associated with tougher or softer items respectively (Teaford and Oyen, 1989b, 1989c). In addition, laboratory experiments have also shown that turnover is a rapid phenomenon, especially when abrasives are present in food, and there is a higher degree of action on crushing than on shearing facets (Teaford and Oyen, 1989b; Teaford and Tylenda, 1991). Other studies have focused on the microwear effect of the properties of food eaten by living primates due to sex, seasonal variability, and habitat range (Teaford and Glander, 1996; Galbany *et al.*, 2011). Finally, *in vivo* microwear studies of human occlusal and buccal postcanine tooth surfaces have shown that the enamel of humans who ate soft, industrialized foods had fewer microwear features than those consuming an induced, abrasive diet, where the rate of wear increased (Teaford and Lytle, 1996; Romero *et al.*, 2007), and showed a different turnover that was lower on the buccal surface than on the occlusal (Pérez-Pérez *et al.*, 1994; Romero and De Juan, 2006; Romero *et al.*, 2007, 2009) (Figure 14.4b). Therefore, although reconstructions of ancient human diet

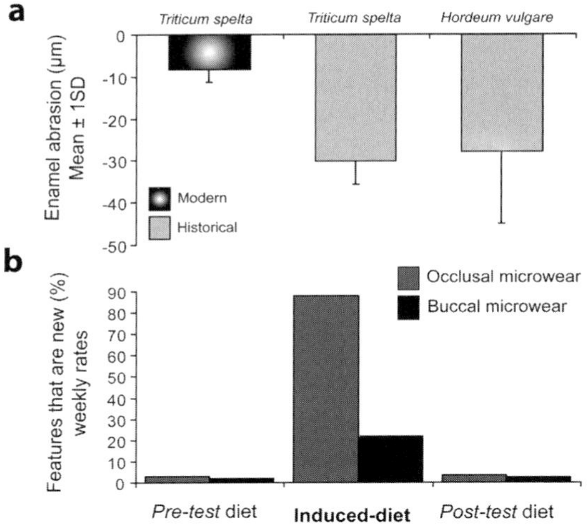

Figure 14.4 a) Molar enamel loss through *in vitro* abrasion by cereal grains processed with modern and historical methods (data from Gügel *et al.*, 2001). b) Differences in weekly rates of *in vivo* occlusal and buccal dental microwear in humans associated with changes in diet (data from Teaford and Lytle, 1996; Romero *et al.*, 2007, 2009). Differences during pre- and post-test *ad libitum* diet are evident with regards to induced abrasive diet (cereal grains processed with stone mortars) controls. Microwear turnovers seem to be less important on buccal than on occlusal surfaces.

behavior based on dental microwear analysis must be understood in terms of abrasiveness, they equally reflect the biomechanics of chewing as well as cultural and environmental factors (Pérez-Pérez *et al.*, 2003; Romero *et al.*, 2003b, 2004; Teaford 2007). In short, more experimental *in vivo* studies on tooth enamel are needed to investigate the relationship between types of abrasive and dental phenomenon.

14.5.3 *Post-mortem* microwear

Trampling processes can affect the teeth obtained from palaeontological or archaeological contexts (Teaford, 1988b; King *et al.*, 1999; Pérez-Pérez *et al.*, 2003; Romero *et al.*, 2004; Smith and Tafforeau, 2008). Nevertheless, experimental and comparative micrograph examinations have demonstrated that *post-mortem* processes produce micro-features that are clearly different from *ante-mortem* microwear affecting tooth surfaces through abrasive particles in food consumed on both occlusal (Teaford, 1988b; King *et al.*, 1999) and non-occlusal surfaces (Pérez-Pérez *et al.*, 2003; Romero *et al.*, 2004) (Figure 14.5). Particles in sediments or erosive processes basically tend to erase or distort *ante-mortem* food microwear (King *et al.*, 1999) and wear patterns caused during chewing, which are located at specific tooth facets (Teaford, 1988b, 2007). Whilst *post-mortem* artifacts on the enamel surfaces can be relatively diagnosed, it is nevertheless still necessary to document *in vivo* microwear patterns in order to compile a database of micro-feature typologies for comparative purposes.

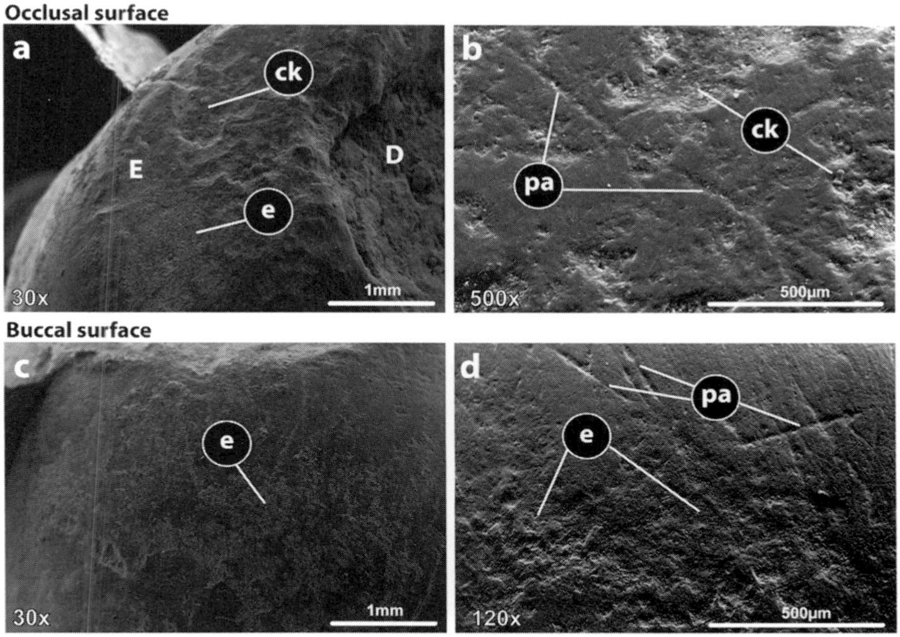

Figure 14.5 SEM micrographs at different magnification levels of *post-mortem* damage on osteoarchaeological molar tooth showing mesial occlusal contact facet (a), occlusal grinding-Phase II facet 9 (b), and buccal surface (c–d) of the protoconid cusp. Physical abrasive and chemical erosive damage can be found on occlusal cusp enamel (E) and dentine (D) facets affecting buccal side (in c) of the tooth. A clear unpreserved enamel surface for microwear analysis is evident. *Post-mortem* enamel cracks (ck), abrasions (pa), and erosion effect (e) that induce enamel prism exposure (see d) due to trampling processes are readily identifiable.

14.6 Dental microwear from a palaeoanthropological perspective

We know that tooth enamel is a tissue subject to wear. If the frequency and morphology of microwear features reflect dietary habits, what can information on dental microwear in extant species tell us about ancient dietary behavior?

14.6.1 Non-human primates and fossil hominins

The dental microwear of living primate species has been examined to determine differences between primate hard-object feeders and leaf-eaters. Species fed a hard diet show a greater frequency of features that are larger in size, whereas those fed a soft diet have a low incidence of small microwear morphologies (Teaford, 1994; Ungar, 1998). Primate hard-object feeders (e.g. *Cebus apella* or *Pongo pygmaeus*) are characterized by large pits on the enamel, whilst leaf-eaters such as *Cebus* or *Colobus* species tend to have fewer pits and more scratches on molar enamel surfaces (Teaford, 1994). In contrast to these extreme models (hard-object feeders and folivorous primates), more frugivorous and omnivorous species (i.e. *Pan troglodytes*) display an intermediate density of features with a higher average scratch size (Ungar *et al.*, 2006). In short, interspecific variability suggests a relationship between dietary hardness and microwear patterns in living primates. Differences between genera may be explained in part by differences in diet composition as well as by differences in incidence of terrestrial feeding events (Teaford, 1994; Ungar and Teaford, 1996; Galbany and Pérez-Pérez, 2004). More frequently, pit density in hard-object feeders has been attributed to the mastication of seeds which may incorporate extraneous grit particles (Ungar and Teaford, 1996; Teaford, 2007).

Buccal dental microwear in extant primates has provided another source of information (Galbany *et al.*, 2005b, 2009). Clearly, interspecific variability related to diet and ecological factors can be found by exploring striation on non-occlusal surfaces of molar teeth (Ungar and Teaford, 1996; Galbany and Pérez-Pérez, 2004; Galbany *et al.*, 2009). Firstly, microwear variability among Cercopithecoidea primates is due to differences in diet; for example, *Colobus* leaf-eater monkeys show a low rate of microwear compared with *Cercopithecus spp.* fruit eaters and *Papio anubis* terrestrial foragers from open savannah environments (Galbany and Pérez-Pérez, 2004; Galbany *et al.*, 2005b). On the other hand, analyses of variability in the African great apes (*Gorilla gorilla* and *Pan troglodytes*) (Galbany *et al.*, 2009) suggest that ecological adaptations constitute an important factor in subspecific level microwear patterns. In fact, although these species are mainly folivorous or frugivorous, their diets vary greatly in terms of ecological conditions and preferences (Fleagle, 1999). These differences in habitat exploitation are significant in research since patterns from fossil hominin specimens can be used to infer ecological diversification. Microwear fabrics can be used to identify broad dietary categories and infer aspects of food item abrasiveness. Therefore, it is important to determine the relationship between dental microwear and feeding behaviors in living primates in order to infer the diet of our ancestors (Galbany *et al.*, 2005b, 2009; Teaford, 1994, 2007).

Dental microwear suggests that after the Miocene, hominins had a wide range of diets (Teaford and Ungar, 2000; Galbany *et al.*, 2005b). Furthermore, the pattern of molar microwear found in the Early Pliocene African hominin *Australopithecus anamensis* (~4.2–3.9 Myr) overlaps with that of the chimpanzee (*Pan troglodytes*) and *Gorilla* species, and similar dietary patterns have been identified in *afarensis* species of the same genera (Grine *et al.*, 2006a, b). When microwear data for *A. afarensis* (Grine *et al.*, 2006a; Estebaranz *et al.*, 2009) are compared with those recorded for several extant primate species with well-documented dietary habits (El-Zaatari *et al.*, 2005; Galbany *et al.*, 2005b; Ungar *et al.*, 2006; Estebaranz *et al.*, 2009), interspecific microwear results for both occlusal and buccal molar surfaces show similar dietary models (Figure 14.6). Clearly, pitting frequency and scratch breadth (μm), as well as buccal microwear density and length (μm), tend to differentiate fossil hominins from the extant primates. *Papio* and *Colobus* species show extreme differences, in contrast to *Pan troglodytes* or *Gorilla gorilla*, which present a similar microwear pattern to those exhibited by fossil forms. The highly folivorous *Colobus* species is clearly distinguished from hard-object eaters such as *Papio ursinus* (El-Zaatari *et al.*, 2005) or *Papio anubis* (Galbany *et al.*, 2005b). Results suggest that the hominin species were capable of processing a fairly wide

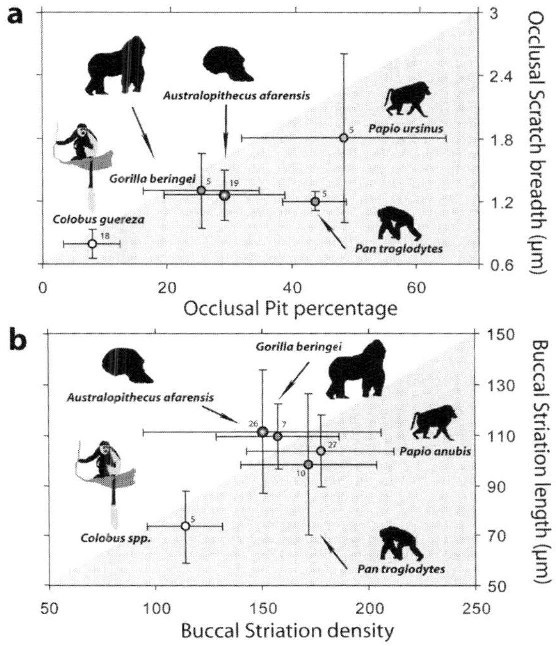

Figure 14.6 Bivariate plots showing occlusal (a) and buccal (b) microwear variables in molar teeth for extant primates and fossil hominin *Australopithecus afarensis*. The numbers indicate sample sizes (mean±SD) (data from El-Zaatari *et al.*, 2005; Galbany *et al.*, 2005b; Ungar *et al.*, 2006; Grine *et al.*, 2006a; Estebaranz *et al.*, 2009). The clear interspecific variability in living primates demonstrates that dietary information is retained in fossil hominins.

range of foods (Grine *et al.*, 2006b), eating fleshy ripe fruits and herbaceous plant materials from both closed-forest and open environments, probably including fallback resources (Estebaranz *et al.*, 2009). Other studies have examined more recent hominin species. For example, molar microwear results for Plio-Pleistocene early *Homo* from Africa suggest the consumption of a varied diet with no excessively hard or tough foods (Ungar *et al.*, 2006). Lastly, comparing buccal microwear patterns from Pleistocene hominins with those from modern hunter-gatherer groups has provided new and different perspectives (Pérez-Pérez *et al.*, 1999, 2003). Results have shown that the higher microwear density upon buccal tooth surfaces in Neanderthals appears to be associated with diets that were much more abrasive than those of modern populations and do not support the hypothesis of a strictly carnivorous diet.

14.6.2 An investigation into human tooth crown microwear

Analyses of dental wear and microwear in humans are of special interest. Compared to other extant primates, humans show a variety of feeding technologies, alimentary, and non-alimentary habits. Gradients of tooth wear and microwear in pre-agricultural and contemporary industrialized populations indicate differences in diet abrasiveness related to the types of resource exploited and food preparation techniques employed (Molnar *et al.*, 1983; Gügel *et al.*, 2001; Kaifu *et al.*, 2003; Romero *et al.*, 2004). Consumption of more refined and processed foods in modern industrialized societies has dramatically decreased the severity of wear. Therefore, study of tooth wear rates from past and present populations has provided information about human behavior.

14.6.2.1 Ancient human populations and diets

Human dental microwear analyses, of both occlusal and buccal surfaces of molar teeth from prehistoric human groups, have provided direct evidence of dietary behaviors (Pérez-Pérez *et al.*, 2003; Ungar *et al.*, 2006; Mahoney, 2007; Romero and De Juan, 2007; Teaford 2007). Firstly, occlusal microwear studies have documented the dietary shift from hunter-gathering to farming subsistence, showing signs of characteristic enamel micro-features (pits and scratches) depending on diet abrasiveness in agricultural economies (Schmidt, 2001; Mahoney, 2007). In fact, the transition from pre-agricultural populations to subsequent agricultural economies in Native American groups and in the Near East has shown a great variety of microwear patterns recording a complicated phenomenon due to differences in habitat and cultural factors (Schmidt, 2001; Organ *et al.*, 2006; Mahoney, 2006b, 2007). For example, food texture has been related to the incidence of pitting, reflecting changes in abrasiveness due to the consumption of exogenous particles, from hunting and gathering subsistence to dependence on agriculture. Secondly, buccal microwear research in humans has shown that variability in intragroup molar microwear values seems to be less significant than inter-individual and intergroup variability (Lalueza *et al.*, 1996; Pérez-Pérez *et al.*, 1999; Romero and De Juan, 2007). Human groups that eat more abrasive items – wild plant foods or milled cereals – usually show higher microwear density values (Pérez-Pérez *et al.*, 1994, 2003; Romero and De Juan, 2005, 2007). Additionally, human populations consuming large amounts of meat in their diet show a low density of striations on non-occlusal teeth surfaces (Lalueza *et al.*, 1996; Ungar and Spencer, 1999; Romero and De Juan, 2007; Polo-Cerdá *et al.*, 2007, 2010; Romero *et al.*, 2010). Although these studies clearly demonstrate the existence of a relationship between dietary habits and interspecific microwear variability, little is yet known about the relationship between dental microwear formation rates, diet, and age. *In vivo* evidence represents one method for clarifying these questions. In recent years and based on pioneering studies (Teaford and Lytle, 1996; Teaford and Tylenda, 1991), analyses of *in vivo* molar crown microwear (occlusal and buccal surfaces) in contemporary humans have been conducted (Romero *et al.*, 2003b, 2007, 2009; Romero and De Juan, 2007) in an attempt to obtain a microwear-related model for comparison with new baseline series of modern hunter-gatherers (Romero *et al.*, 2010) and ancient human groups (Romero and De Juan, 2007).

14.6.2.2 Occlusal-buccal microwear interaction

When occlusal and buccal microwear variables are compared with selected tooth microwear patterns recorded in previous studies of ancient and modern human groups from different environments and presenting different dietary habits (Lalueza *et al.*, 1996; Mahoney, 2007; Romero and De Juan, 2007; Alrousan and Pérez-Pérez, 2008; El-Zaatari, 2008; Polo-Cerdá *et al.*, 2010) (Figure 14.7), a significant number of similar patterns are observed despite the different abrasive effects on occlusal and non-occlusal enamel surfaces. As we have shown in primates (Figure 14.6), pitting incidence and scratch breadth (μm) for occlusal surfaces (Mahoney, 2007; El-Zaatari, 2008) and microwear density and length (μm) for buccal surfaces tend to discriminate human groups by dietary habits (Pérez-Pérez *et al.*, 2003; Romero *et al.*, 2004; Polo-Cerdá *et al.*, 2007). Data obtained for contemporary humans (*in vivo* sample) (Romero and De Juan, 2007) show that there is no food in the contemporary diet capable of causing the microwear patterns observed among hunter-gatherer and farming populations. Occlusal-enamel pit frequency and low striation density in buccal surfaces are a common pattern among contemporary humans as a product of tooth-to-tooth interaction and softer non-abrasive diets. Dental microwear in the groups compared indicated different diets and food processing technologies. The meat-eaters lived between latitudes 50° and 70°. Alaskan Eskimos (Ipiutak and Aleuts) (El-Zaatari, 2008), or Inuits, and Fueguians (Tierra del Fuego) (Lalueza *et al.*, 1996) hunt almost exclusively marine animals for subsistence. By contrast, prehistoric populations are represented by Natufian hunter-gatherers (~12–10 000 BC) from the Near East (Mahoney, 2007; Alrousan and Pérez-Pérez, 2008) and Neolithic and Chalcolithic human groups from the Near East (~10–5000 BC) (Mahoney, 2007) and southern Spain (~5–2000 BP) (Romero and De Juan, 2007; Polo-Cerdá *et al.*, 2010). The prehistoric sample based their subsistence economies on hunting, animal husbandry, and plant cultivation, which reflects their specific environmental conditions and adaptations. Nevertheless, changes in the percentages of wild or cultivated products and food processing techniques are evident across periods (Mahoney, 2007; Romero and De Juan, 2007). Occlusal and buccal microwear patterns in meat-eaters show a frequency and morphology of microwear features related to a lesser consumption of plant foods (Figure 14.7). Meat is not hard enough to damage the enamel, and meat-eating human groups clearly show dental microwear variability compared to the other groups. However, in these groups, food grit derived from meat dried or roasted for conservation, and large bite forces during chewing created the larger features found on occlusal enamel (Figure 14.7a) and the longer ones on buccal enamel (Figure 14.7b). Finally, the variation in occlusal-buccal microwear patterns clearly reflects tough food consumption factors related to chewing processes and tooth-to-tooth contact (Lucas and Peters, 2000), due to the high occlusal pit density in meat-eaters or the contemporary human sample. The microwear patterns of the Natufian, Neolithic, and Chalcolithic populations are characterized by a high density of small features, seeming to indicate a more abrasive diet than that of contemporary humans. Ancient hunter-gatherers (Natufian sample) are characterized by a diet in which tough foods predominated, requiring substantial pressure to chew, since large-grained abrasives could significantly impact microwear patterns. An exception to this is found in the high frequency of pits and wide scratches in the hunter-gatherers from the Natufian period (Mahoney, 2007), plotted with data obtained from the Eskimo sample (El-Zaatari, 2008). From the Neolithic and

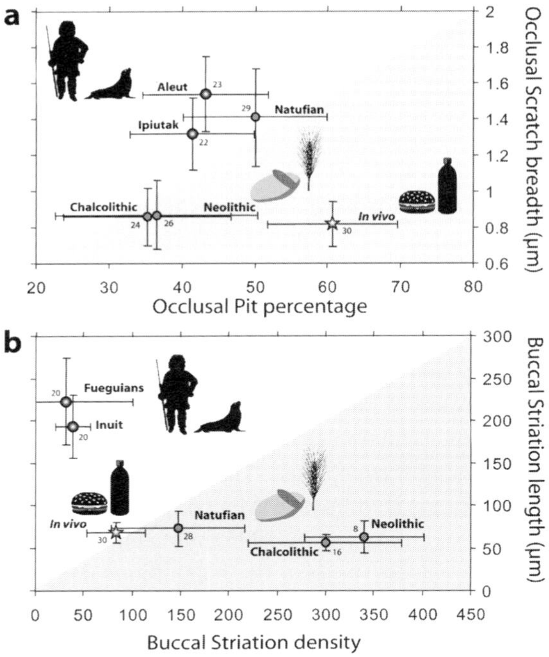

Figure 14.7 Bivariate plots of measurements of molar occlusal (a) and buccal (b) microwear in modern humans from different environments and presenting different dietary habits. The numbers indicate sample sizes (mean±SD) (data from Lalueza *et al.*, 1996; Mahoney, 2007; Romero and De Juan, 2007; Alrousan and Pérez-Pérez, 2008; El-Zaatari, 2008; Polo-Cerdá *et al.*, 2010).

Chalcolithic onwards, a mixed farming economy has been documented (Mahoney, 2007), with a reduction in the frequency of animal hunting and an increase in the husbandry of domestic animals, with a predominance of ovicaprids (sheep and goats) (Romero and De Juan, 2007). Differences among prehistoric groups are probably due to different sized contaminants and meat intake in the diet. Phytolith plant species or exogenous grit may have had an impact on the enamel during chewing, producing different microwear feature morphologies (Gügel *et al.*, 2001). Although some foods do not contain abrasive items during chewing, occlusal microwear can be affected. In a non-abrasive diet, repetitive loading during chewing can induce a high percentage of pits associated with small microwear morphologies. Finally, the inferences about molar microwear patterns of the populations examined have led researchers to consider significant differences in food processing techniques and environmental exploitation methods rather than general dietary regimes in human populations. Significant differences with the *in vivo* contemporary sample suggest that exogenous grit, together with the effects of occlusal forces during mastication, are the principal factors behind the differences obtained. The results across the different groups examined clearly suggest changes in the abrasiveness of food related to food processing techniques and the biomechanisms of chewing, since a high microwear density pattern can be caused by factors other than dietary-related ones.

14.7 Conclusions and future implications

The objective of this research was to analyze the probable interaction between, and significance of, occlusal and buccal-dental microwear and differences between diet in ancient and contemporary humans. As coincident results were found, dissimilar microwear patterns showed similar diet-related signs. Despite the fact that microwear patterns on occlusal and buccal enamel surfaces are related to their specific formation processes, dental microwear patterns can be attributed to changes in dietary texture. When the authors compared primate and fossil hominin microwear patterns with hunter-gatherer or farming populations no simple interspecific models were found, as diet microwear shows signs of variability according to a series of specific ecological niche, morphology, and climate factors (Pérez-Pérez *et al.*, 2003; Romero and De Juan, 2007; Teaford, 2007; Galbany *et al.*, 2009). In turn, contemporary Western population models might not be representative of ancient human diets. Probably, the significance of dental microwear in ancient human groups is relevant when considering cultural dietary factors and the amount of gritty particles in diets. In particular, food processing procedures have been clearly related to changes in nutritional characteristics of diets since the Neolithic period and later during the Industrial era (Cordain *et al.*, 2005). Therefore, the role of food processing techniques in humans undoubtedly has a potential effect on the abrasiveness of the diet and microwear and wear formation rates. Reconstruction of past human diets through dental microwear must depend on a comparatively large sample of known dietary habits identified in modern hunter-gatherers and industrialized people, in which microwear, diet, and intrapopulation differences by sex and age are represented. Future research is needed to clarify these questions.

14.8 References

Alrousan, M. and Pérez-Pérez, A. (2008). Non-occlusal microwear of the last hunter-gatherer from Near East and Europe. In: *Genes, ambiente y enfermedades en poblaciones humanas*, ed. J. L. Nieto, J. A. Obón, and S. Baena. Zaragoza, Prensa Univ. Zaragoza, pp. 45–59.

Brothwell, D. (1969). The study of archaeological materials by means of the scanning electron microscope: an important new field. In: *Science in Archaeology*, ed. D. Brothwell and E. Higgs. London, Thames and Hudson, pp. 564–66.

Calcagno, J. M. and Gibson, K. R. (1991). Selective compromise: evolutionary trends and mechanisms in hominid tooth size. In: *Advances in Dental Anthropology*, ed. M. A. Kelley and C. S. Larsen. New York, Wiley-Liss, pp. 59–76.

Cordain, L., Eaton, S. B., Sebastian, A., *et al.* (2005). Origins and evolution of the Western diet: health implications for the 21st century. *American Journal of Clinical Nutrition*, **81**, 341–54.

Covert, H. H. and Kay, R. F. (1981). Dental microwear and diet – implications for determining the feeding behaviors of extinct primates, with a comment on the dietary pattern of *Sivapithecus*. *American Journal of Physical Anthropology*, **55**, 331–6.

El-Zaatari, S. (2008). Occlusal molar microwear and the diets of the Ipiutak and Tigara populations (Point Hope) with comparisons to the Aleut and Arikara. *Journal of Archaeological Science*, **35**, 2517–22.

El-Zaatari, S., Grine, F. E., Teaford, M. F., and Smith, H. F. (2005). Molar microwear and dietary reconstruction of fossil Cercopithecoidea from the Plio-Pleistocene deposits of South Africa. *Journal of Human Evolution*, **51**, 297–319.

Estebaranz, F., Martínez, L. M., Galbany, J., Turbón, D., and Pérez-Pérez, A. (2009). Testing hypotheses of dietary reconstruction from buccal dental microwear in *Australopithecus afarensis*. *Journal of Human Evolution*, **57**, 739–50.

Fleagle, J. G. (1999). *Primate Adaptation and Evolution*. London, Academic Press.

Fiorenza, L., Benazzi, S., and Kullmer, O. (2009). Morphology, wear and 3D digital surface models: materials and techniques to create high-resolution replicas of teeth. *Journal of Anthropological Sciences*, **87**, 211–18.

Galbany, J. and Pérez-Pérez, A. (2004). Buccal enamel microwear variability in Cercopithecoidea Primates as a reflection of dietary habits in forested and open savanna environments. *Anthropologie*, **42**, 13–19.

Galbany, J., Martínez, L. M., and Pérez-Pérez, A. (2004). Tooth replication techniques, SEM imaging and microwear analysis in primates: methodological obstacles. *Anthropologie*, **42**, 5–12.

Galbany, J., Martínez, L. M., López-Amor, H. M., *et al.* (2005a). Error rates in buccal-dental microwear quantification using scanning electron microscopy. *Scanning*, **27**, 23–9.

Galbany, J., Moyà-Solà, S., and Pérez-Pérez, A. (2005b). Dental microwear variability on buccal tooth enamel surfaces of extant Catarrhini and Miocene fossil *Dryopithecus laietanus* (Hominoidea). *Folia Primatologica*, **76**, 325–41.

Galbany, J., Estebaranz, F., Martínez, L. M., *et al.* (2006). Comparative analysis of dental enamel polyvinylsiloxane impression and polyurethane casting methods for SEM research. *Microscopy Research and Technique*, **69**, 246–52.

Galbany, J., Estebaranz, F., Martínez, L. M., and Pérez-Pérez, A. (2009). Buccal dental microwear variability in extant African Hominoidea: taxonomy versus ecology. *Primates*, **50**, 221–30.

Galbany, J., Altmann, J., Pérez-Pérez, A., and Alberts, S. C. (2011). Age and individual foraging behavior predict tooth wear in Amboseli baboons. *American Journal of Physical Anthropology*, **144**, 51–9.

Gordon, K. D. (1982). A study of microwear on chimpanzee molars: implications for dental microwear analysis. *American Journal of Physical Anthropology*, **59**, 195–215.

Gordon, K. D. (1984). Hominoid dental microwear: complications in the use of microwear analysis to detect diet. *Journal of Dental Research*, **63**, 1043–6.

Gordon, K. D. (1988). A review of methodology and quantification in dental microwear analysis. *Scanning Microscopy*, **2**, 1139–47.

Gordon, K. D. and Walker, A. C. (1983). Playing "possum": a microwear experiment. *American Journal of Physical Anthropology*, **60**, 109–12.

Grine, F. E., Ungar, P. S., and Teaford, M. F. (2002). Error rates in dental microwear quantification using scanning electron microscopy. *Scanning*, **24**, 144–53.

Grine, F. E., Ungar, P. S., Teaford, M. F., and El-Zaatari, S. (2006a). Molar microwear in *Praeanthropus afarensis*: evidence for dietary stasis through time and under diverse paleoecological conditions. *Journal of Human Evolution*, **51**, 297–319.

Grine, F. E., Ungar, P. S., and Teaford, M. F. (2006b). Was the early Pliocene hominin '*Australopithecus*' *anamensis* a hard object feeder? *South African Journal of Science*, **102**, 301–10.

Gügel, I. L., Grupe, G., and Kunzelmann, K. H. (2001). Simulation of dental microwear: characteristic traces by opal phytoliths give clues to ancient human dietary behavior. *American Journal of Physical Anthropology*, **114**, 124–38.

Henke, W. (1998). Current aspects of dental research in Paleoanthropology. In: *Dental Anthropology: Fundamentals, Limits, Prospects*, ed. K. W. Alt, F. W. Rösing, and M. Teschler-Nicola. Stuttgart, Gustav Fischer, pp. 179–200.

Hillson, S. W. (1996). *Dental Anthropology*. Cambridge, Cambridge University Press.

Irish, J. D. and Nelson, G. C. (2008). *Technique and Application in Dental Anthropology*. Cambridge, Cambridge University Press.

Kaifu, Y., Kasai, K., Townsend, G. C., and Richards, L. C. (2003). Tooth wear and the "design" of the human dentition: a perspective from evolutionary medicine. *Yearbook of Physical Anthropology*, **46**, 47–61.

Kay, R. F. and Hiiemae, K. M. (1974). Jaw movement and tooth use in recent and fossil primates. *American Journal of Physical Anthropology*, **40**, 227–56.

Kay, R. F. and Covert, H. H. (1983). True grit: a microwear experiment. *American Journal of Physical Anthropology*, **61**, 33–8.

King, T., Andrews, P., and Boz, B. (1999). Effect of taphonomic processes on dental microwear. *American Journal of Physical Anthropology*, **108**, 359–73.

Koppe, T., Meyer, G., and Alt, K. W. (ed.) (2009). *Comparative Dental Morphology. Frontiers of Oral Biology*, vol. **13**. Basel, Karger.

Kullmer, O., Benazzi, S., Fiorenza, L., *et al.* (2009). Occlusal fingerprint analysis: quantification of tooth wear pattern. *American Journal of Physical Anthropology*, **139**, 600–5.

Krueger, K. L., Scott, J. R., Kay, R. F., and Ungar, P. S. (2008). Technical note: Dental microwear textures of "Phase I" and "Phase II" facets. *American Journal of Physical Anthropology*, **137**, 485–90.

Lalueza, C., Pérez-Pérez, A., and Turbón, D. (1996). Dietary inferences through buccal microwear analysis of Middle and Upper Pleistocene human fossils. *American Journal of Physical Anthropology*, **100**, 367–87.

Larsen, C. S. (2003). Animal source foods and human health during evolution. *Journal of Nutrition*, **133**, 3893–7.

Larsen, C. S. (2006). The agricultural revolution as environmental catastrophe: implications for health and lifestyle in the Holocene. *Quaternary International*, **150**, 12–20.

Lucas, P. W. (2004). *Dental Functional Morphology: How Teeth Work*. New York, Cambridge University Press.

Lucas, P. W. and Peters, C. R. (2000). Function of postcanine tooth shape in mammals. In: *Development, Function and Evolution of Teeth*, ed. M. F. Teaford, M. M. Smith, and M. W. J. Ferguson. Cambridge, Cambridge University Press, pp. 282–9.

Lucas, P. W., Constantino, P. J., and Wood, B. A. (2008). Inferences regarding the diet of extinct hominins: structural and functional trends in dental and mandibular morphology within the hominin clade. *Journal of Anatomy*, **212**, 486–500.

Mahoney, P. (2006a). Inter-tooth and intra-facet dental microwear variation in an archaeological sample of modern humans from the Jordan Valley. *American Journal of Physical Anthropology*, **129**, 39–44.

Mahoney, P. (2006b). Dental microwear from Natufian hunter-gatherers and early Neolithic farmers: comparisons between and within samples. *American Journal of Physical Anthropology*, **130**, 308–19.

Mahoney, P. (2007). Human dental microwear from Ohalo II (22, 500–23 500 cal. BP), Southern Levant. *American Journal of Physical Anthropology*, **139**, 489–500.

Molnar, S., McKee, J. K., Molnar, I. M., and Przybeck, T. R. (1983). Tooth wear rates among contemporary Australian aborigines. *Journal of Dental Research*, **62**, 562–5.

Newesely, H. (1993). Abrasion as an intrinsic factor in palaeodiet. In: *Prehistoric Human Bone. Archaeology at the Molecular Level*, ed. J. B. Lambert and G. Grupe. Berlin Heidelberg, Springer-Verlag, pp. 293–308.

Olsen, S. L. (1988). *Scanning Electron Microscopy in Archaeology*. Oxford, BAR Int. Series.

Organ, J. M., Teaford, M. F., and Larsen, C. S. (2006). Dietary inferences from dental occlusal microwear at Mission San Luis de Apalachee. *American Journal of Physical Anthropology*, **128**, 801–11.

Pérez-Pérez, A. (2004). Why buccal microwear? *Anthropologie*, **42**, 1–3.

Pérez-Pérez, A., Lalueza, C., and Turbón, D. (1994). Intraindividual and intragroup variability of buccal tooth striation pattern. *American Journal of Physical Anthropology*, **94**, 175–87.

Pérez-Pérez, A., Bermúdez de Castro, J. M., and Arsuaga, J. L. (1999). Nonocclusal dental microwear analysis of 300 000-year-old *Homo heidelbergensis* teeth from Sima de los Huesos (Sierra de Atapuerca, Spain). *American Journal of Physical Anthropology*, **108**, 433–57.

Pérez-Pérez, A., Espurz, V., Bermúdez de Castro, J. M., de Lumley, M. A., and Turbón, D. (2003). Non-occlusal dental microwear variability in a sample of Middle and Late Pleistocene human populations from Europe and the Near East. *Journal of Human Evolution*, **44**, 497–513.

Peters, C. R. (1982). Electron-optical microscopic study of incipient dental microdamage from experimental seed and bone crushing. *American Journal of Physical Anthropology*, **57**, 283–301.

Polo-Cerdá, M., Romero, A., Casabó, J., and De Juan, J. (2007). The Bronze Age burials from Cova Dels Blaus (Vall d'Uixó) Castelló, Spain: an approach to palaeodietary reconstruction through dental pathology, occlusal wear and buccal microwear patterns. *Journal of Comparative Human Biology*, **58**, 297–307.

Polo-Cerdá, M., García Prósper, E., and Romero, A. (2010). Bioantropología y Paleopatología. Herramientas para la investigación histórico-arqueológica. In: *Restos de vida, restos de muerte*, ed. A. Pérez Fernández and B. Soler Mayor. Valencia, Museu de Prehistòria de València, pp. 95–116.

Romero, A. and De Juan, J. (2003). Scanning electron microscopy in paleoanthropological research. In: *Science, Technology and Education of Microscopy: an Overview*, ed. A. Mendez-Vilas. Badajoz, Spain, Formatex, pp. 420–30.

Romero, A. and De Juan, J. (2005). Scanning microscopy exam of hominoid dental enamel surface: exploring the effect of abrasives in the diet. In: *Science, Technology and Education of Microscopy: an Overview. Current Issues in Multidisciplinary Microscopy Research and Education*, ed. A. Méndez-Vilas and L. Labajos-Broncano. Badajoz, Spain, Formatex. pp. 1–17.

Romero, A. and De Juan, J. (2006). Análisis de microdesgaste dentario-vestibular en humanos bajo dieta *ad libitum*. *Revista Española de Antropología Física*, **26**, 103–8.

Romero, A. and De Juan, J. (2007). Intra- and interpopulation human buccal tooth surface microwear analysis: inferences about diet and formation processes. *Anthropologie*, **45**, 61–70.

Romero, A., Martínez-Ruiz, N., and De Juan, J. (2003a). Dental microwear correlation rates on occlusal and non-occlusal molar surface. *International Journal of Dental Anthropology*, **4**, 16–22.

Romero, A., Martínez-Ruiz, N., Amorós, A., and De Juan, J. (2003b). Microdesgaste dental *in vivo*: modelo preliminar para interpretar su formación y variabilidad en grupos humanos antiguos. *Revista Española de Antropología Física*, **24**, 5–18.

Romero, A., Martínez-Ruiz, N., and De Juan, J. (2004). Non-occlusal dental microwear in a Bronze-Age human sample from East Spain. *Anthropologie*, **42**, 65–9.

Romero, A., Galbany, J., Pérez-Pérez, A., and De Juan, J. (2007). Microwear formation rates in human buccal tooth enamel surfaces: an experimental *in vivo* analysis under induced-diet. In: *New Perspectives and Problems in Anthropology*, ed. É. B. Bodzsár and A. Zsákai. Newcastle, Cambridge Scholars Publishing, pp. 135–46.

Romero, A., Galbany, J., Martínez-Ruiz, N., and De Juan, J. (2009). *In vivo* turnover rates in human buccal dental-microwear. *American Journal of Physical Anthropology, Supplement*, **48**, 223–4.

Romero, A., Ramírez-Rozzi, F. V., Froment, A., De Juan, J., and Pérez-Pérez, A. (2010). Buccal dental-microwear analysis among Pygmy hunter-gatherers from Western Central Africa. *American Journal of Physical Anthropology, Supplement*, **50**, 200–1.

Sanson, G. D., Kerr, S. A., and Gross, K. A. (2007). Do silica phytoliths really wear mammalian teeth? *Journal of Archaeological Science*, **34**, 526–31.

Schmidt, C. W. (2001). Dental microwear evidence for a dietary shift between two nonmaize-reliant prehistoric human populations from Indiana. *American Journal of Physical Anthropology*, **114**, 139–45.

Smith, T. M. and Tafforeau, P. (2008). New visions of dental tissue research: tooth development, chemistry, and structure. *Evolutionary Anthropology*, **17**, 213–26.

Strait, S. G. (1997). Tooth use and the physical properties of food. *Evolutionary Anthropology*, **5**, 199–211.

Teaford, M. F. (1988a). A review of dental microwear and diet in modern mammals. *Scanning Microscopy*, **2**, 1149–66.

Teaford, M. F. (1988b). Scanning electron microscope diagnosis of wear patterns versus artifacts on fossil teeth. *Scanning Microscopy*, **2**, 1167–75.

Teaford, M. F. (1994). Dental microwear and dental function. *Evolutionary Anthropology*, **3**, 17–30.

Teaford, M. F. (2000). Primate dental morphology revisited. In: *Development, Function and Evolution of Teeth*, ed. M. F. Teaford, M. M. Smith, and M. W. J. Ferguson. Cambridge, Cambridge University Press, pp. 290–304.

Teaford, M. F. (2007). What do we know and not know about dental microwear and diet? In: *Evolution of the Human Diet: The Known, the Unknown and the Unknowable*, ed. P. S. Ungar. New York, Oxford University Press, pp. 106–31.

Teaford, M. F. and Glander, K. E. (1996). Dental microwear and diet in a wild population of mantled howling monkeys (*Alouatta palliata*). In: *Adaptive Radiations of Neotropical Primates*, ed. M. A. Norconk, A. L. Rosenberger, and P. A. Garber. New York, Plenum Press, pp. 433–49.

Teaford, M. F. and Lytle, J. D. (1996). Brief communication: diet-induced changes in rates of human tooth microwear: a case study involving stone-ground maize. *American Journal of Physical Anthropology*, **100**, 143–7.

Teaford, M. F. and Oyen, O. J. (1989a). Live primates and dental replication: new problems and new techniques. *American Journal of Physical Anthropology*, **80**, 73–81.

Teaford, M. F. and Oyen, O. J. (1989b). *In vivo* and *in vitro* turnover in dental microwear. *American Journal of Physical Anthropology*, **80**, 447–60.

Teaford, M. F. and Oyen, O. J. (1989c). Differences in rate of molar wear between monkeys raised on different diets. *Journal of Dental Research*, **68**, 1513–18.

Teaford, M. F. and Tylenda, C. A. (1991). A new approach to the study of tooth wear. *Journal of Dental Research*, **70**, 204–7.

Teaford, M. F. and Ungar, P. S. (2000). Diet and the evolution of the earliest human ancestors. *Proceedings of the National Academy of Sciences*, **97**, 13506–11.

Teaford, M. F. and Ungar, P. S. (2007). Dental adaptations of African apes. In: *Handbook of Paleoanthropology*. Volume 1: *Principles, Methods, and Approaches*, ed. W. Kenke, W. Rothe, and I. Tattersall. Heidelberg, Springer-Verlag, pp. 1107–32.

Ungar, P. S. (1998). Dental allometry, morphology and wear as evidence for diet in fossil primates. *Evolutionary Anthropology*, **6**, 205–17.

Ungar, P. S. (2004). Dental topography and diets of *Australopithecus afarensis* and Early *Homo*. *Journal of Human Evolution*, **46**, 605–22.

Ungar, P. S. (2007). Dental topography and human evolution: with comments on the diets of *Australopithecus africanus* and *Paranthropus robustus*. In: *Dental Perspectives on Human Evolution: State of the Art Research in Dental Anthropology*, ed. S. Bailey and J. J. Hublin. New York, Springer-Verlag, pp. 321–44.

Ungar, P. S. and Bunn, J. (2008). Primate dental topographic analysis and functional morphology. In: *Technique and Application in Dental Anthropology*, ed. J. D. Irish and G. C. Nelson. Cambridge, Cambridge University Press, pp. 253–65.

Ungar, P. S. and M´Kirera, F. (2003). A solution to the worn tooth conundrum in primate functional anatomy. *Proceedings of the National Academy of Sciences*, **100**, 3874–7.

Ungar, P. S. and Spencer, M. A. (1999). Incisor microwear, diet, and tooth use in three Amerindian populations. *American Journal of Physical Anthropology*, **109**, 387–96.

Ungar, P. S. and Teaford, M. F. (1996). Preliminary examination of non-occlusal dental microwear in anthropoids: implications for the study of fossil primates. *American Journal of Physical Anthropology*, **100**, 101–13.

Ungar, P. S., Brown, C. A., Bergstrom, T. S., and Walker, A. (2003). Quantification of dental microwear by tandem scanning confocal microscopy, and scale sensitive fractal analyses. *Scanning*, **25**, 185–93.

Ungar, P. S., Grine, F. E., Teaford, M. F., and El-Zaatari, S. (2006). Dental microwear and diets of African early *Homo*. *Journal of Human Evolution*, **50**, 78–95.

Ungar, P. S., Scott, R. S., Scott, J. R., and Teaford, M. F. (2008). Dental microwear analysis: historical perspectives and new approaches. In: *Technique and Application in Dental Anthropology*, ed. J. D. Irish and G. C. Nelson. Cambridge, Cambridge University Press, pp. 389–425.

Ungar, P. S., Scott, R. S., Grine, F. E., and Teaford, M. F. (2010). Molar microwear textures and the diets of *Australopithecus anamensis* and *Australopithecus afarensis*. *Philosophical Transactions of the Royal Society London B, Biological Science*, **365**, 3345–54.

Wrangham, R. W., Jones, J. H., Laden, G., Pilbeam, D., and Conklin-Brittain, N. (1999). The raw and the stolen: Cooking and the ecology of human origins. *Current Anthropology*, **40**, 567–94.

Index

Figures and Tables are indicated in boldface

Numbers
3-D imaging, *see* stereo imaging

agar-string fracturing, 39–42, **40**
angiogenesis, 23–24, 63
artifacts
  in chromosomes, 161
  in sample preparation, 5–6, 34, 35, 73, 213, 223, 224
  with in vivo cryo-fixation, 196–197, **197**
attachment virus stage, **101**, 101–103, **102**

backscattered electron microscopy, *see also* field emission SEM
  and chromosomes, 148–152, 154–157
  in correlative microscopy, 85
  in dental research, 226–229
  detectors, 34, 142, 217–220
  in SEM, 2, 3, **3**
biology
  future of SEM in, 1–2
  importance of expertise in for SEM, 6–7
bleaching, 213
block face imaging, 10–11, *see also* stereo imaging
blood vessels, *see also* erythrocytes
  corrosion casting in, 17, 18–24, **24**
  in dental research, 214
  in organs, 201–208
BSE, *see* backscattered electron microscopy
budding virus stage, **101**, 107–112, **107**, **109**, **111**, 121–126

capillary sprouts, *see* angiogenesis
casting media, *see also* corrosion casting; sample preparation
  history of, 16–18
  problems with, 16–19
  properties of, 19
  resin, 213–214, **214**, 219, 239–240
cathodoluminescence, **3**, 214

cell cycle
  chromosome replication, 138, 143, 159
  Herpes Simplex virus, 117
  in mitochondria, 59–60
  viral infection, **101**, 101–107, **102**, **104**, **105**
cerebellar nerves
  mossy and climbing fibers, 180–184, **181**, **182**, **183**, 189–191
  Purkinje cells, 183–188, **185**, **186**, **187**, **188**, 191
  scanning electron microscopy in, 171–180, 189
  stellate neurons, 187, **187**
chromomeres, 143–146
chromosomes
  dynamic matrix model in, 158–160, **159**
  focused ion beam milling in, 139, 154–158, 160, 161–167
  history of visualization, 138–139
  influence of fixation additives, 145–148, **147**, **148**
  isolated SEM data collection, 143–146, **144**, **145**, 160
  resolution, 160–161
  SEM techniques, 139–143
  in situ SEM data collection, 143, **144**
  staining and labeling, 148–154, 160
  stereo imaging, 155–158, **158**, 160–161
  structure, 138, 157–158
cleaved glass fracturing, 36–37, **36**
CLSM, *see* confocal laser scanning microscopy
coating, *see also* labeling; marker; sample preparation
  in corrosion casting, 20–21, 239
  in dental research, 212
  in sample preparation, 4, 6
colloidal gold, *see* labeling
color anaglyph images, 47–48, **48**, *see also* stereo imaging
conductive bridges, 20–21
confocal laser scanning microscopy, 180, 191
congugate
  marker process, 94–96
  synthesis, 96

conventional SEM
in cerebral tissue, 179, 180, 181, 186–191
in dental research, 215–216, **216**, **217**
epitopes, 84–87
introduction to, 2, **3**
on viruses, 101–103, 110
correlative microscopy, 84–96
corrosion casting, *see also* casting media
applications, 26–28
history, 16–18
image interpretation, 21–24, **22**
procedure, 18–21
quantitative analysis, 23–24
vascular studies, 25–27
CPD, *see* critical point drying
critical point drying, 4, 5–6, 20, **73**, 142, 212, *see also* sample preparation
cross fractures, **182**, 213, **214**, 227–229
cross sections, 213, **214**
cryo-fixation, *see also* fixation; freezing methods; *in vivo* cryotechnique
in chromosomes, 139
drop/cryo technique, 141, 159
Herpes Simplex virus, 117–119
introduction to, 4
new technologies in, 10, **76**
in *toxoplasma*, 7

dehydration
in dental research, 212–213
high resolution low voltage SEM, 34
in vivo cryotechnique, **197**, 201
dendrites, **52**, 57, **58**, 181, **181**, 182, **182**, **183**, 190
dental research, *see also* paleoanthropology research
composite restorations in, 211–212, **212**
conventional SEM, 214–216, **217**
environmental SEM, 229–231, **231**
fractography in, 227–231, **228**
high resolution field emission SEM, 217–220, **218**, **219**
importance of SEM in, 211, 231
sample preparation, 212–214, **213, 214**
x-ray analysis, 220–227, **221**, **222**, **224**, **226**, **227**
detergent treatment, 34–36, 46
diet-caused teeth wear, 245–251
drop/cryo technique, 141, **147**, 158
dynamic matrix model, 158–160

ECM, *see* extracellular matrix
EDS, *see* x-rays
EDX, 86, 89
EELS, *see* energy filtering imaging
electron spectroscopic imaging, 84–87, **87**
enamel, *see* teeth
endoplasmic reticulum, **52**, **53**, **57**, 57, **58**, 59
energy dispersive spectroscopy, *see* x-rays
energy filtering imaging, 85–86, 90
entry virus stage, **101**, 103–104, **104**
environmental SEM
in dental research, 229–231
in paleoanthropological research, 239
epitopes
of chromosomes, 152–154
as markers, 84–87, **85, 88**
erythrocytes, *see also* blood vessels
in aorta and vena cava, 204–206, **205**
in liver, **205**, 206
ESEM, *see* environmental SEM
extracellular matrix, *see also* networks
fixation, 167–169
mucus, 167, **171**, **173**, 173–174
role of, 165–166
sample preparation, 169–170
zona pellucida, 166–167,170, **170**, **171**, 172–175

FESEM, *see* field emission SEM
FIB, *see* focused ion beam milling
field emission SEM, *see also* backscattered electron microscopy
in cell mitosis, 10
in cerebral tissue, 179–180, 183–184, 191
in chromosomes, 142, 161
in correlative microscopy, 87–89, **89**
Herpes Simplex virus, 118
fixation, *see also* cryo-fixation; sample preparation
agar-string fracturing, 39–41, **40**
in chromosomes, 145–147, **147**, 148, 160
cleaved glass fracturing, 36–37, **36**
de-embedding semi-thin sections, 41–44, **43**
in dental research, 212–213
extracellular matrix, 167–169
high resolution low voltage SEM, 34–44, **35**
tape-ripping, 37, **38**
wet-ripping, 37–39, **39**
flourescent second antibody, 88–89
focused ion beam milling
of chromosomes, 140, 154–157,160
new technology in, 1
new technology of, 10, 11
of viral structures, 112–113, **113**
fractography, 227–229
fracturing
agar-string, 39–41, **40**
cleaved glass, 36–37, **36**
freeze fracture, *see* freezing methods
freezing methods, 7–9, 44–46, 119, 173, *see also* cryo-fixation; *in vivo* cryotechnique

GAGs, *see* glycosaminoglycans
glycosaminoglycans, 165–166
Golgi apparatus, **52**, **53**, **57**, **58**, 59, 116, 126
granule cells, 64, 65
gylcoprotein matrices, *see* extracellular matrix

Herpes Simplex virus
and advances in HR-FE-SEM, 103–108, 117, 119, 127
budding, 121–126
cryo-fixation, 117–118, 119
field emission SEM, 118–119
infection of cells with, 117
nuclear pore changes, 126–132
nuclear surface in, 119–120, **121**
nuclei expansion, 120, **121**
hexamethyldisilazane, 213
high resolution field emission SEM
in chromosomes, 139
in dental research, 217–220
Herpes Simplex virus, 103–108, 117, 118–119, 127
in mitochondria, 51–66, **52**
sample preparation, 66–67, 170
high resolution low voltage SEM
dehydration, 34
extracellular matrix, 170–171
fixation, 34–44
freezing methods, 44–46, **47**
introduction to, 9–10
of mitochondrial networks, 51–67
stereo image creation in, 46–48, **48**
HMDS, *see* hexamethyldisilazane
HR-FE-SEM, high resolution field emission SEM
HRLVFESEM, *see* high resolution low voltage SEM
hyperfusion, 64–65

IA, *see* ion abrasion
imaging methodology, 85, 86, **86**, 90
immunolabeling, *see* labeling
in situ SEM data collection, 142, 143, **144**, 160
in vitro studies
cryotechnique, 196–208
large cells, 63–64
microwear
mitochondrial networks, 59–61
in vivo cryotechnique, *see also* cryo-fixation; freezing methods
application to organs, **199**, 201–208, **204, 205**
dehydration, 197, 201
fixation, **197**, 200
how to perform, 197–199, **198**
and prevention of artifacts, 196–197, **197**
sample preparation, 199–200, **199**
in vivo studies
Herpes Simplex virus, 117, 118, 119, 127, 133
microwear, 243–245
mitochondrial networks, 59–65
ion abrasion, 112–113, **113**
IVCT, *see in vivo* cryotechnique

Keimmethode, 92–93, **93**
kidney blood flow, 200–203, **203**

labeling, *see also* coating; marker
in chromosomes, 139, 141, **149**
in correlative microscopy, 84–96, 86, 88, **86**, 89
immuno, 37, **85**, 152–154, 180, 217–218
light microscopy, 63, 66, 84–89, 90, 139, 143, 159
limitations of SEM techniques, 33, 112, 139, 160, 203, 215, 225
LM, *see* light microscopy

marker, *see also* coating; labeling
conjugation, 94–96
shapes, 93–94
synthesis, 90–93
mastication, 240–243, **240**, *see also* teeth
microcomputed tomography, 24
microcorrosion casting, *see* corrosion casting
micro-CT, *see* microcomputed tomography
microwear
analysis, 238–242, **253**, **254**
from an anthropological perspective, 245,–251, **250**
formation, 242–245
mitochondria
during apaptosis, 62–63
cellular structure, 52–53, **52**
continuous nature of, 59–61, 65
in dendrites, 57, **58**, 64–65
endoplasmic reticulum, 52, 53, 57, 58, 59
Golgi apparatus, **53**, **57**, **58**, 59
imaging methodology, 65–66
mitochondria networks
classification, **53**, 55–57, **57**
diversity, **53**, 62
formation in large cells *in vivo*, 63, 65
formation *in vitro* cells, 64–65
history of study, 51–52
size, 61
morphometry, 23–24
mucus, 167, 172, **172, 173**, 173–174
muscle cells, **23, 44**, 63–64

nanocrystals, 89–90
networks, *see also* extracellular matrix
mitochondria, 51–66, 54, 57
vascular, 24–27
neurons
and mitochondria networks, 64
stellate, 188

ODO method, 65
OTO staining method, 6

paleoanthropology research, *see also* dental research
importance of SEM in, 236–237, 238–239
microwear analysis, 238–242
microwear formation process, 242–245
microwear from an anthropological perspective, 245–251, **247, 250**

paleoanthropology research (cont.)
  tooth shape and function, 237, 238
particle
  shapes, 93–94
  synthesis, 90–93, **184**, **185**
plant
  chromosome imaging, 139–140, 142–144
  fixation, 37, **38**
  freezing methods, 44–45, **45**
  sample preparation, 7
Platinum Blue staining, **47**, 148–152, **151, 171**
plunge freezing, 7
polishing, 211, **218**, **224**
propane jet freezing, 8
protein labeling, 90
Purkinje cells, 181, 183–188, **185, 186, 187**, 191

replication virus stage, **101**, 103–104, **105**, 116–117
resin casts, 17, 239–240, **239**
resolution, 160–161, 225–227, **225**, *see also* stereo imaging
ruthenium red dye, 168, 169, 170, 174, 175, *see also* saponin

sample preparation, *see also* casting media; coating; critical point drying; fixation
  artifacts in, 5–6, 34, 35, 73, 213, 223, 224
  conventional, 4–7, 73, **73**, 79, 169–170
  in dental research, 212–214
  high resolution field emission SEM, 66–67, 170–172
  high voltage STEM, 78
  low voltage STEM, 78
  new developments in, 10–11
  by prefixant extraction, **73**, 73–75, **74**, 79–80
  in vivo cryotechnique, 199–200
saponin, 168, 169, 170, 174, 175, *see also* ruthenium red dye
scanning electron microscopy, *see also* scanning transmission electron microscopy; transmission electron microscopy
  conventional, 3, **3**, 85, 101–103, 110, 179, 180, 181, 187–191, 214–215
  cryo-fixation, 75, **76**, 80–81, 117, 118, 121, 122
  environmental, 1, 4, 229–230, 239
  field emission, 10, 88, **88**, 118, 142, 161, 179, 183–187, **186**, 191
  high resolution field emission, 51–67, 103–108, 117, 118, 127, 217–218
  high resolution low voltage, 9–10, 34–44, **35**, **36**, 45, **45**, 46–48, **47, 48**, 74–75, **74**, 170–172, **172, 173**
scanning helium ion microscopy, 11
scanning transmission electron microscopy, *see also* scanning electron microscopy; transmission electron microscopy
  in correlative microscopy, 86–87, **86, 87, 89**
  high voltage, **77**, 78, 81
  low voltage, **74**, 77–78, 80–81
SE, *see* secondary electrons
secondary electrons, **3**

SEM, *see* scanning electron microscopy
senescent cells, 65
SHIM, *see* scanning helium ion microscopy
slam freezing, 7–8
stages
  chromosome cell, 138–139
  viral infection, 100–112, 101
STEM, *see* scanning transmission electron microscopy
stereo imaging, *see also* block face imaging; color anaglyph images; resolution
  of cells, **47, 74, 76**
  of chromosomes, **147**, 155–156, 161
  extracellular matrix, 170, **171, 172**
  high resolution low voltage SEM, 9–10, 46–48, **48**
  by stereo pairs, 217, **217**
  of teeth microwear, 237
  transmission electron microscopy, 33

tape-ripping fixation, 37, **38**, *see also* wet-ripping fixation
teeth, *see also* mastication
  and diet, 245–246, 250
  microwear, 237, **241**
  mold making, **239**
  structure of, **212**, 217
TEM, *see* transmission electron microscopy
three-dimensional imaging, *see* stereo imaging
Toxoplasma, 7
trafficking virus stage, **101**, 104–105, **214**
transmission electron microscopy, *see also* scanning electron microscopy; scanning transmission electron microscopy
  and cerebellar bells, 179, 183, 187, 189
  and chromosomes, 139, 142, 143, 160
  in correlative microscopy, 87, **87**
  in dental research, **213**, 217–220, 225–226
  and Herpes Simplex virus, 117, 132–133
  imaging using, 1, 2, 31, **35**, 51
  internal cell structures, 44–46, **47**
  mitochondrial networks, 63, 66
  resolution of, 72–73
tumors, 25–27, **27**
Type-1 mitochondrial network, **53**, 54–56
Type-2 mitochondrial network, **53**, 54–56
Type-3 mitochondrial network, **53**, 54, 57
Type-4 mitochondrial network, **53**, 57, **57**

uncoating virus stage, 101, 103–104, **105**, 119

variable pressure SEM, *see* environmental SEM
vascular system, *see* blood vessels
virus infection
  Herpes Simplex, 116–133
  stages, **101**, 102–112, **104**, **105**, **106**, **107**, **109**, **111**, 113

wavelength dispersive spectronometers, *see* X-rays
WDS, *see* X-rays
wet-ripping fixation, 37–38, **39**, *see also* tape-ripping fixation

X-rays
  in dental research, 220–226, **221**, **222**, **223**, **224**, **226**, 227
  in elemental analysis, 2, **3**
  microanalysis, **197**, 201, 206–208, **207**

zona pellucida, 166–167, 170–171, **171**, 172–175
ZP, *see* zona pellucida